ONE HEALTH ATLAS

Books on One Health

One Health, une seule santé.
Théorie et pratique des approches intégrées de la santé
J. Zinsstag, E. Schelling, D. Waltner-Toews, M.A. Whittaker, M. Tanner, eds.
2020, 584 p.
[One Health: the theory and practice of integrated health approaches,
2nd edition, CABI, 2021]

Sortir des crises.
One Health en pratique
S. Gardon, A. Gautier, G. Le Naour, S. Morand, eds.
2022, 264 p.

Une seule santé.
S'ouvrir à d'autres savoirs
N. Lainé
2024, 80 p.

This publication was funded through LabEX AGRO 2011-LABX-002, project number 2200211, under the I-Site Muse programme, coordinated by Agropolis Fondation. It received additional support from CIRAD. Additional backing was provided by the PREZODE initiative.

To cite this book:
Roger F., Olive M.-M., Peyre M., Pfeiffer D., Zinsstag J., eds. 2025. One Health Atlas, Quæ, CABI, 208 p., https://doi.org/10.1079/9781836992578.0000

Éditions Quæ
RD 10, 78026 Versailles Cedex, France
www.quae.com
www.quae-open.com

CABI
Nosworthy Way, Wallingford,
Oxfordshire OX10 8DE, United Kingdom
www.cabi.org
www.cabidigitallibrary.org

ISBN paperback: 978 1 83699 255 4
ISBN PDF: 978 1 83699 256 1
ISBN ePub: 978 1 83699 257 8

Edited by
François Roger • Marie-Marie Olive • Marisa Peyre • Dirk Pfeiffer • Jakob Zinsstag

ONE HEALTH ATLAS

Table of contents

Foreword

The One Health approach embodies a long-standing concept: that human, animal, plant and environmental health are interdependent and bound to the health of the ecosystems that sustain them.

It is a collaborative, whole-of-society, whole-of-government approach to understanding, anticipating and addressing risks to global health at the interface between these sectors.

While global actors have long acknowledged the benefits of One Health and, more recently, have formalized partnerships to improve health governance at the international level (such as the Quadripartite Collaboration on One Health, comprising the FAO, WHO, WOAH and UNEP), its implementation remains a significant challenge.

As global health threats become more complex, the need to operationalize One Health at national and local levels has never been more urgent. The COVID-19 pandemic, a public health crisis caused by a virus of possible anim al origin, underlined the validity of the One Health approach in understanding and addressing such threats.

Often used to coordinate multisectoral efforts for prevention, preparedness and response to zoonotic diseases, this approach is critical for the control of priority zoonotic diseases such as rabies, avian influenza, Ebola or vector-borne diseases. Furthermore, numerous cross-cutting issues, such as antimicrobial resistance, food safety, climate change and weak health infrastructure, need to be addressed from a multisectoral and multidisciplinary perspective, which the One Health approach guarantees.

Risk drivers such as climate and land-use change, unsustainable agricultural practices, globalization and the unregulated wildlife trade provide multiple opportunities for pathogens to evolve into new forms, making cross-species spillover events more frequent and intense.

Tackling these major global health risks cannot be done in isolation. It requires the full cooperation of the animal, human, plant and environmental health sectors. By making One Health accessible and actionable, we can build resilient systems that protect human, animal and environmental health alike.

But beyond the concept itself, what we need now is to speak about "One Health in Practice" in a concrete narrative.

This atlas, which illustrates the rationale behind this approach through case studies, illustrations and thoughtful analyses, serves as a valuable reference and a practical guide for a range of stakeholders, including researchers, students, policymakers and practitioners, to better understand and engage with these critical issues. It demonstrates the benefits of breaking down silos to create practical, collaborative solutions that integrate public health, veterinary medicine, agriculture and environmental science as well as economics and social science. As highlighted in this atlas, public-private partnerships, One Health education, networks, governance and the science-policy interface are particularly effective in supporting the multisectoral collaboration needed for impactful One Health initiatives.

I am pleased that the World Organisation for Animal Health (WOAH) has contributed to several chapters of this project, reflecting how our organization is helping policymakers and stakeholders to envision a future in which human, animal and environmental health systems are mutually beneficial and supportive. We achieve this by setting standards, developing guidelines, gathering scientific expertise, sharing quality data and working with a strong network of partners across sectors.

The full potential of the One Health approach remains untapped because of significant gaps in its operationalization. It must evolve into a systematic way of approaching health governance, enabling the sharing of resources and knowledge, and facilitating coordinated policies and investment.

Governments have a key role to play in embedding One Health in national policies and programmes. But One Health must also be accessible to local communities, smallholder farmers and front-line health workers. Bridging these gaps requires inclusive communication, education and stakeholder engagement.

Operationalizing One Health is not an end in itself; it must answer today's challenges and lead to measurable improvements in health outcomes, economic resilience and environmental sustainability. We share a common goal across sectors: to reduce the risks of zoonotic diseases and pandemics, address priority health threats such as antimicrobial resistance and vector-borne diseases, mitigate the effects of climate change and support development goals.

Governments, international organizations, civil society and the private sector must work together to turn the promise of One Health into a reality that benefits everyone, everywhere.

Illustrations of practical examples of how the One Health approach is being used across sectors worldwide, and its added value to health, the economy and society as presented in this atlas, are important to scale up its uptake.

Together, we can achieve a more integrated, equitable and effective health system that secures our shared future.

Emmanuelle Soubeyran

Director General of the
World Organisation for Animal Health (WOAH)

Preface

The idea for this One Health Atlas emerged from the recognition that, while a wealth of articles, books and studies exists on the subject of One Health, there is still a need for a visual, accessible resource that clearly captures the multifaceted nature of this evolving approach. One Health is inherently multidimensional: it encompasses scientific, political, geographical, economic and social perspectives. A graphical and structured representation, such as this atlas, offers an intuitive way to navigate and understand the interconnected complexities of this integrated framework.

This atlas is divided into four sections, each focusing on a specific dimension of One Health. Supported by theoretical foundations, case studies and practical examples, these sections aim to provide readers with both foundational knowledge and applied insights. Each double-page spread (chapter) is designed to function independently, allowing for flexible use in courses, workshops and various educational settings. The book's structure is further enriched by an introductory chapter that sets the stage for understanding One Health, and a concluding chapter that examines the critical interface between science and society.

One Health is a constantly evolving approach that thrives within a dynamic ecosystem of projects, publications, conferences and collaborative initiatives. It is continually shaped by the dedication of researchers, policymakers, practitioners and communities who are actively pushing the boundaries of this interdisciplinary field. This vibrant environment reflects a growing global acknowledgment of the interconnectedness of human, animal and environmental health.

Governments and institutions around the world are increasingly embedding One Health principles into policies and practices, recognizing its potential to tackle pressing global health challenges such as zoonotic disease emergence, antimicrobial resistance, food security and the profound impacts of biodiversity and climate change on health.

Each part of this atlas reflects this ongoing expansion, showcasing how One Health principles are not only conceptual but also being applied to real-world challenges. By highlighting the contributions of diverse actors—scientists, decision makers and field practitioners alike—this atlas aspires to inform, inspire and advance the global discourse on One Health.

The concepts of impermanence and interdependence in Buddhist philosophy tangibly resonate with the One Health approach. Impermanence is evident in the constant changes affecting ecosystems, pathogens and societal dynamics. Ecosystems evolve under the influence of natural or human-induced factors, such as climate change or urbanization, while pathogens continually adapt by mutation. These transformations highlight the fragility of local and global health equilibria, requiring proactive and adaptive approaches.

Interdependence underscores the close connections between the domains of life and health—humans, animals, plants and ecosystems. This interdependence also extends to the essential cooperation among communities, governments, scientists and organizations to address health challenges. One Health enhances our ability to anticipate crises, adopt systemic perspectives, strengthen resilience and promote international collaboration—key elements for tackling the growing complexity of global health in an ever-changing world.

François Roger
Hanoi—Bangkok, 2025

Traditional glass mosaic from Wat Xieng Thong temple (Luang Prabang, Laos), depicting daily life and human–animal interactions in rural landscapes. These artistic scenes reflect a long-standing cultural awareness of the relationships between people, domestic animals and wildlife.

Acknowledgements

We would like to express our deepest gratitude to all those who contributed to the development of this atlas.

We extend our heartfelt thanks to all the authors for their collaboration and invaluable contributions. Their expertise and dedication have been essential in bringing this work to fruition, enriching it with the scientific and interdisciplinary depth that defines its value.

We also wish to acknowledge the generous financial support of the Agropolis Foundation, whose contribution made this project possible, as well as CIRAD for its additional financial support. Thank you as well to the LabEx AGRO and its joint research unit SELMET, which greatly facilitated the administrative aspects.

We extend our appreciation to the cartography studio AFDEC for their work in creating detailed and informative maps. We would also like to recognize Daan Vink and Thierry Lefrançois from CIRAD for their valuable support.

We are particularly grateful to Teri Jones-Villeneuve for her meticulous copyediting of the English to ensure clarity and accessibility. Our sincere thanks also go to Laetitia Perotin-Meslay for the graphic design, which has given this atlas a cohesive and visually appealing look.

Finally, we would also like to offer a very special thank you to Christelle Fontaine and Anne Dievart at Éditions Quæ for their unwavering support and guidance throughout the editing and publication process.

François Roger
Marie-Marie Olive
Marisa Peyre
Dirk Pfeiffer
Jakob Zinsstag

Introduction to One Health

Jakob Zinsstag, Lisa Crump

One Health is an integrative and systemic concept to understand and improve public health. Its focus extends beyond humans and encompasses the health and well-being of animals (including pets, livestock and wildlife), plants and key ecosystem services. Humans, animals and the environment are all closely linked, and failing to recognize this interconnectedness hinders effective primary prevention of diseases at their source. This interconnectedness is reflected in the spread of zoonotic diseases (infectious diseases transmitted between wildlife or domestic animals and humans), antimicrobial resistance, the effects of climate change on health and the global COVID-19 pandemic. One Health aims to demonstrate the added value of having stakeholders from different disciplines and fields working together to produce new knowledge that would not be possible separately (Zinsstag *et al.* 2015). The added value gained from a One Health approach can be leveraged on various levels depending on the contributors involved. One level is interdisciplinary, where professionals from different disciplines (e.g. human and veterinary medicine and other related disciplines) collaborate to produce additional systems knowledge. This knowledge can then be applied to improve human and animal health, generate financial savings or enhance environmental services through solutions that would not be possible without collaboration. Another level of added value is the co-production of transformational knowledge between academic and non-academic actors (e.g. businesses and communities) in transdisciplinary processes (Zinsstag *et al.* 2023).

An example of the added value that a One Health approach can offer is joint human and animal vaccination services for mobile pastoralists in Chad, where

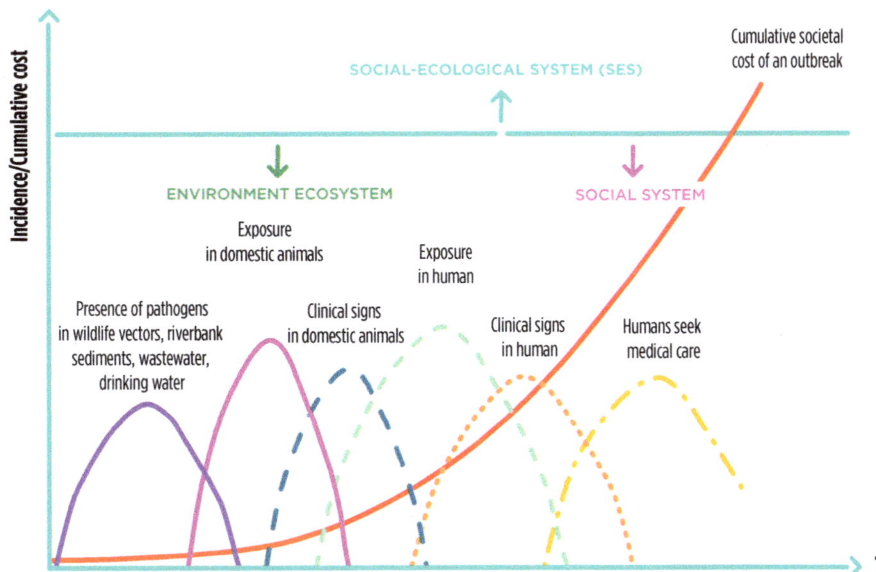

Figure 1. Reported clinical signs allegedly associated to natural SARS-CoV-2 infection in the different animal hosts worldwide.

populations that would otherwise be excluded gain access to healthcare. Because human and animal health services share a cold chain and transport costs, they are able to save costs (Schelling *et al.* 2007). In another example, the benefits of brucellosis control in Mongolia for public health alone were not enough to justify the cost of a mass livestock vaccination campaign to prevent the disease in humans. But when all the benefits of vaccinating livestock for this disease are added up across the health and agricultural sectors, the overall societal benefits of mass vaccination are three times higher than the intervention cost (Roth *et al.* 2003). Shared infrastructure can also produce savings. For instance, the World Bank estimates that the Canadian Science Centre in Winnipeg for Human and Animal Health, which hosts laboratories for highly contagious human and animal diseases under one roof, is able to reduce its operations costs by 26% (World Bank 2012; Zinsstag *et al.* 2018).

Doctors alone can no longer solve all the health problems the world faces today, which range from pandemics to antibiotic resistance and food security. They must join forces with veterinarians and professionals from other

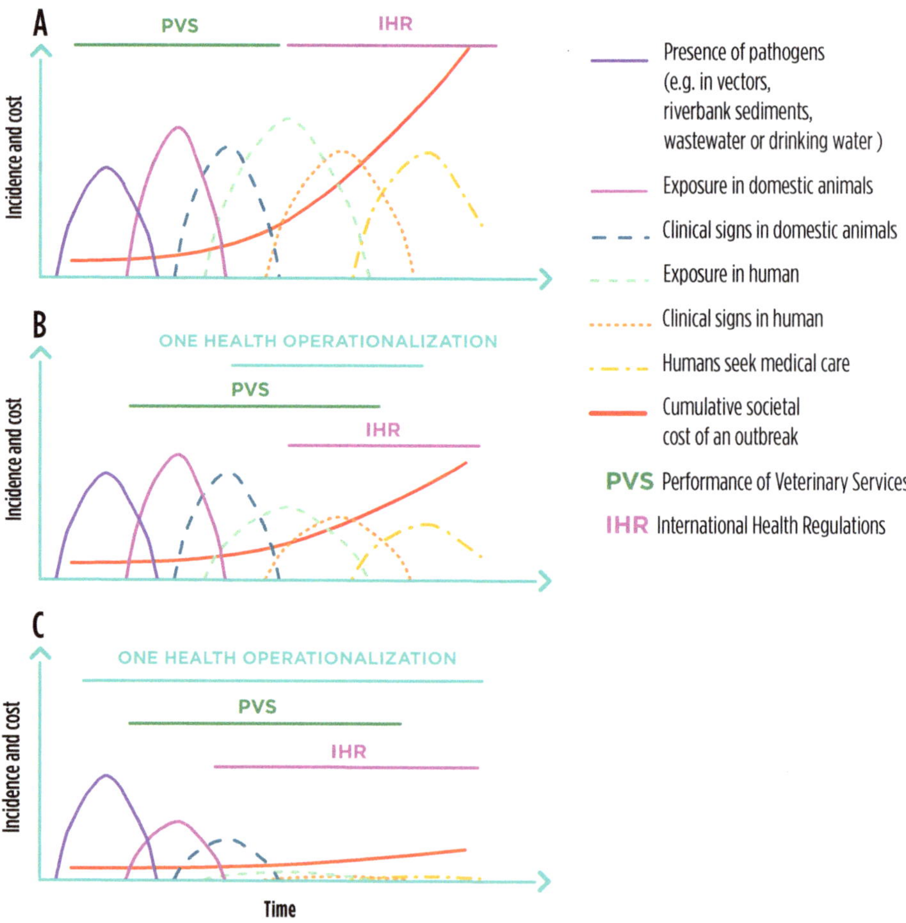

Figure 2. Vision of One Health governance for global health security. From Zinsstag *et al.* 2023.
To better understand interactions between antibiotic resistance in humans, animals and their environment, integrated One Health surveillance–response systems must be expanded to include antibiotic resistance. As soon as antibiotic resistance is detected in one location, all other areas should be quickly informed. This can prevent exposure to antibiotic resistance and reduce the emergence of new resistance. Several countries already have such systems in place, with the Canadian Integrated Program for Antimicrobial Resistance Surveillance (CIPARS) at the forefront. In this One Health atlas, we provide examples of the benefits of a geographically-based One Health approach with a view to informing scientists, health and government authorities and practitioners. The atlas also shows gaps where the potential of One Health is not yet been leveraged worldwide.

disciplines. In 2012, the World Bank showed how a One Health approach could be implemented (Figure 1; Zinsstag *et al.* 2020). When emerging diseases are observed in wildlife, the associated costs (red line) are low, but they begin to rise as soon as emerging diseases appear in livestock, peaking when humans become ill and infect each other. The COVID-19 pandemic is a good illustration of this type of situation. Preventing or mitigating outbreaks of emerging diseases is possible if we link disease surveillance and response systems to effectively communicate information about the environment, animals and humans. Aside from a few initiatives like the integrated West Nile virus surveillance programme in Emilia-Romagna, Italy (Paternoster *et al.* 2017), most countries still have separate surveillance-response systems for humans and animals. National governments and international organizations play an important role in promoting integrated human-animal-environment surveillance-response systems in their countries and networking to connect them internationally. International collaboration can help prevent or mitigate future pandemics to save lives and curb financial losses. A recent study summarized the evidence of what One Health could do for global health security (Zinsstag *et al.* 2023b). While One Health governance, operationalized at different levels, may not prevent future outbreaks of emerging diseases, it can significantly lower the effects on human and animal health and related costs (Figure 2).

Antibiotic resistance—considered a "silent pandemic"—is another global problem today. Bacterial infections that no longer respond to antibiotics are killing more and more people and animals around the globe, leading to both high treatment costs and significant follow-up costs due to loss of labour. Reasons for this escalating antibiotic resistance are manifold and still mostly unknown. However, scientists believe the frequent misuse of antibiotics, such as when patients do not complete their full course of treatment, and the overuse of antibiotics to promote livestock growth drive antibiotic resistance in humans.

References

Paternoster G., Martins S.B., Mattivi A., Cagarelli R., Angelini P. *et al.* 2017 Economics of One Health: Costs and benefits of integrated West Nile virus surveillance in Emilia-Romagna. *PLoS One,* 12(11), e0188156. https://doi.org/10.1371/journal.pone.0188156

Roth F., Zinsstag J., Orkhon D., Chimed-Ochir G., Hutton G. *et al.* 2003. Human health benefits from livestock vaccination for brucellosis: Case study. *Bulletin of the World Health Organization,* 81(12), 867–876.

Schelling E., Bechir M., Ahmed M.A., Wyss K., Randolph T.F., Zinsstag J. 2007. Human and animal vaccination delivery to remote nomadic families, Chad. *Emerging Infectious Diseases,* 13(3), 373–379. https://doi.org/10.3201/eid1303.060391

World Bank. 2012. *People, Pathogens and Our Planet: Volume 2: The Economics of One Health.* Report No. 69145-GLB.

Zinsstag J., Crump L., Schelling E., Hattendorf J., Maidane Y.O. *et al.* 2018. Climate change and One Health. *FEMS Microbiology Letters,* 365(11), fny085. https://doi.org/10.1093/femsle/fny085

Zinsstag J., Kaiser-Grolimund A., Heitz-Tokpa K., Sreedharan R., Lubroth J. *et al.* 2023b. Advancing One human-animal-environment Health for global health security: what does the evidence say? *The Lancet,* 401(10376), 591–604. https://doi.org/10.1016/S0140-6736(22)01595-1

Zinsstag J., Pelikan K., Berger Gonzalez M., Kaiser-Grolimund A., Crump L. *et al.* 2023a. Value-added transdisciplinary One Health research and problem solving, in *Handbook of Transdisciplinarity: Global Perspective* (Lawrence R.J., ed.). Edward Elgar Publishing, Cheltenham, United Kingdom.

Zinsstag J., Schelling E., Waltner-Toews D., Whittaker M., Tanner M. 2015. *One Health: The Theory and Practice of Integrated Health Approaches.* CABI.

Zinsstag J., Utzinger J., Probst-Hensch N., Shan L., Zhou X.-N. 2020. Towards integrated surveillance-response systems for the prevention of future pandemics. *Infectious Diseases of Poverty,* 9(1), 140. https://doi.org/10.1186/s40249-020-00757-5

1

Foundations and recent history

One Health's foundations trace back to ancient integrative thinking, spanning from the philosophies of Hippocrates to the wisdom of Indigenous knowledge systems. These roots have been enriched by pivotal milestones, such as the One World, One Health Conference in Manhattan in 2004 and the One Planet, One Health, One Future Conference in 2019 in Berlin, which shaped One Health principles and broadened its scope. This section explores the theoretical and historical underpinnings of One Health, highlighting its evolution through key pandemics, from the plague to COVID-19, and its relevance to contemporary zoonotic threats.

Concepts through time: a brief history from Hippocrates to COVID-19

François Roger

A fundamental principle of the One Health concept is that all living organisms —including humans, animals, plants and the planet itself—are interconnected and their existence is mutually beneficial.

In the fifth century BCE, Hippocrates, known as the "father of medicine", acknowledged the impact of the environment on human health. This view has evolved throughout the years in reaction to growing health concerns (Figure 1). Veterinarian Claude Bourgelat, who established the world's first veterinary school in Lyon, France, in the mid-1700s and is considered the "father of veterinary medicine", was a prominent figure in this school of thought. In his pioneering work, Bourgelat linked human and animal health. Modern veterinary medicine owes a great debt to his seminal work, which highlighted the need for a more comprehensive approach to health. Epidemiologist Calvin Schwabe revived this notion in the twentieth century, calling it "One Medicine". To better understand and manage disease, Schwabe called for a more holistic approach to public health that considers the interdependence of all forms of life (see Box).

The spread of zoonotic diseases in recent decades, such as the H5N1 avian flu, SARS, MERS-CoV, Ebola and, most recently, COVID-19, has increased the importance of the One Health approach. Due to the risks associated with altered interactions between humans, animals and the environment, these health crises have confirmed the need for a collaborative approach to fight them. Today, One Health is recognized and promoted by international organizations including the World Health Organization (WHO), World Organisation for Animal Health (WOAH), Food and Agricultural Organization (FAO) and United Nations Environment Programme (UNEP). In an increasingly globalized world, interdisciplinary and multi-sectorial efforts should be implemented to tackle public health issues, with the ultimate goal of improving overall health and well-being.

NATURE AND (ONE) HEALTH

Although "One Health" is a somewhat recent concept, it draws upon the ideas of many thinkers who have explored the interconnectedness of nature and humans throughout history. **Hippocrates** (c. 460–370 BCE) examined the influence of the environment on human health; **Aristotle** (384–322 BCE) later explored the interconnectedness of living beings through his concept of "telos", which suggested a natural purpose for all life forms. **Aldo Leopold** (1887–1948), a key figure in ecology, proposed a land ethic that highlighted the need to respect and preserve ecosystems, an idea echoed by the One Health philosophy of today. Other important thinkers may have indirectly influenced the development of holistic approaches: **Hans Jonas** (1903–1993) discussed the ethical obligations humans have towards nature and future generations, emphasizing environmental stewardship, in his work *The Imperative of Responsibility*. With his theory of complexity, **Edgar Morin** (born 1921) advocated for an interdisciplinary and systemic approach to addressing global challenges. The work of **Gregory Bateson** (1904–1980) in systems theory underscored the importance of understanding biological, social and ecological systems as interconnected entities. More contemporary philosophers, like **Michel Serres** (1930–2019) in *The Natural Contract*, have called for redefining the relationship between humans and nature, recognizing that human survival depends on the health of the planet.

References

Ancheta J., Fadaak R., Anholt R.M., Julien D., Barkema H.W., Leslie M. 2021a. The origins and lineage of One Health, Part I. *The Canadian Veterinary Journal*, 62(8), 883–885.

Ancheta J., Fadaak R., Anholt R.M., Julien D., Barkema H.W., Leslie M. 2021b. The origins and lineage of One Health, Part II. *The Canadian Veterinary Journal*, 62(10), 1131–1133

Zinsstag J., Schelling E., Waltner-Toews D., Tanner M. 2011 From "one medicine" to "one health" and systemic approaches to health and well-being. Preventing Veterinary Medecine, 101(3-4), 148–156. https://doi.org/10.1016/j.prevetmed.2010.07.003

Epidemiological Transitions

Agricultural Revolution
Settlement, population density, agriculture and livestock, emergence of zoonoses

Industrial Revolution
Improved nutrition, antibiotics, etc. Development of cancer, diabetes, cardiovascular diseases.

Since the mid-20th century
Antimicrobial resistance, emerging diseases, new pandemics

Hippocrates (460–370 BCE)
Treatise on Airs, Waters, and Places. Health depends on a clean environment.

Claude Bourgelat (1712–1779)
Founded the first veterinary school in the world in **1762**, highlighted the importance of animal health for public health, and encouraged an interdisciplinary approach to solving public health problems.

William Osler (1849–1919)
Canadian physician who emphasized the importance of studying animal diseases for human health in the early 20th century.

Rudolf Virchow (1821–1902)
1855 Coined the term "zoonosis".

1855 · 1965 · 2004 · 2007 · 2008 · 2009 · 2010 · 2011 · 2018 · 2019

1965 Calvin Schwabe coins "One Medicine" in his textbook *Veterinary Medicine and Human Health.*

2004 The Wildlife Conservation Society conference releases the 12th Manhattan Principles leading to the concept of One World, One Health.

2007 Contributing to One World, One Health. A strategic framework for reducing risks of infectious diseases at the animal-human-ecosystems interface by WHO, OIE and FAO.

2008 International Ministerial Conference in New Delhi recommends One Health approach to 111 countries and 29 international organisations attending.

2009 One Health office established at the CDC.

2009 USAID establishes the Emerging Pandemic Threats Program using a One health approach.

2010 The EU reaffirms commitment to using One Health in the "Global Response to the Avian Influenza Crisis" report.

2010 The Hanoi Declaration recommending One Health is adopted by 71 countries after the H1N1 pandemic and avian influenza outbreak.

2013 The first international One Health Congress in Melbourne, Australia.

2018 One Health European Joint Programme (OHEJP) launched.

2019 First OHEJP Annual Scientific Meeting in Dublin, Ireland.

CALVIN SCHWABE (1927–2006)

An epidemiologist and veterinary parasitologist, described and promoted the concept of "One Medicine". He proposed a unified human and veterinary approach to zoonoses in his 1964 work, Veterinary Medicine and Human Health. "There is no paradigm difference between human medicine and veterinary medicine. Both sciences share a common body of knowledge in anatomy, physiology, pathology, and the origins of diseases across all species."

Figure 1. Timeline. Based on Brown H.L., Passey J.L., Getino M., Pursley I., Basu P. *et al.* 2020. The One Health European Joint Programme (OHEJP), 2018–2022: an exemplary One Health initiative. *Journal of Medical Microbiology*, 69(8), 1037-1039. https://doi.org/10.1099/jmm.0.001228

From Manhattan to Berlin: how the principles of One Health have evolved

Lucy Keatts,
Mathieu Pruvot

In 2004, the Wildlife Conservation Society (WCS) hosted a symposium on global health challenges linked to the crossover of human, animal and environmental health. The resulting recommendations, set out in the "Manhattan Principles", underpinned the multisectoral, interdisciplinary and collaborative "One World, One Health" approach.

Many global entities have since adopted this approach, now known simply as One Health. Over the next decade, the One Health movement was often too narrowly focused on a few topics, such as emerging diseases and zoonoses, with a heavy emphasis on human health and agricultural impacts. While these are important issues, such a limited One Health approach cannot deliver on its full global potential.[1]

WCS recognized this limitation and the urgent need for a coordinated, more extensive One Health approach. It partnered with the Climate and Environmental Foreign Policy Division at the German Federal Foreign Office in 2019 to organize the One Planet, One Health, One Future conference, which marked a clear shift from trying to better understand the underlying principles of One Health to improving real-world implementation. With more scientists, policymakers and practitioners from the Global South leading the agenda, the 2019 participants reflected a much broader range of stakeholders and included experts from government, academia and civil society representing policy, sociology, philosophy, economics, ecology, and human and veterinary medicine. These experts produced the Berlin Principles, which updated the Manhattan Principles and recommended integrating ecosystem health into One Health along with pressing issues such as non-communicable diseases, climate change and antimicrobial resistance. They emphasized the need for human, animal and environmental health sectors to be bolder, work together and strengthen the key role of biodiversity and One Health across all policies.

The Berlin Principles warned of the dire consequences of ignoring the interconnected health issues. Just months after the 2019 meeting, the global community got a devastating wake-up call: the COVID-19 pandemic. Unless we fully integrate biodiversity conservation and ecosystem health into future One Health approaches, we dramatically increase risks for another calamitous pandemic of zoonotic origin. Intact and functioning environments must be considered in an expanded social-ecological model of health going forward, to reduce pandemic risks as well as deleterious health impacts from climate change and biodiversity loss. There remains a need for the health sector to better embrace environmental integration and related One Health actions in international health regulations and national action plans. While governments, multilateral organizations and donors have all voiced a wide-ranging commitment to One Health since the COVID-19 pandemic, operationalization remains elusive and often challenging.

1. oneworldonehealth.wcs.org

References

Building Interdisciplinary Bridges to Health in a "Globalized World": https://oneworldonehealth.wcs.org/About-Us/Mission/The-Manhattan-Principles.aspx

The Berlin Principles: https://oxfordinberlin.eu/the-berlin-principles-on-one-health#:~:text=The%20Berlin%20Principles%20(below)%20are,climate%20change%20and%20antimicrobial%20resistance.&text=intrinsically%20connected%20and%20profoundly%20influenced%20by%20human%20activities

The Manhattan Principles: https://oneworldonehealth.wcs.org/About-Us/Mission/The-Manhattan-Principles.aspx

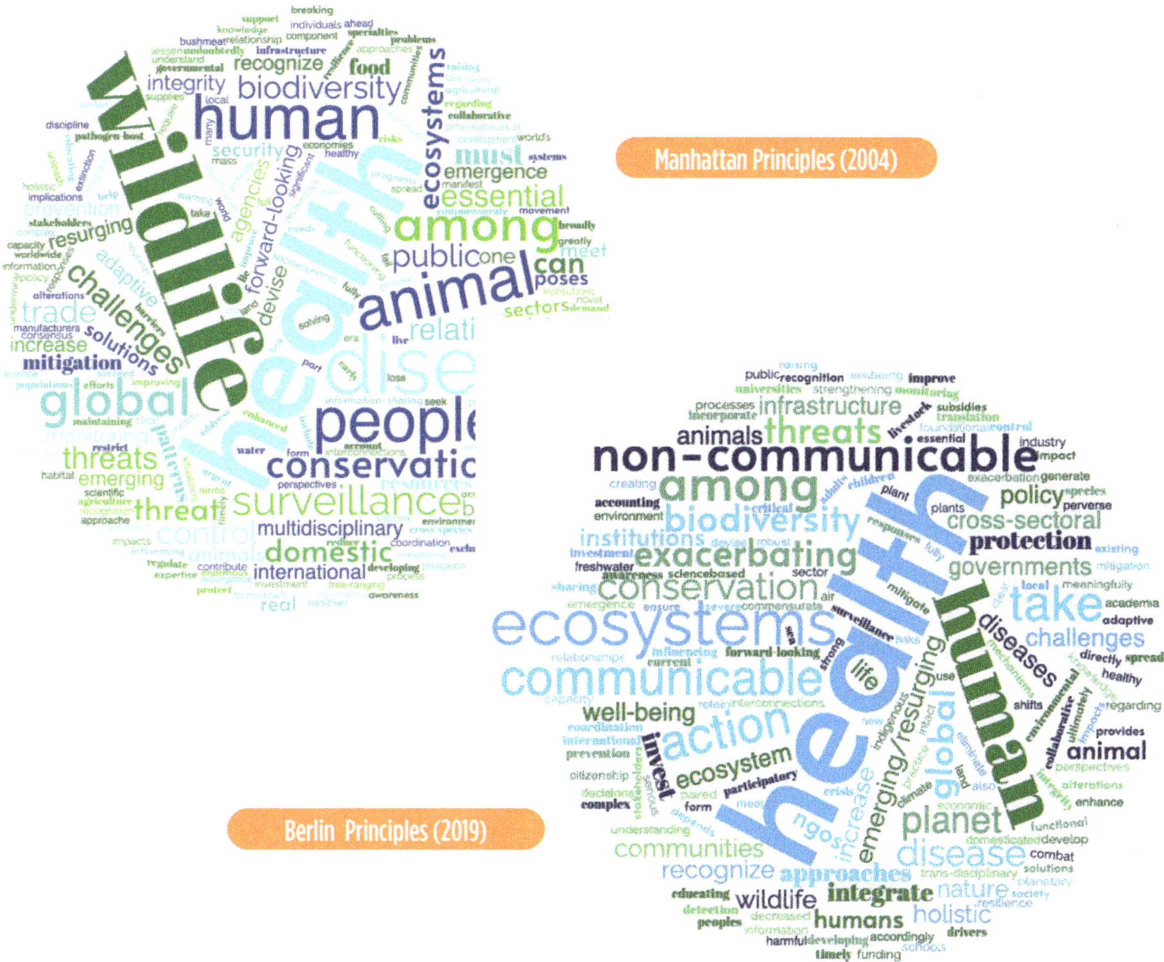

Figure 1. Word clouds from the Manhattan Principles (1) and Berlin Principles (2).

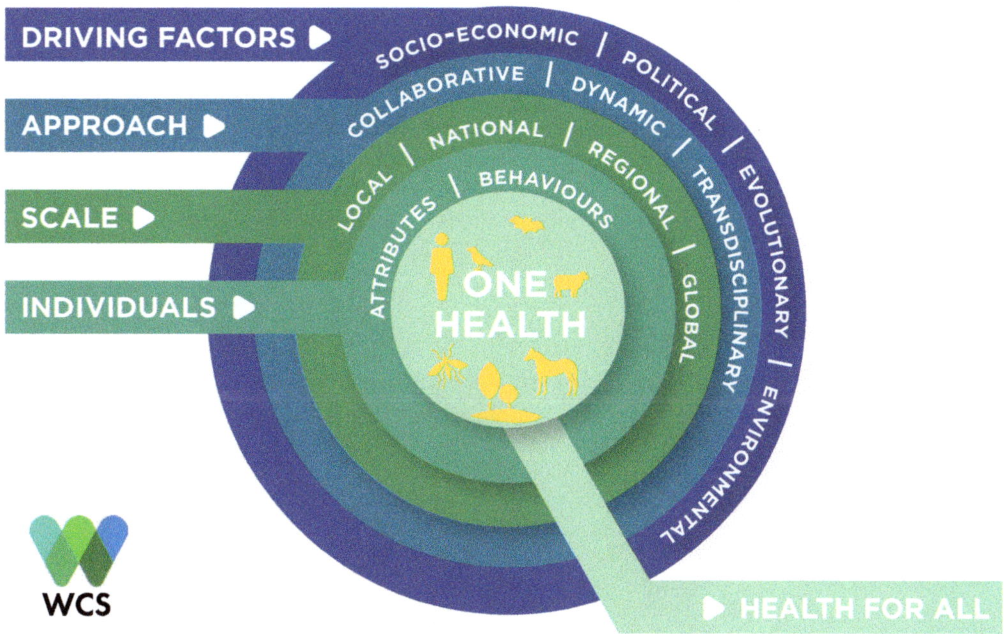

Figure 2. Wildlife Conservation Society (WCS) One Health implementation approaches.

Indigenous people and (One) Health

François Roger

Indigenous community health traditions, which recognize the interdependence of all living things, laid the groundwork for a kind of integrated approach to health. These practices span generations and embody a holistic view of health that includes environmental preservation and caring for one's mental, spiritual, emotional and psychological health.

Living in harmony has given Indigenous people a unique perspective on how the environment affects human well-being. For better ecosystem management and disease prevention, Indigenous knowledge must be incorporated into global health policies, and this traditional paradigm can teach us a lot about how to do just that.

According to Dr. Carol Zavaleta-Cortijo, a Peruvian medical doctor and socio-environmental scientist, Indigenous peoples in Peru, such as the Shawi, were able to endure the COVID-19 pandemic with remarkable strength. These communities successfully navigated the crisis by utilizing traditional knowledge and practising isolation. This approach is in line with the principles of One Health since it draws on Indigenous knowledge to address global health issues and stresses the interdependence of human, animal and environmental health. More inclusive and resilient public health approaches can be achieved by including Indigenous perspectives in broader health strategies.

Indigenous Canadian viewpoints that value the inextricable link between humans and their natural surroundings are consistent with One Health's multidisciplinary approaches. Nevertheless, there is a clear lack of comprehensive Indigenous knowledge integration in One Health studies. Developing culturally appropriate health interventions that blend traditional ways of knowing with modern health practices is essential for addressing climate change, zoonoses, and the sociocultural relationships between humans and animals through a One Health lens. This requires genuine engagement with Indigenous communities.

When thinking about human health, the One Health paradigm must not ignore the impact of social and political factors. Disparities in human health are, in fact, the product of a complex network of political and economic factors. Among other things, this means that other knowledge systems have not been sufficiently integrated and that One Health frameworks fail to adequately acknowledge or represent Indigenous knowledge.

References

Blume A.W. 2024. One Health models are lacking an Indigenous perspective. *BMJ*, 386, q2015. https://doi.org/10.1136/bmj.q2015

Hillier S.A., Taleb A., Chaccour E., Aenishaenslin C. 2021. Examining the concept of *One Health* for indigenous communities: A systematic review. *One Health*, 12, 100248. https://doi.org/10.1016/j.onehlt.2021.100248

Pollowitz M., Allick C., Campbell K.B., Ellison N.L.K., Perez-Aguilar G. *et al.* 2024. One Health, many perspectives: Exploring Indigenous and Western epistemologies. *CABI One Health*, 3(1), 0015. https://doi.org/10.1079/cabionehealth.2024.0015

Riley T., Anderson N.E., Lovett R., Meredith A., Cumming B., Thandrayen J. 2021. One Health in indigenous communities: A critical review of the evidence. *International Journal of Environmental Research and Public Health*, 18(21), 11303. https://doi.org/10.3390/ijerph182111303

Sudlovenick E., Jenkins E.J., Loseto L.L. 2024. Comparative review of One Health and Indigenous approaches to wildlife research in Inuit Nunangat. *One Health*, 18, 100620. https://doi.org/10.1016/j.onehlt.2024.100846

Zavaleta-Cortijo C. What can indigenous people teach the world About One Health? https://onehealthtrust.org/news-media/podcasts/what-can-indigenous-people-teach-the-world-about-one-health/

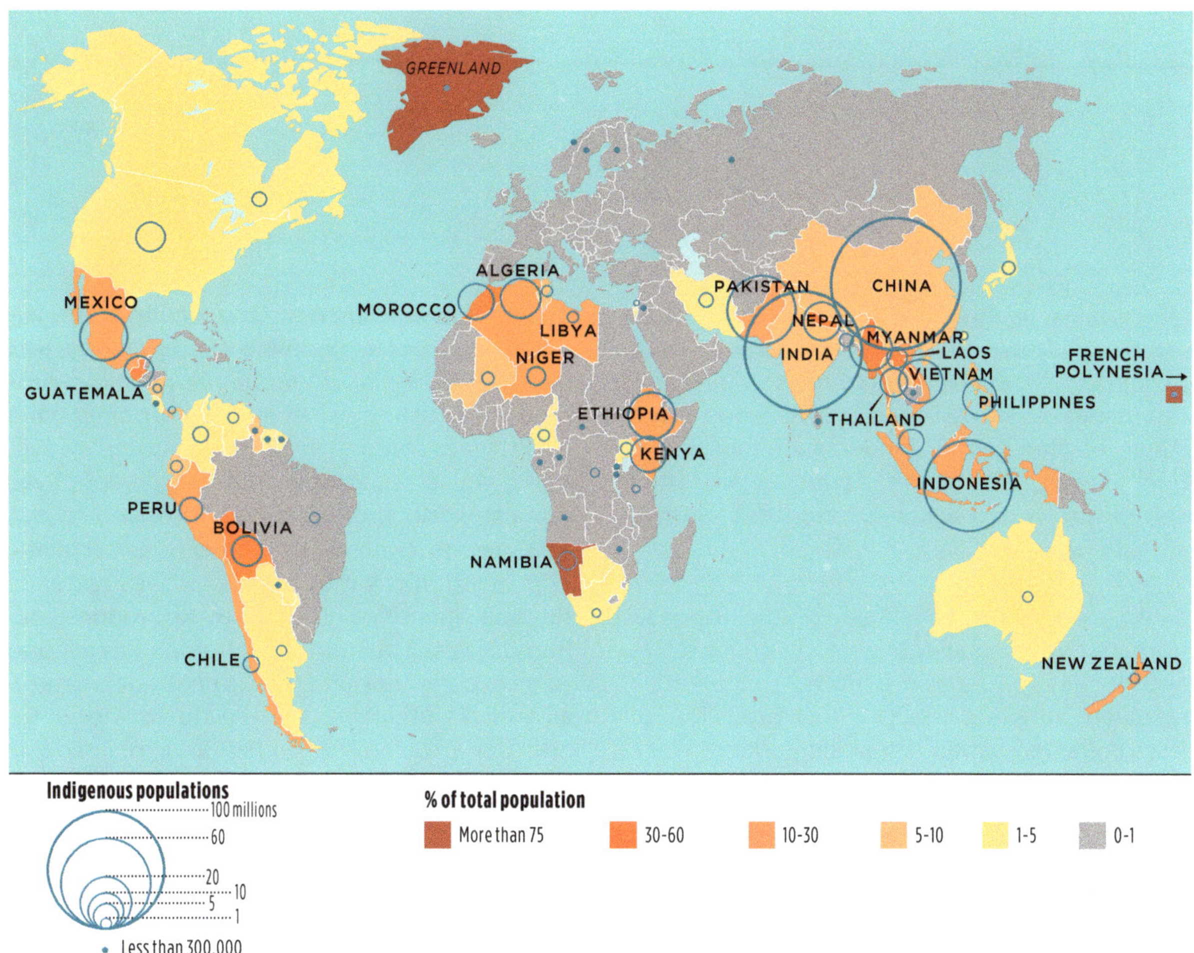

Figure 1. Global map of lands managed and/or controlled by Indigenous peoples. Source: IWGIA (https://iwgia.org/en/) and World Bank (https://www.worldbank.org/en/topic/indigenouspeoples).

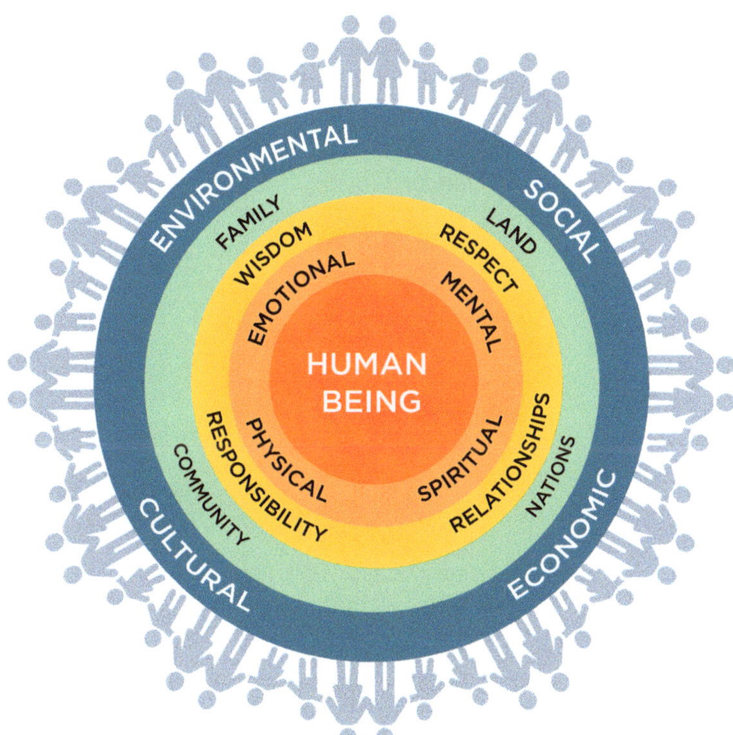

Figure 2. First Nations perspective on health and wellness.

This illustration was developed by the First Nations Health Authority (FNHA) of British Columbia, Canada. It aims to visually represent FNHA's vision: "Healthy, self-determining, and vibrant First Nations children, families, and communities in British Columbia". This figure illustrates a holistic approach to health. This representation is designed as a tool for FNHA and First Nations communities to create a shared understanding of holistic well-being. It is adaptable and can be freely customized to reflect the needs and perspectives of each community. Source: www.fnha.ca.

Epistemology of One Health: bridging disciplines and integrating knowledge

Nicolas Antoine-Moussiaux, Stéphane Leyens, Séverine Thys, Simon Rüegg, Aurélie Binot

Complex health problems require collaboration between different scientific disciplines. But deep divergences in the way these disciplines conceive and value knowledge (their epistemology) hinder such collaboration. Projects combining disciplines are often split into work packages and thus remain siloed (Figure 1).

An often-cited obstacle is the opposition between qualitative and quantitative approaches to research (Figure 2). Qualitative research entails subjective interpretation of data collected within a specific context, while quantitative approaches require representative samples to generalize results at the population level. However, necessary dialogue is impeded by myriad practices, judgement biases and epistemologies. Social sciences often (although not exclusively) rely on a constructivist, inductive and interpretative approach. Biomedical research establishes experimental facts to test hypotheses within controlled conditions to decontextualize knowledge, but it also gains knowledge from epidemiological studies. Meanwhile, modelling calls for mathematical translations of reality to create the object of analysis. As a result, this highly diverse landscape is one with wide divergences, exacerbated by technical jargon that further hampers collaborations.

The epistemology of One Health acknowledges that each discipline sheds unique and valuable light on a complex reality and promotes the dialogue between viewpoints. One Health is based on the theory of complex systems, which recognizes multiple perspectives on real-world problems and the need to act and decide even when there is uncertainty. This need for various perspectives extends beyond scientifically validated knowledge and harnesses the full range of human knowledge (e.g. experiential and traditional knowledge), as well as other ways to relate to the world and our problems (e.g. the arts, philosophy, spirituality). This approach, called transdisciplinary research, entails the broad participation of stakeholders and negotiation between divergent values, engaging political and intercultural dialogue (Figure 1). One Health is thus built on strong communication, translation and mediation activities (Figures 3 and 4). For example, One Health practitioners must often bridge and balance anthropocentric and biocentric ethics, where collaborations sometimes hinge on how terms—such as "nature" and "environment"—translate our worldviews. From a systems thinking standpoint, they must also continuously go back and forth between holism and reductionism. Indeed, the need to see the "big picture" does not eliminate the need to identify detailed mechanisms.

References

Antoine-Moussiaux N., Janssens de Bisthoven L., Leyens S., Assmuth T., Keune H. *et al.* 2019. The good, the bad and the ugly: Framing debates on nature in a One Health community. *Sustainability Science,* 14(6), 1729–1738. https://doi.org/10.1007/s11625-019-00674-z

Antoine-Moussiaux N., Leyens S. 2023. Harnessing concepts for sustainability: A pledge for a practice. *Sustainability Science*, 18, 2441–2451. https://doi.org/10.1007/s11625-023-01375-4

Cash D.W., Clark W.C., Alcock F., Dickson N.M., Eckley N. *et al.* 2003. Knowledge systems for sustainable development. *PNAS*, 100(14), 8086–8091. https://doi.org/10.1073/pnas.1231332100

Duboz R., Echaubard P., Promburom P., Kilvington M., Ross H. *et al.* 2018. Systems thinking in practice: Participatory modeling as a foundation for integrated approaches to health. *Frontiers in Veterinary Science,* 5, 303. https://doi.org/10.3389/fvets.2018.00303

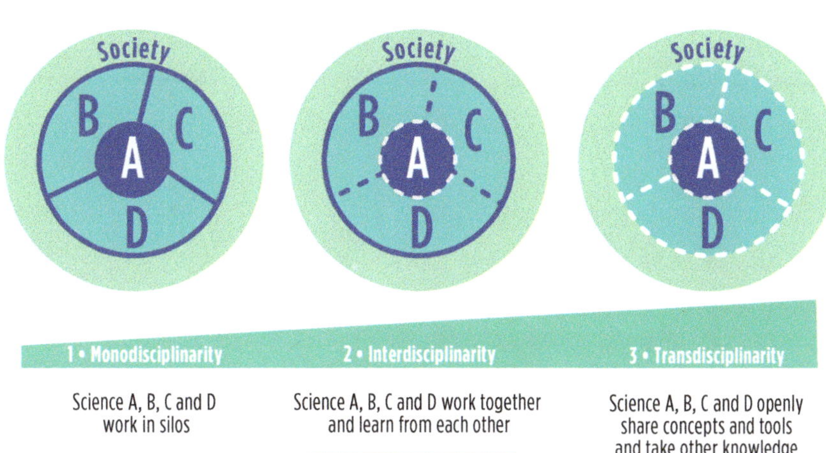

Figure 1. Mono-, Inter- and Transdisciplinarity represent distinct forms of work along a continuum of openness to a variety of knowledge forms. Importantly, all three forms are needed, in an iterative and adaptive way throughout the solving of complex health problems.

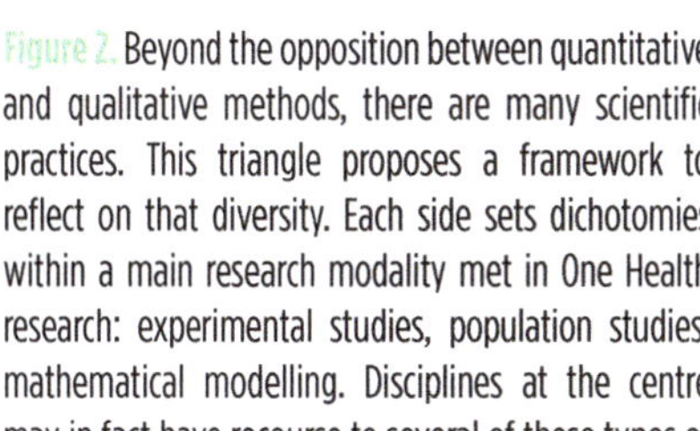

Figure 2. Beyond the opposition between quantitative and qualitative methods, there are many scientific practices. This triangle proposes a framework to reflect on that diversity. Each side sets dichotomies within a main research modality met in One Health research: experimental studies, population studies, mathematical modelling. Disciplines at the centre may in fact have recourse to several of these types of

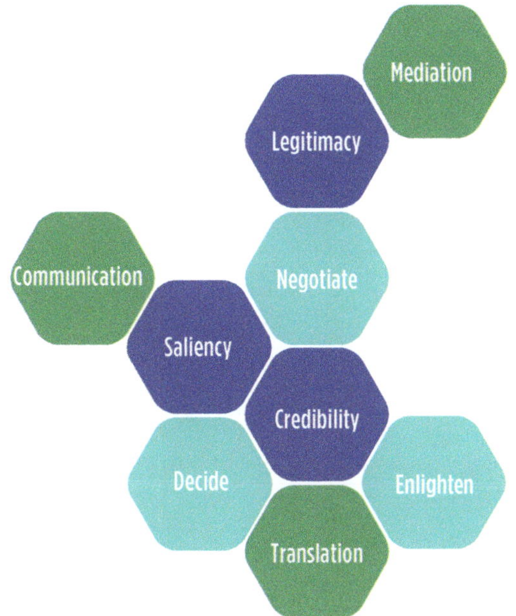

Figure 3. Knowledge in transdisciplinary research may be valued based on three criteria (dark blue hexagons). Hexagons in the figure are arranged to connect criteria to their goals (verbs in light blue hexagons) and to the needed practices (green hexagons) (adapted from Cash *et al.* 2003).

Figure 4. Knowledge sharing is facilitated by objects that are meaningful to all collaborators and that all can manipulate. These so-called "boundary objects" are key to One Health implementation and may be of many different types (e.g. models, species, games, art). Since One Health aims to manage health, health management concepts may themselves act as boundary objects. Source: AI-produced illustration (DALL-E).

The historical context of the human plague: from miasmas to One Health

François Roger,
Marie-Marie-Olive

Human plague, caused by the bacterium *Yersinia pestis*, is an example of an ancient zoonotic disease. Pandemics occurred from the tenth century (Figure 1). Ancient European and Chinese ideas about disease transmission were sketchy at best and lacked the precise data on germs and vector transfer that modern science is able to provide. For example, scientists demonstrated in the nineteenth century that fleas spread the plague from infected animals to humans after the disease initially appeared in Central Asian rodent populations (Figure 2).

China had always prioritized public health in its response to epidemics. The Chinese may not have understood the plague specifically, but they were familiar with the general principles of disease transmission. Traditional philosophical ideas, such as the harmony of yin and yang and qi (meaning "life force energy"), had a significant impact on their health-care practices. Despite not knowing how the plague spreads from animals, they instituted quarantines and other health measures to limit infections more generally.

A third of the European population perished during the second plague pandemic. The epidemic was blamed on miasmas, or polluted air emissions, or seen as a divine punishment. Since transmission by vectors and animal reservoirs remained unclear, the miasmatic idea prevailed. Neither the significance of the environment nor the possibility of transmission from animals was considered during the plague's early stages.

Human plague cases still occur around the world (Figure 3). The current understanding of plague dynamics shows how the human-animal-environment interfaces are essential for disease development and transmission. Environmental variations, such as changes in natural habitats, biodiversity and climate, influence rodent populations and the behaviours of vectors like fleas, which in turn could increase the risk of transmission to humans. Thus, a One Health strategy is necessary to manage and prevent the plague, and includes monitoring rodent populations, studying disease carriers and educating communities. Understanding the environmental factors that affect transmission of the plague is also crucial for preventing future epidemics of the same type.

References

Audoin-Rouzeau F. 2003. *Les Chemins de la peste. Le rat, la puce et l'homme*. Presses universitaires de Rennes. https://doi.org/10.4000/books.pur.8382

Benedict C. 1996. *Bubonic Plague in Nineteenth-Century China*. Stanford University Press.

D'Ortenzio E., Lemaître N., Brouat C., Loubet P., Sebbane F. *et al.* 2018. Plague: Bridging gaps towards better disease control. *Médecine et Maladies Infectieuses*, 48(5), 307–317. https://doi.org/10.1016/j.medmal.2018.04.393

Lachenal G., Thomas G. 2023. *Atlas historique des épidémies*. Autrement.

Mollaret H.H. 1989. Le cas de la peste. *Annales de Démographie Historique*, 1989(1), 131–138. https://doi.org/10.3406/adh.1989.1733

Nicoud M. 2023. À l'épreuve de la peste : Médecins et savoirs médicaux face à la pandémie (XIVe-XVe siècles). *Annales Histoire, Sciences Sociales*, 78(3), 505–541. https://doi.org/10.1017/ahss.2023.105

Sikora M., Canteri E., Fernandez-Guerra A., Oskolkov N., Ågren R. *et al.* 2025. The spatiotemporal distribution of human pathogens in ancient Eurasia. *Nature*, 643(8021), 1011–1019. https://doi.org/10.1038/s41586-025-09192-8

Slack P. 2021. *Plague: A Very Short Introduction*. Oxford University Press.

First pandemic: 6th–8th centuries	Second pandemic: 14th–18th centuries	Third pandemic: 1855–mid-20th
The Plague of Justinian began in 541 CE, primarily affecting the Byzantine Empire and Mediterranean port cities. It killed millions of people before subsiding in the 700s.	The second plague pandemic started with the Black Death in 1347. This pandemic overwhelmed Europe and Asia (and possibly Africa), resulting in the death of 75 to 200 million people. Recurrences of this plague were reported until the 18th century.	Originating in China before spreading to India and then other parts of the world, the third plague pandemic stimulated advances in scientific understanding (reservoirs and vectors). It led to significant casualties, particularly in India and China, until its decline in the mid-20th century.

Figure 1. Timeline covering the three major human plague pandemics. Two major advances were made during the third plague pandemic: first, in 1894, when Alexandre Yersin, considered by some to be a forefather of One Health, discovered the pathogen responsible for epidemics; and second, in 1898, when Paul-Louis Simond discovered the role of the rat flea, *Xenopsylla cheopis*, as a vector of *Yersinia pestis*.

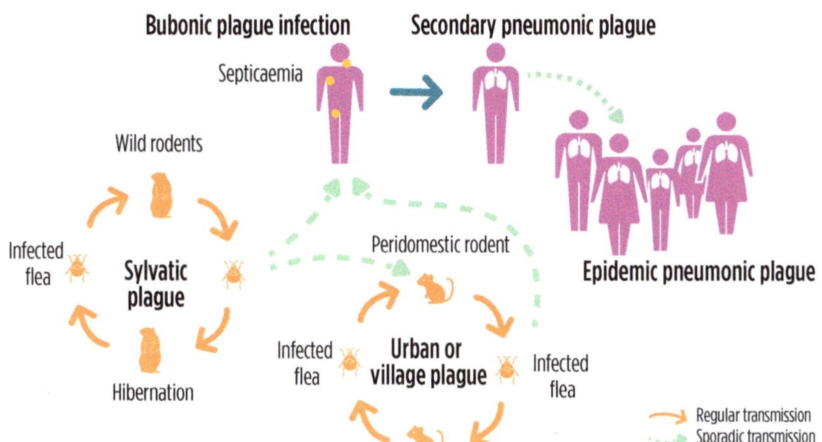

Figure 2. Transmission cycle.
Primary hosts are black and brown rats. Sylvatic plague circulates in wild rodents via fleas, which can hibernate, sustaining the infection cycle. Urban plague spreads to domestic rodents, transmitting to nearby humans through fleas, leading to outbreaks. If human bubonic plague evolves, it can become pneumonic plague, spreading directly between humans via respiratory droplets. From: Lachenal and Thomas 2023.

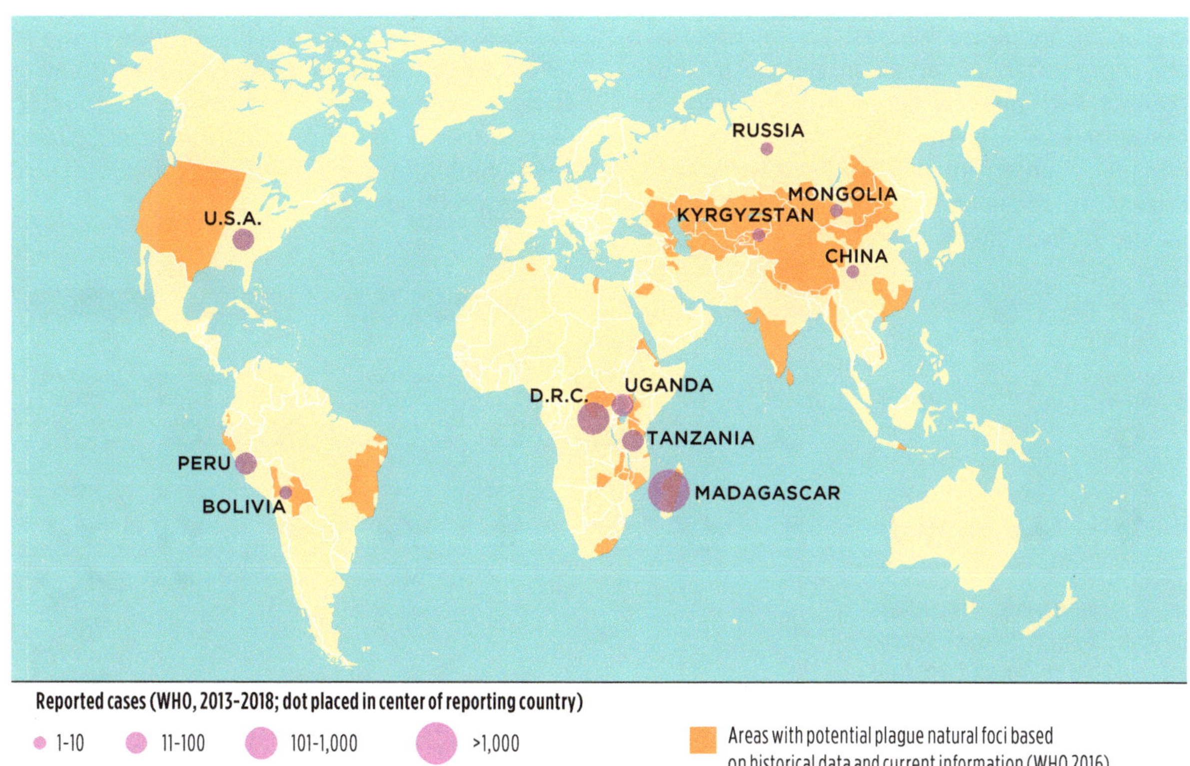

Figure 3. Plague remains endemic in parts of South and North America, Africa, and Central Asia, where rodent reservoirs persist. Outbreaks still occur, notably in Madagascar, Peru, and the D.R. of Congo. Sources: WHO and CDC.

From past pandemics to recent threats: the ongoing zoonotic risks of influenza

Claire Hautefeuille,
Marisa Peyre

Prior to the emergence of coronavirus diseases, the four major pandemics of the twentieth century were all caused by animal influenza viruses. The Spanish influenza of 1918, which killed between 50 and 100 million people, was linked to an H1N1 serotype virus that emerged as a result of an adaptation of a low pathogenic avian influenza virus to mammals. The Asian influenza of 1957 (H2N2), which caused 70,000 deaths, and the Hong Kong influenza of 1968 (H3N2), which caused 56,000 deaths, both emerged as a result of recombination between human and low pathogenic avian influenza viruses. Finally, the H1N1pdm (2009) killed 200,000 people and developed from pig viruses (Figure 1).

Increased surveillance in humans and animals and improved sequencing made it possible to identify human infection with swine and avian influenza viruses. These infections are more often observed in people in close contact with animals (e.g. people working on farms or in live animal markets or slaughterhouses). Human infection with animal influenza viruses mainly causes clinical signs similar to the human flu and is often not detected. That said, since 1997 over 900 cases of human infection were reported for H5N1 and over 1,500 for H7N9 avian influenza viruses. Wild aquatic migratory birds from the Anatidae family are known to play a role in the emergence of new strains and the global spread of avian influenza viruses. Climate change has an impact on migratory patterns and thus on the occurrence of avian influenza outbreaks. Since 2021, the emergence of a new H5N1 strain caused a large number of cases in various wild bird species (including non-Anatidae birds) on all continents - even in South America, which was previously relatively free of avian influenza. This H5N1 strain also caused many cases in mammals, with viral transmission observed between mammals. In domestic animals, the most critical situation was the high number of outbreaks in cattle in the United States, with more than 130 infected dairy herds in 12 states identified in June 2024. A wide range of wild mammal species were also affected, with transmission between mammals suspected (e.g. within a sea lion colony in Peru). These cases highlight the potential zoonotic risk of this strain (Figure 2).

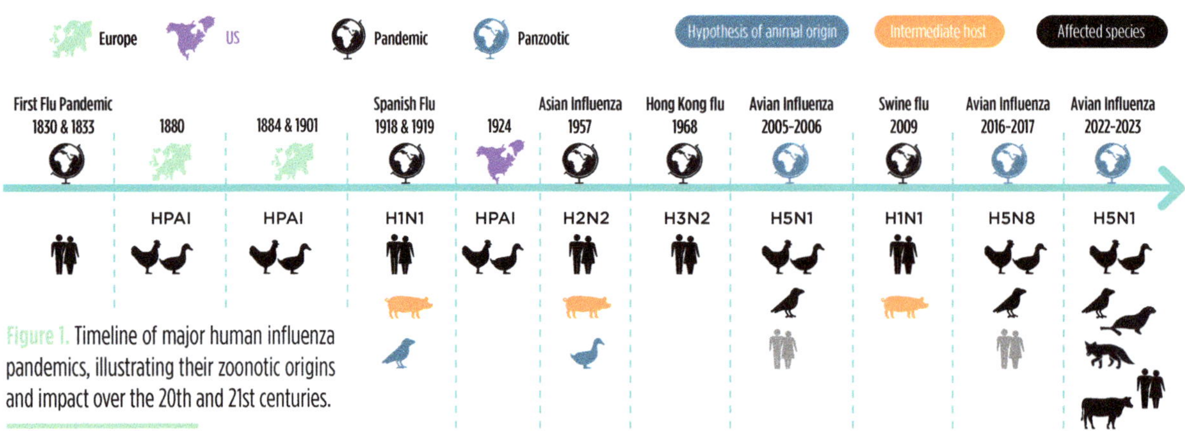

Figure 1. Timeline of major human influenza pandemics, illustrating their zoonotic origins and impact over the 20th and 21st centuries.

References

Adlhoch C., Fusaro A., Gonzales J.L., Kuiken T., Marangon S. *et al.* 2023. Avian influenza overview December 2022 - March 2023. *EFSA Journal*, 21(3), e07917. https://doi.org/10.2903/j.efsa.2023.7917

Alexakis L., Fusaro A., Kuiken T., Mirinavičiūtė G., Ståhl K. *et al.* 2024. Avian influenza overview March - June 2024. EFSA Journal, 22(7), e8930. https://doi.org/10.2903/j.efsa.2024.8930

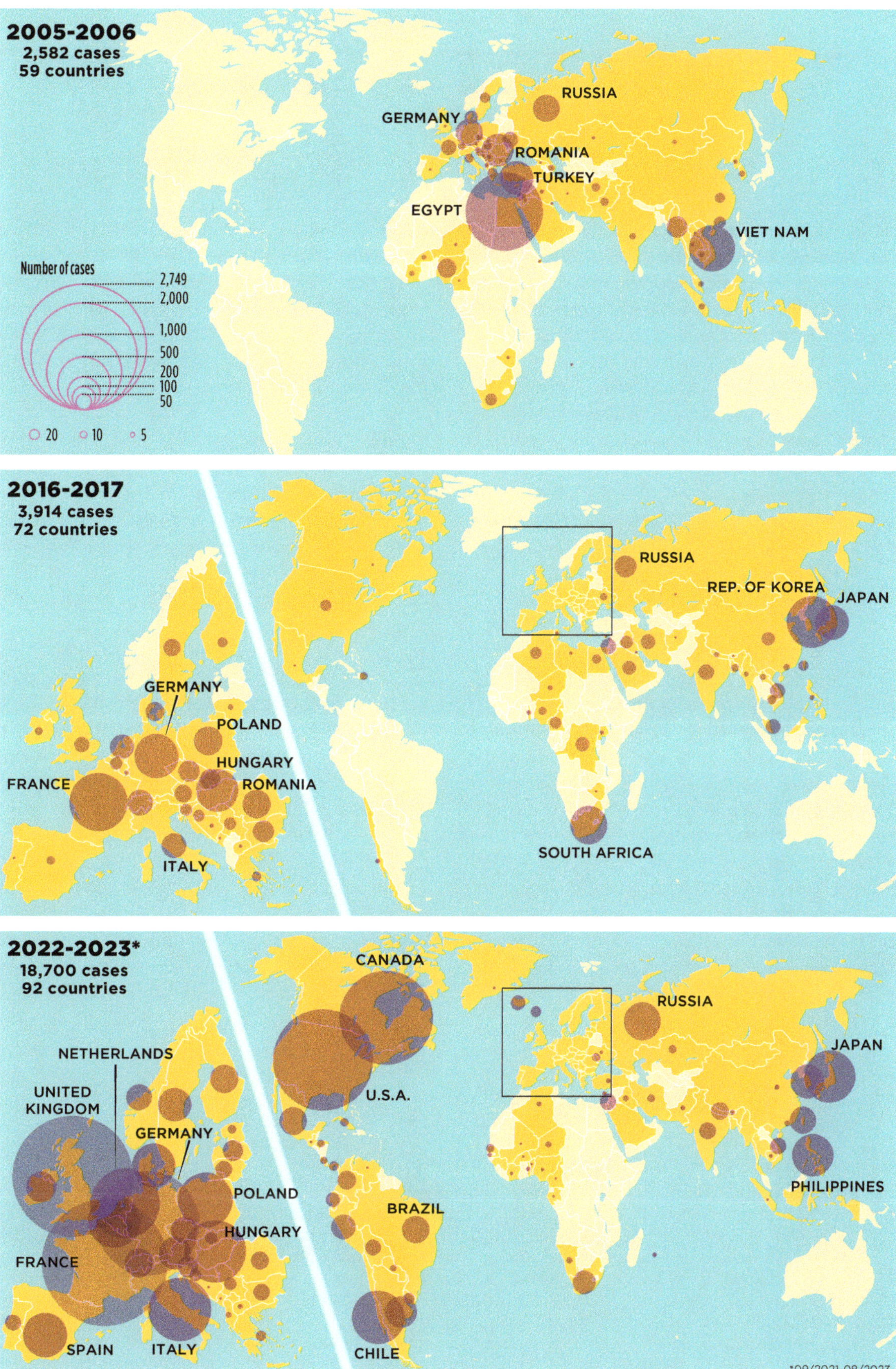

Figure 2. Global distribution of recent avian and swine influenza outbreaks, highlighting their spread and zoonotic risk.

Unveiling filovirus reservoirs in Africa and Asia: the necessary role of One Health

François Roger,
Mathieu Bourgarel

Filoviruses, which include the Ebola and Marburg viruses, represent significant global public health challenges due to their high case fatality rates and potential to cause epidemics. Despite extensive research, their natural history remains poorly understood. Natural reservoirs for these viruses are elusive, although fruit bats are the primary suspects for virus maintenance. The ecological, socio-economic and environmental factors that drive spillover events—when these viruses are transmitted to humans—are still generally unclear (Figure 1).

Identifying the natural reservoirs of filoviruses is a multidimensional challenge. Suspected reservoirs like bats often harbour asymptomatic infections, complicating detection and understanding of viral maintenance mechanisms. While the Marburg virus has been isolated from *Rousettus aegyptiacus,* only fragments of Ebola virus genomes have been found in fruit bats, despite the analysis of hundreds of thousands of wildlife samples across Africa. Furthermore, African tropical rainforests, where these viruses are believed to circulate, host exceptional biodiversity, with many species potentially acting as reservoirs or vectors. Filoviruses likely circulate among multiple reservoirs and intermediary hosts, including bats, non-human primates and other terrestrial mammals. The genetic diversity of filoviruses also suggests that strains vary geographically, requiring context-specific surveillance in both Africa and Asia.

Spillover events are rare and intermittent, often occurring in remote areas with limited access to healthcare and research infrastructure, which hinders detailed investigations. These transmission dynamics are further exacerbated by socioecological factors such as habitat destruction and hunting, which amplify the risk of zoonotic transmission.

Identifying filovirus reservoirs remains critical for preventing and controlling these deadly diseases (Figure 2). The One Health approach offers an integrated and effective framework to address these challenges. This strategy emphasizes integrated surveillance, interdisciplinary collaboration among virologists, ecologists, epidemiologists and anthropologists, and community engagement to strengthen disease reporting and prevention. Advanced molecular research, including next-generation sequencing and artificial intelligence, can further aid in identifying reservoirs and intermediary hosts.

References

Caron A., Bourgarel M., Cappelle J., Liégeois F., De Nys H.M., Roger F. 2018. Ebola virus maintenance: If not (only) bats, what else? *Viruses*, 10(10), 549. https://doi.org/10.3390/v10100549

De Nys H.M., Kingebeni P.M., Keita A.K., Butel C., Thaurignac G. et al. 2018. Survey of Ebola viruses in frugivorous and insectivorous bats in Guinea, Cameroon, and the Democratic Republic of the Congo, 2015-2017. *Emerging Infectious Diseases*, 24(12), 2228-2240. https://doi.org/10.3201/eid2412.180740

Lee-Cruz L., Lenormand M., Cappelle J., Caron A., De Nys H. et al. 2021. Mapping of Ebola virus spillover: Suitability and seasonal variability at the landscape scale. *PLOS Neglected Tropical Diseases,* 15(9), e0009683. https://doi.org/10.1371/journal.pntd.0009683

Olival K.J., Hayman D.T.S. 2024. Filoviruses in bats: Current knowledge and future directions. *Viruses, 6,* 1759-1788. https://doi.org/10.3390/v6041759

Semancik C.S., Cooper C.L., Postler T.S., Price M., Yun H. et al. 2024. Prevalence of human filovirus infections in sub-Saharan Africa: A systematic review and meta-analysis protocol. *Systematic Reviews*, 13(218). https://doi.org/10.1186/s13643-024-02626-w

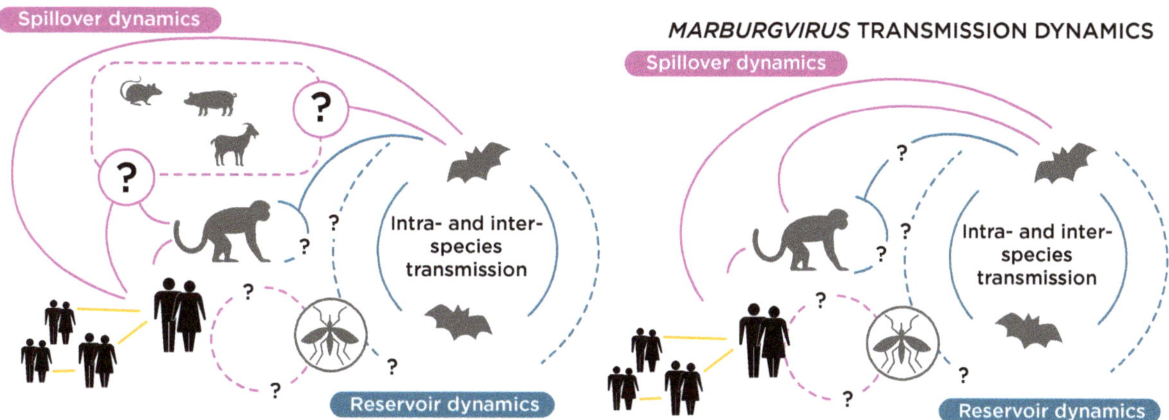

Figure 1. Diagrams depicting the hypothesized transmission cycle involving suspected reservoirs (e.g. fruit bats), incidental hosts (e.g. non-human primates), and humans. Source: Olivial and Hayman 2024.

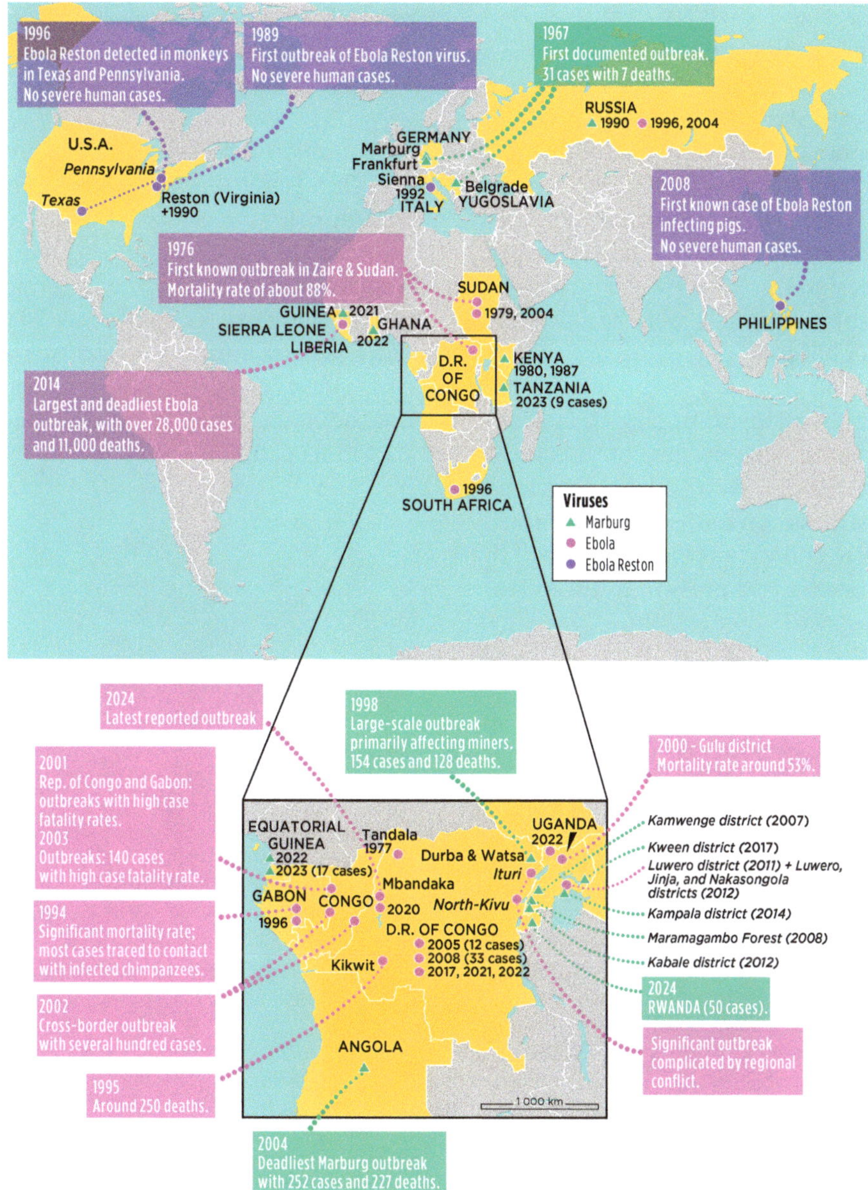

Figure 2. Geographical distribution of filovirus outbreaks and epidemics. Sources: CDC (https://www.cdc.gov/ebola/outbreaks/index.html) and WHO (https://www.who.int/emergencies/disease-outbreak-news).

One Health risk assessment: lessons from SARS, MERS-CoV and COVID-19

François Roger,
Flavie Goutard

The emergence of SARS, MERS-CoV and COVID-19 has underscored the interconnection of human and animal compartments—including domestic animals and wildlife—in diverse ecosystems, shaping disease dynamics (Figure 1). However, these many dimensions have yet to be fully integrated into a cohesive One Health approach for managing health risks. For example, MERS-CoV, a zoonotic pathogen with a high fatality rate, is primarily transmitted from dromedary camels to humans (Figure 2). Scientific understanding of how ecological dynamics and species interactions shape MERS-CoV transmission is still limited, which hampers effective prevention and control efforts. Similarly, COVID-19 demonstrated that without systematic One Health integration at each stage of response, timely and effective intervention is challenging.

A proactive One Health approach is essential during the earliest stages of crises like COVID-19 and MERS-CoV to clarify pathogen origins and transmission pathways. Such clarity can significantly enhance efforts to control the spread of outbreaks and prevent future ones.

One Health relies on long-term collaboration across human, animal and environmental health sectors and extends beyond times of crisis. Effective implementation would deepen our understanding of interspecies transmission and support more robust prevention strategies. For instance, early integrated surveillance during the initial COVID-19 outbreak could have helped identify SARS-CoV-2 reservoirs and shed light on transmission pathways, thus enhancing response efforts.

A unified risk assessment framework under the One Health umbrella—referred to as "One Health risk assessment"—could also be implemented (Figure 3). Doing so would enable the anticipation of future pandemics and make health systems more flexible. Such a framework would improve responses at key points of transmission, promote interdisciplinarity, and advance integrated health governance on a global scale.

References

Azhar E.I., El-Kafrawy S.A., Farraj S.A., Hassan A.M., Al-Saeed M.S. *et al.* 2014. Evidence for camel-to-human transmission of MERS coronavirus. *New England Journal of Medicine*, 370(26), 2499–2505. https://doi.org/10.1056/NEJMoa1401505

Delahay R.J., de la Fuente J., Smith G.C., Sharun K., Snary E.L. *et al.* 2021. Assessing the risks of SARS-CoV-2 in wildlife. *One Health Outlook*, 3, 7. https://doi.org/10.1186/s42522-021-00039-6

Funk A.L., Goutard F.L., Miguel E., Bourgarel M., Chevalier V. *et al.* 2016. MERS-CoV at the animal-human interface: Inputs on exposure pathways from an expert-opinion elicitation. *Frontiers in Veterinary Science*, 3, 88. https://doi.org/10.3389/fvets.2016.00088

Guan Y., Zheng B.J., He Y.Q., Liu X.L., Zhuang Z.X. *et al.* 2003. Isolation and characterization of viruses related to the SARS coronavirus from animals in southern China. *Science*, 302(5643), 1351–1354. https://doi.org/10.1126/science.1087139

Irving A.T., Ahn M., Goh G., Anderson D.E., Wang L.F. 2021. Lessons from the host defences of bats, a unique viral reservoir. *Nature*, 589(7842), 363–370. https://doi.org/10.1038/s41586-020-03128-0

WHO, FAO, WOAH. 2020. Joint risk assessment operational tool (JRA OT): An operational tool of the tripartite zoonoses guide. Taking a multisectoral, one health approach: A tripartite guide to addressing zoonotic diseases in countries. World Health Organization. https://www.who.int/publications/i/item/9789240015142

Worobey M., Levy J.I., Malpica Serrano L., Crits-Christoph A., Pekar J.E. *et al.* 2024. The Huanan seafood wholesale market in Wuhan was the early epicenter of the COVID-19 pandemic. *Science*, 377(6609), 951–959. https://doi.org/10.1126/science.abp8715

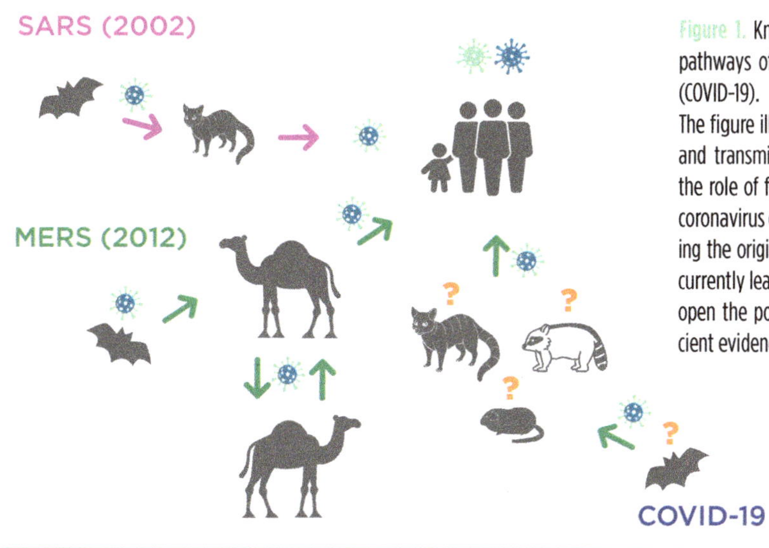

Figure 1. Knowledge and hypotheses on the origins and spillover pathways of SARS-CoV (SARS), MERS-CoV (MERS) and SARS-CoV-2 (COVID-19).
The figure illustrates potential animal reservoirs, intermediate hosts and transmission routes leading to human infection, emphasizing the role of free and captive wildlife as well as domestic animals in coronavirus emergence. In the absence of definitive evidence concerning the origin of COVID-19, the majority of scientists and institutions currently lean toward a zoonotic origin hypothesis, while still leaving open the possibility of a laboratory-related incident, given insufficient evidence to definitively rule it out.

Infected dromedary camels

	Direct contact	Indirect contact	Milk	Raw meat	Urine	Mean confidence	Kendall's W
Number of experts	*14*	*11*	*9*	*7*	*8*	*3.8**	-
Relative importance	14	4	2.5	1.2	1	3.8	0.56*

Camel workers

Figure 2. Example of MERS-CoV transmission pathways from infected dromedaries to camel workers based on expert opinion. The table shows the relative importance of different exposure routes as assessed by a panel of experts, with each route evaluated for its perceived risk. The number of experts assigning importance to each route and the mean confidence scores reflect the level of consensus on these risks. Direct contact was deemed the highest risk route, with a strong mean confidence rating of 3.8, indicating a moderate agreement among experts (Kendall's W = 0.56). From: Funk *et al.* 2016.

FORMATION OF A MULTISECTORAL TECHNICAL TEAM

Gather experts from human health, animal health and environmental sectors. These experts should include epidemiologists, veterinarians, wildlife specialists and public health professionals. This team forms the core of the risk assessment process.

RISK FRAMING AND DEFINITION

Identify and define the specific hazard to be assessed (e.g. a zoonotic virus such as COVID-19). It is essential to frame the risks clearly in terms of scope, objectives and geographical context (national or regional). This step guides the risk assessment to ensure it is relevant for management decisions.

RISK PATHWAY IDENTIFICATION AND DIAGRAMMING

Develop a risk pathway diagram that outlines the logical progression of the hazard from its source to the potential infection of a human or animal host. This diagram helps visualize critical points where interventions can be implemented.

FORMULATION OF RISK ASSESSMENT QUESTIONS

Develop a risk pathway diagram that outlines the logical progression of the hazard from its source to the potential infection of a human or animal host. This diagram helps visualize critical points where interventions can be implemented.

RISK CHARACTERIZATION AND MANAGEMENT

Develop a risk pathway diagram that outlines the logical progression of the hazard from its source to the potential infection of a human or animal host. This diagram helps visualize critical points where interventions can be implemented.

Figure 3. Comprehensive approach to One Health risk assessment. From: WHO *et al.* 2020.

COVID-19: institutionalizing One Health

Thierry Lefrançois

The World Health Organization (WHO) defines health as "a state of complete physical, mental and social well-being and not merely the absence of disease or infirmity". However, health is still viewed through the prism of human diseases, and the WHO's definition does not include activities that promote the long-term health of animals, the environment, or a territory. For a more comprehensive vision of health, we need to rethink our approach to One Health by more effectively integrating the environment to account for the health of all living organisms in a given ecosystem. There is an urgent need to move on from a limited vision of health to a more integrated, holistic vision.

The COVID-19 crisis revealed the need for integrated approaches, and especially the One Health approach, which entails multi-sectoral, multi-disciplinary and multi-stakeholder organization on all levels (local, national, regional and global).

One Health emphasizes the need for better communication between researchers and policy makers, which the COVID-19 pandemic made clear. Communication can be improved by:

- Rethinking world governance of health to incorporate the One Health approach;
- Supporting the establishment of regional One Health networks, with a focus on emergence zones, to conduct concrete research on and surveillance of new diseases;
- Organizing, at a national level, inter-ministerial collaboration through a comprehensive and concerted approach to prevent, detect early and manage crises (Figure 1). Joint efforts must be made to prevent and monitor major emergence risks and connect all sectors and stakeholders from the outset to manage health crises by lifting administrative barriers and facilitating information and data sharing (Figure 2). Cooperation between scientists and decision-makers is an essential link in the chain, as is better coordination between the French Ministries of Health, Agriculture, Ecological Transition and Foreign Affairs.

A One Health approach calls for a paradigm shift when it comes to training health professionals and decision-makers on complex issues to give them actionable skills in a variety of contexts (biodiversity, climate change, ecological transition, etc.). The One Health approach could become the standard for many other medium- and long-term societal challenges.

The pandemic showed the necessity of coming together and implementing institutional changes, transdisciplinary research and concrete actions in the field, drawing on the social fabric and developing new training and education methods for a variety of stakeholders (including decision-makers) to enhance preparedness for future emerging infectious diseases and adopting a One Health approach.

References

de Garine-Wichatitsky M., Binot A., Morand S., Kock R., Roger F. *et al.* 2020. Will the COVID-19 crisis trigger a One Health coming-of-age? *Lancet Planet Health*, 4, e377–e378.

Ghai R.R., Wallace R.M., Kile J.C., Shoemaker T.R., Vieira A.R. *et al.* 2022. A generalizable one health framework for the control of zoonotic diseases. *Scientific Report*, 12, 8588. https://doi.org/10.1038/s41598-022-12619-1

Jakab Z., Selbie D., Squires N., Mustafa S., Saikat S. 2021. Building the evidence base for global health policy: The need to strengthen institutional networks, geographical representation and global collaboration. *BMJ Glob Health*, 6, e006852. https://doi.org/10.1136/bmjgh-2021-006852

Lefrançois T., Malvy D., Atlani-Duault L., Benamouzig D., Druais P.L. *et al.* 2023. After two years of pandemic, translating One Health into action is urgent. *The Lancet*, 401 (10378), 789–794. https://doi.org/10.1016/S0140-6736(22)01840-2

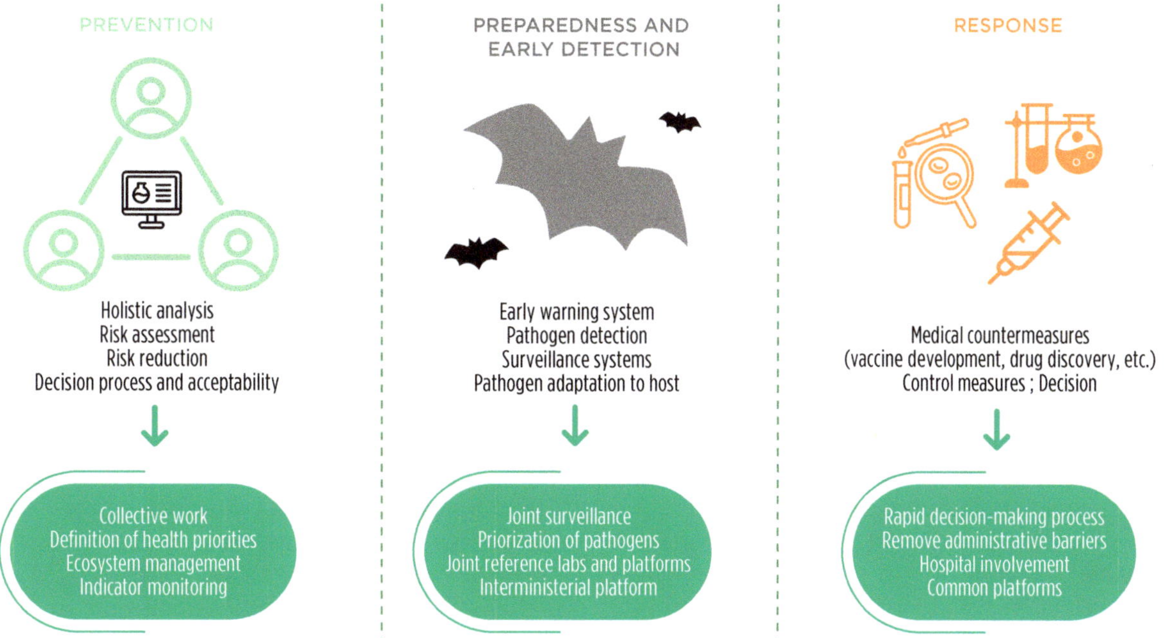

Figure 1. An ambitious roadmap for each step of a pandemic crisis (© Patricia Doucet and Delphine Guard-Lavastre).
The origin of SARS-CoV-2 remains uncertain. Two main hypotheses are considered: a natural spillover from wildlife, supported by the presence of susceptible animals at the Wuhan market and parallels with past zoonoses; and an accidental laboratory leak, suggested by the virus's unique features and proximity to the Wuhan Institute of Virology. While no conclusive evidence supports either theory, most scientific assessments still favour a natural origin.

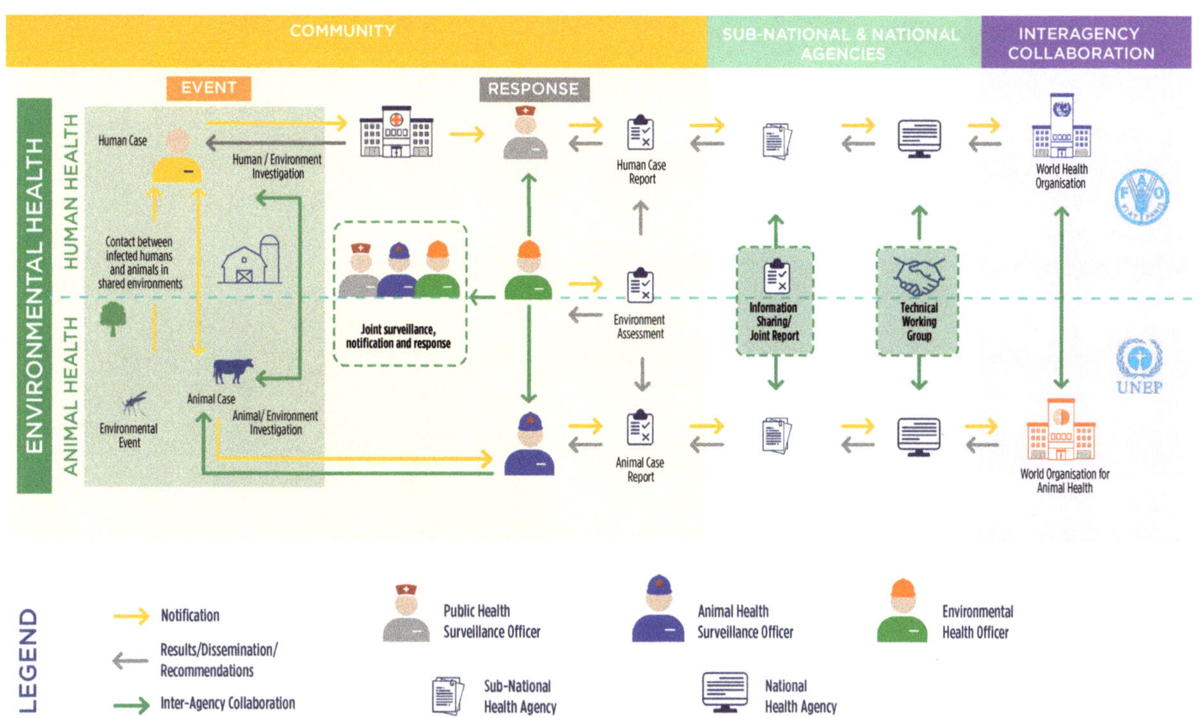

Figure 2. Operationalizing event-based surveillance through a One Health approach.
This diagram shows how human, animal and environmental health sectors can coordinate to detect, investigate and respond to health events. It emphasizes timely data sharing, joint risk assessment and interagency collaboration across governance levels to improve early warning and crisis response, as demonstrated during the COVID-19 pandemic. Adapted from Ghai *et al.* 2022.

Publishing One Health: evolving global patterns

François Roger,
Marie-Christine Lambert,
Émilie Chirouze

An analysis of scientific publications related to One Health between 2004 and 2022 was carried out on a data set from the Web of Science (WoS) and MEDLINE databases (query date: 22 February 2023). A total of 3,498 articles were selected following keyword searches by authors, Medical Subject Headings (MeSH) terms from MEDLINE, and keywords generated by WoS algorithms. The selected terms were "One Health", "EcoHealth" and "Planetary Health". Figure 1 shows the annual evolution of the number of publications.

Several major pandemics that occurred during this period are highlighted: the H1N1 influenza pandemic (2009-2010), the Ebola outbreak (2014-2016 in West Africa), MERS-CoV (since 2012) and the COVID-19 pandemic (since late 2019). Key milestones are also featured: the first international One Health conference in 2004, commonly known as the "Manhattan Conference", and the 2011 Berlin conference, which brought together scientists and policymakers. Since 2011, the World One Health Congress (https://globalohc.org/) has emerged as a major event, with its eighth edition in 2024.

The results show a notable increase in publications mentioning the terms "One Health" or "Planetary Health" over the past two decades. Major pandemics caused by zoonotic diseases, international conferences and decisions by international organizations have had a decisive impact on the scientific community and its research in these areas, with post-pandemic funding also likely having an effect. Before this period, the SARS (2002-2003) and H5N1 avian influenza (starting in 2003) crises were major catalysts that demonstrated the need for an integrated approach to address global health challenges. These crises underscored the importance of interdisciplinary collaboration and laid the groundwork for the adoption of the One Health concept and its institutionalization by scientists.

Figure 2 illustrates the geographical distribution of One Health publications. An analysis of co-publications by continent shows that Europe and North America are the most prolific regions publishing on these topics, along with several of the BRICS countries (Brazil, India, China and South Africa). Europe, as a whole, stands out, likely due to research policies oriented towards an integrated vision of global health and development cooperation in the Global South. Figure 3 shows a high concentration of publications in Global North countries, with significant interactions between the North and South regions, but relatively fewer studies exclusively focused on the Global South.

This brief analysis—non-exhaustive and trend-focused—highlights not only the evolution of research in the One Health domain and related concepts but also the geographical and economic disparities in scientific production. It draws attention to the importance of strengthening research capacities in under-represented regions.

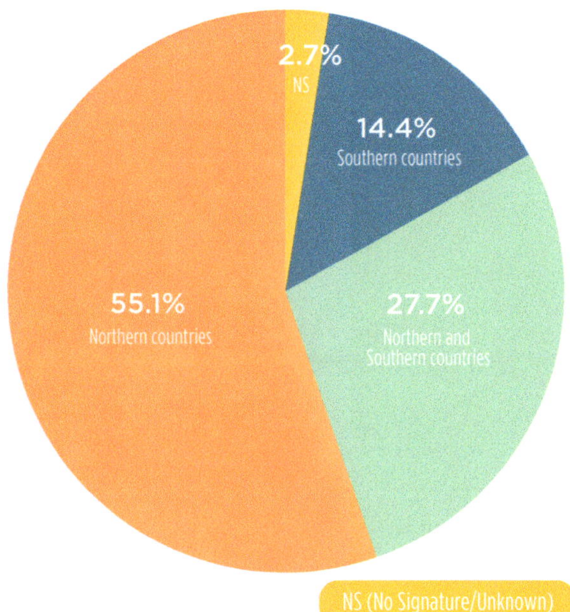

Figure 3. Distribution between Global North and Global South countries (based on the affiliations of the authors of the papers).

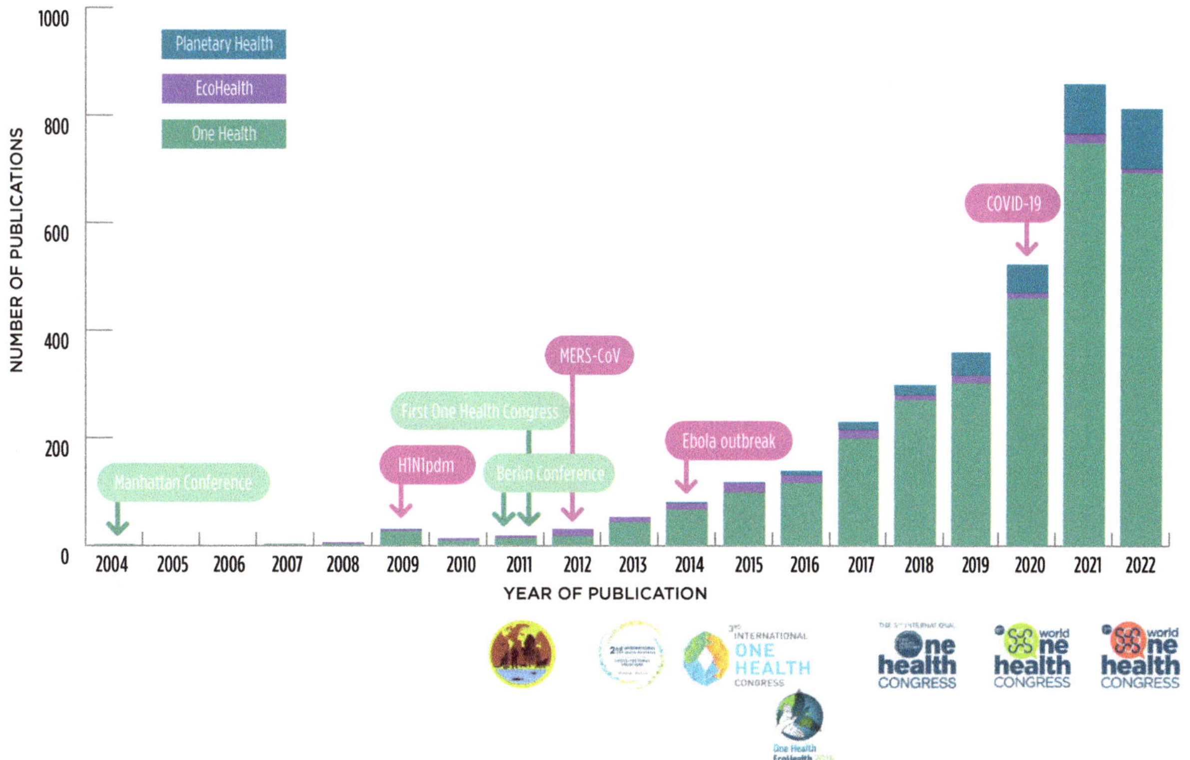

Figure 1. Evolution of scientific publications from 2004 to 2022 for the keywords "One Health", "EcoHealth" and "Planetary Health".
The decline observed in 2022 for One Health is likely not due to a lack of interest in these topics but rather a combination of several factors: delays in the publication process (data extracted in early 2023), a redeployment of research resources as the health crisis subsided, and the natural variability of scientific output.

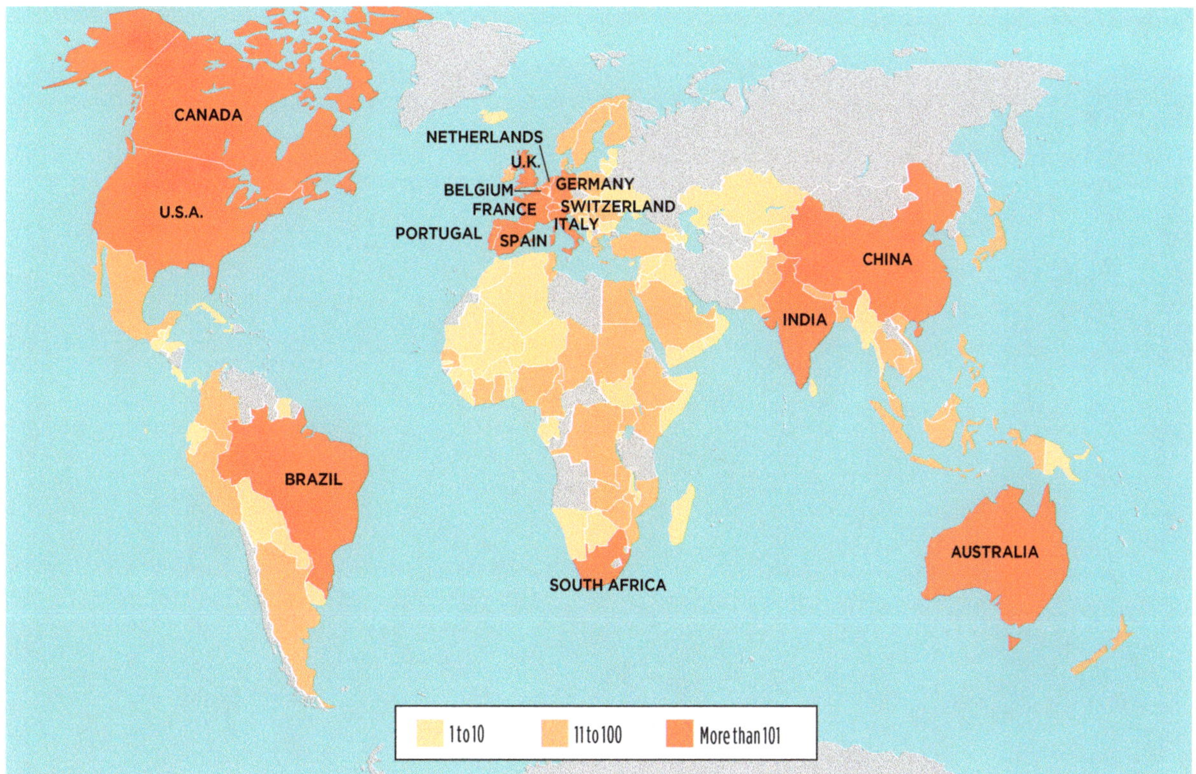

Figure 2. Geographical distribution of publications with "One Health" as a keyword.
Many scientific articles are international collaborations, meaning that a single article may be connected to several countries depending on the authors' affiliated institutions. Each country associated with an article has been counted, which means the total number of publications per country exceeds the total number of articles analysed.

Global trends in #OneHealth: insights from online searches and hashtags

François Roger

An analysis of Google Trends (analysis active from May 2006) and X (formerly Twitter, active from March 2006) activity linked to the hashtag #OneHealth provides significant insights into how interest and engagement regarding global health issues evolved over time (Figures 1 and 2).

Overall, these data reveal a shift in the One Health approach and its growing adoption in public and scientific discourse, especially during health crises. Global health alerts clearly trigged increased awareness of the importance of an integrated health approach that connects human, animal and environmental aspects. This phenomenon is apparent on Google Trends as well as in tweets, where relevant hashtags mirror the diversity of concerns and commitments around One Health (Figure 3).

While this analysis does not illustrate it graphically, the concept of One Health has evolved to include global issues such as antimicrobial resistance and biodiversity, which were previously underrepresented in global health discussions The continued discussions around these topics, even after the immediate crises subsided, suggest a lasting integration of the concept into global public health strategies and collective consciousness.

Information from X and Google Trends reflects more than just the public's interest in One Health. It could influence how One Health strategies are developed, communicated and implemented. By integrating this type of data into decision-making processes, policymakers can create more responsive, informed and effective One Health policies.

Google Trends (trends.google.com): comparison between the search terms "One Health" and "zoonoses" from 2004 to present, worldwide. Category: all categories; search type: Web search.
Tweets via @FEDICA (fedica.com): #OneHealth from 1 July 2006, to 31 December 2023. Tweets: 558K; contributors: 136K; countries: 221; cities: 7.2K; maximum reach: 500M + impressions: 10.3M.

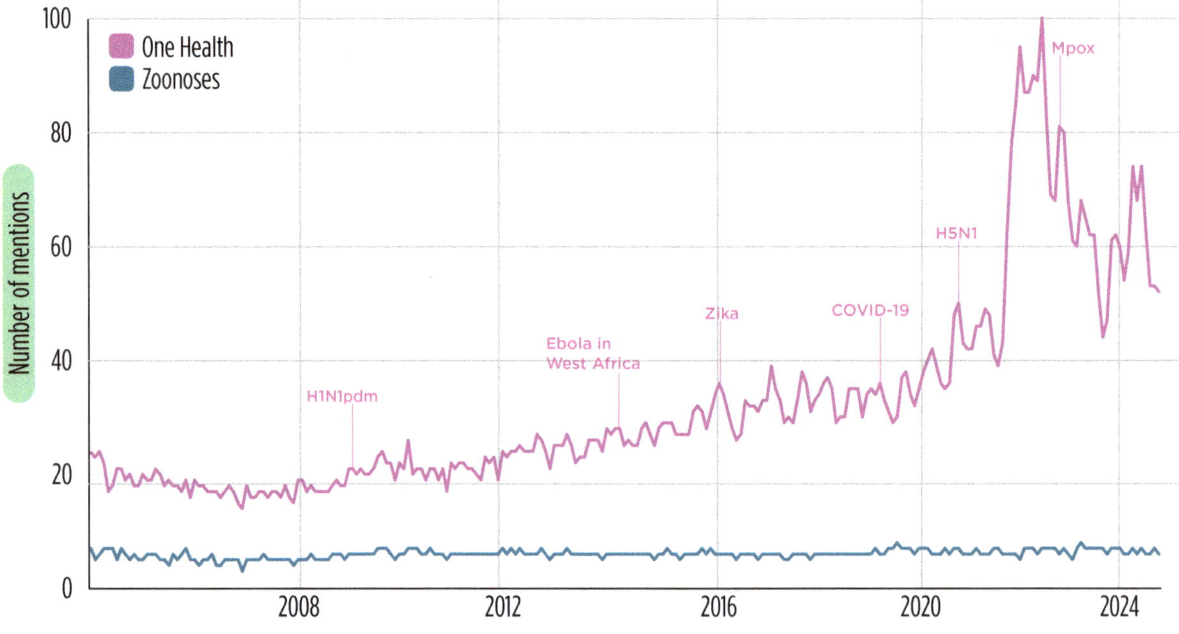

Figure 1. Trends in Google searches for "One Health" and "Zoonoses" over time, showing spikes in interest during major health crises (H1N1pdm 2009, Ebola, COVID-19). Search frequency is scaled from 0 to 100, with "Zoonoses" serving as a kind of baseline.

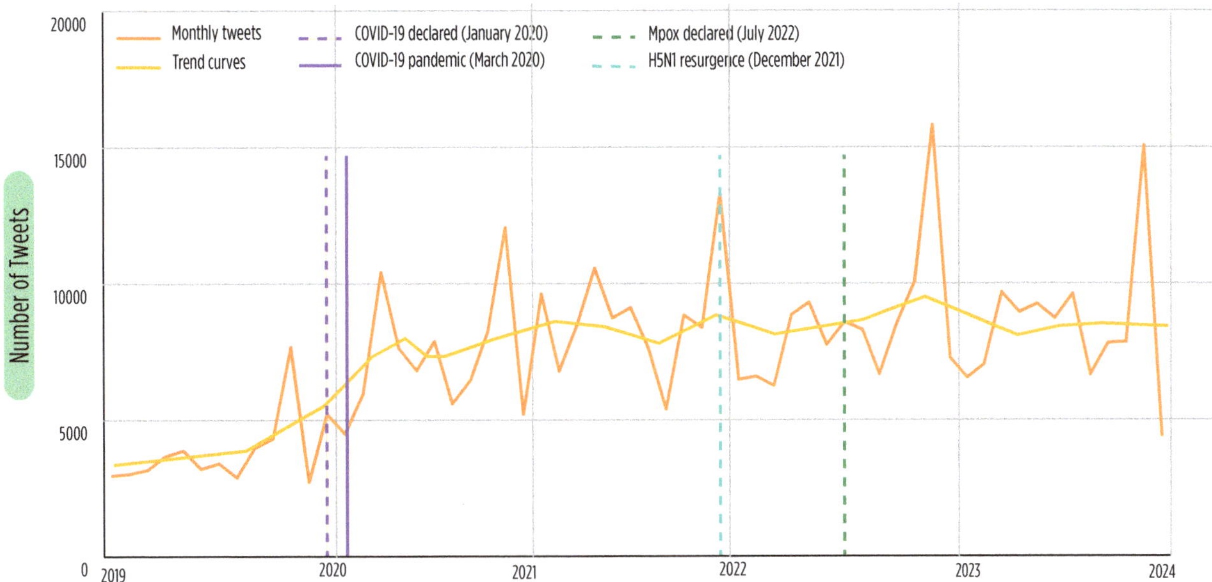

Figure 2. Number of tweets per month since 2019 with trend curve.
The number of tweets about One Health has steadily increased since 2019, with notable peaks correlating with significant public health events. For example, the official announcement of the COVID-19 pandemic in March 2020 led to a significant spike in the number of tweets, reaching an all-time high. Other significant peaks were observed during the announcement of the H5N1 resurgence in December 2021 and the declaration of the Mpox outbreak in July 2022.

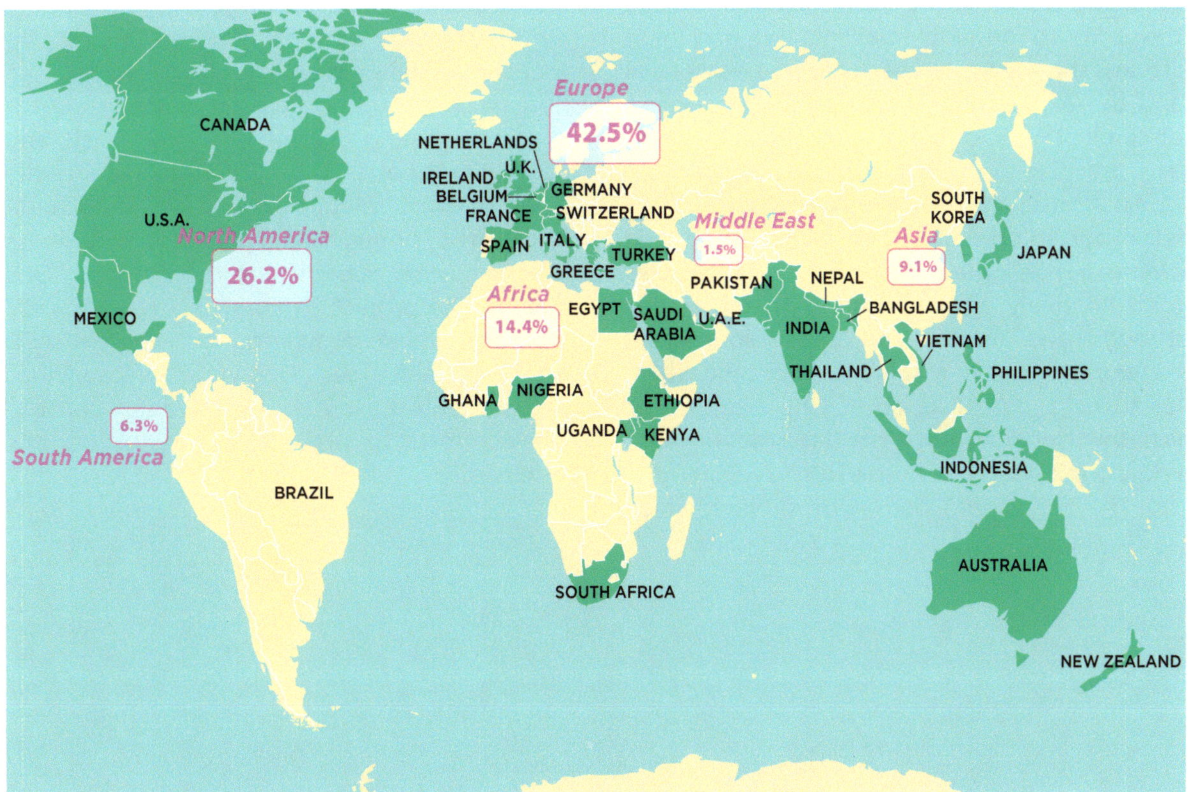

Figure 3. Geographical distribution of #OneHealth activity and scientific publications.
This map shows the proportion of tweets and publications related to "One Health" across world regions, highlighting disparities in global engagement. North America and Europe account for the majority of online and scientific activity, while Africa, Asia and Latin America remain underrepresented despite their vulnerability to emerging zoonoses. The data underline the need to strengthen One Health visibility, research capacity and digital participation in the Global South.

Engaging with professional media to raise public awareness of One Health

Daan Vink, Brieuc Lahellec, Tu Tu Zaw Win, Flavie Goutard

Public awareness of the One Health concept remains superficial, with a lack of understanding of its practical applications and benefits. The scientific community has made little concerted effort to engage with a wide audience. Promotion of One Health by implementing agencies and research organizations tends to be overly technical and scientific, and communication with policy- and decision makers is not targeted at a broad public. However, public engagement and education are essential for effective implementation of One Health activities at community level. Public-facing, professional media (including print, TV, internet and radio) along with journalists play a critical role in raising public awareness. Such media have the skills to develop relevant and engaging stories, the means to reach diverse audiences and the expertise to maximize the impact of disseminating information (Figure 1).

An Australian study exploring how journalists reported on avian influenza and pandemic planning highlighted challenges and strategies for effectively communicating health issues through the media. The Earth Journalism Network provides resources and case studies to help journalists cover One Health topics. However, few journalists have the requisite knowledge and skills to perform such reporting. The "Media for One Health" project, which is led by Canal France International (CFI), aims to enhance the capacity of media outlets in Vietnam, Cambodia, Laos and the Philippines to produce and disseminate information on the One Health approach. The project supports 55 experienced journalists working in diverse media (Figure 2). It provides training in innovative techniques for the production of thematic content on health-related topics, incorporates mentoring and technical support, and encourages cooperation between the journalists. Specific topics of focus include ensuring accuracy and trustworthiness of reporting, avoiding misinformation and building effective relationships with trusted experts to develop relevant and popularized stories (Figure 3). The project aims to result in around a hundred One Health-focused media items.

By facilitating ties between journalists and scientific experts and strengthening interactions between media outlets and their audiences, the project aims to raise awareness among a wider public of the importance of One Health. The project aligns with the PREZODE initiative, which, in synergy with the GREASE platform, enhances One Health capacities across South-East Asia by fostering collaborative efforts to prevent emerging zoonotic diseases.

References

CFI Media Development. Media for One Health. https://cfi.fr/en/project/media-one-health

Leask J., Hooker C., King C. 2010. Media coverage of health issues and how to work more effectively with journalists: A qualitative study. *BMC Public Health,* 10, 535. https://doi.org/10.1186/1471-2458-10-535

Robinson D. 2022. A journalist's guide to covering and implementing the One Health approach in reporting. https://earthjournalism.net/resources/tipsheet/a-journalists-guide-to-covering-and-implementing-the-one-health-approach-in

Figure 1. Information on the One Health approach from scientific and technical agencies plus policy- and decision makers has so far appealed only to a small public audience. This project aims to forge a link between scientists and technical experts on the one side and journalists and their media organizations on the other to widen interest among the general public.

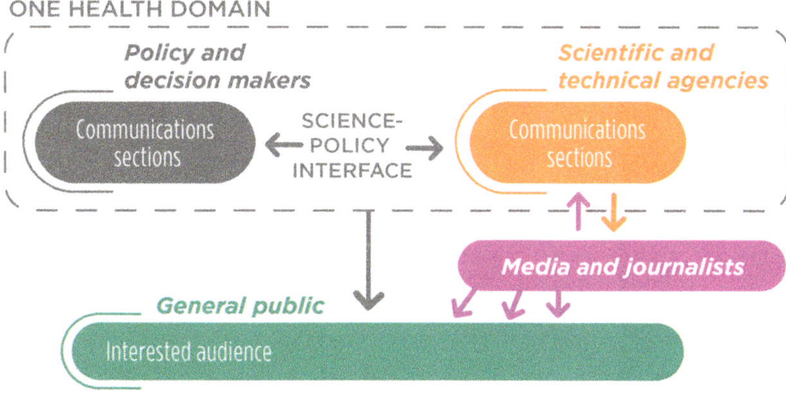

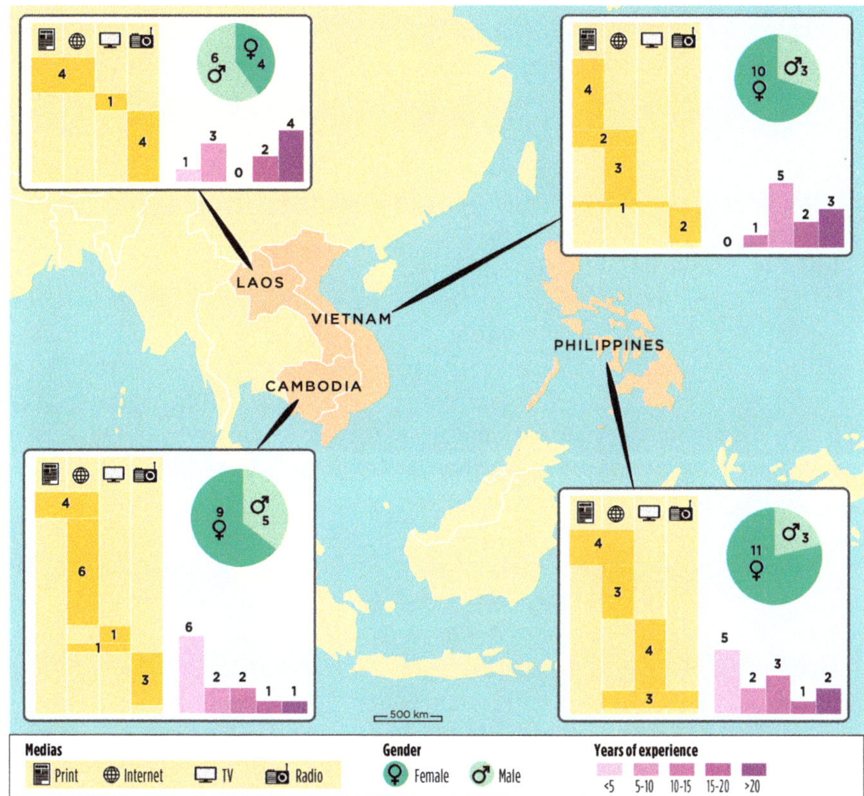

Figure 2. Demographics of the participating journalists in four South-East Asian countries, including gender, journalistic experience and the numbers of different types of media represented.

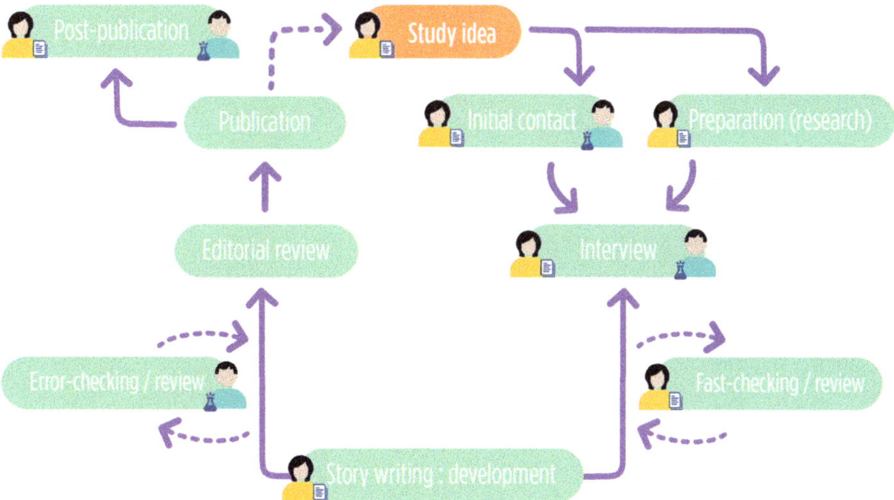

Figure 3. Steps in the story development cycle, from conception of a One Health story idea to publication.

Integrated Health: one goal, many approaches

François Roger

Integrated health approaches—such as One Health, EcoHealth, Planetary Health and Global Health—represent interdisciplinary frameworks that highlight the interconnectedness of human health with broader environmental factors. Although these concepts share similar goals, they differ in their theoretical foundations, approaches and levels of intervention (Figure 1). One Health focuses on the interrelation of human, animal, plant and environmental health, promoting collaboration across the medical, veterinary and environmental sciences to address zoonotic diseases and health threats like pandemics and antimicrobial resistance. EcoHealth adopts a broader ecosystem-centred view, integrating social, cultural and economic factors to enhance community sustainability and resilience by understanding human-environment interactions. Planetary Health expands the focus to the global biosphere, addressing issues like climate change, pollution and soil degradation. It emphasizes preserving ecosystems to safeguard health, drawing on interdisciplinary knowledge to tackle ecological crises with holistic solutions. Together, these approaches offer complementary frameworks for addressing complex, interconnected health challenges.

Global Health and Public Health approaches concentrate on international health challenges, including transboundary health issues and inequalities in healthcare access. These approaches aim to improve healthcare systems and strengthen international cooperation to respond to major health crises, particularly in resource-limited countries. In French, the term "global health" can be translated in different ways to reflect a shift in focus. *Santé mondiale* mainly deals with transboundary health issues and public health challenges at the international level, addressing infectious diseases, healthcare and access to services, with a focus on health inequities and socioeconomic factors. Meanwhile, *santé globale* emphasizes a more humanistic approach within a global governance context, and involves multilevel and multisector interventions to address health issues that transcend national borders.

One Health has emerged as an effective approach to addressing contemporary public health challenges due to its simplicity, institutional recognition and adaptability. Today, it stands as a pillar in the prevention and management of global health crises (Figure 2).

References

Amuasi J.H., Lucas T., Horton R., Wincler A.S. 2020. Reconnecting for our future: The Lancet One Health Commission. *The Lancet*, 395(10235), 1469–1471. https://doi.org/10.1016/S0140-6736(20)31027-8

Lerner H., Berg C. 2017. A comparison of three holistic approaches to health: One Health, EcoHealth, and Planetary Health. *Frontiers in Veterinary Science*, 4, 163. https://doi.org/10.3389/fvets.2017.00163

Myers S., Frumkin H. (Eds.). 2020. *Planetary Health: Protecting Nature to Protect Ourselves*. Island Press.

Roger F., Caron A., Morand S., Pedrono M., de Garine-Wichatitsky M. *et al.* 2016. One Health and EcoHealth: the same wine in different bottles? *Infection Ecology & Epidemiology*, 6(1), 30978. https://doi.org/10.3402/iee.v6.30978

Zinsstag J. 2012. Convergence of EcoHealth and One Health. *EcoHealth*, 9(4), 371–373. https://doi.org/10.1007/s10393-013-0812-z

		One Health		EcoHealth	Planetary Health	
		Narrow	Wide		Narrow	Wide
Core contributing science	Human	Public health	Public health Human medicine Molecular and microbiology Health economics Social sciences	Public health Human medicine Rural and urban development and planning Social sciences Anthropology	Public health Human medicine	Human medicine Economy Energy Natural resources
	Animal	Veterinary medicine	Veterinary medicine	Veterinary medicine	-	Agricultural sciences (including plant and animal production sciences)
	Ecosystem	-	Environmental health Ecology	Conservation and ecosystem management	-	Ecology Other environmental sciences (including climate and biodiversity research) Marine sciences
Knowledge system		Western scientific	Western scientific	Western scientific Indigenous knowledge	Western scientific	Western scientific
Core values	Health	Individual health	Individual and population health	Population health	Individual and population health	Individual and population health
	Groups	Humans Animals	Humans Animals Ecosystems	Humans Animals Ecosystems	Humans	Humans
	Other			Biodiversity Sustainability (for humans, animals, ecosystems)	Sustainability (for humans)	Sustainability (for humans)

Figure 1. One Health, EcoHealth and Planetary Health: interconnected approaches addressing health through collaboration, ecosystem resilience, and global environmental sustainability. From Lerner and Berg 2017.

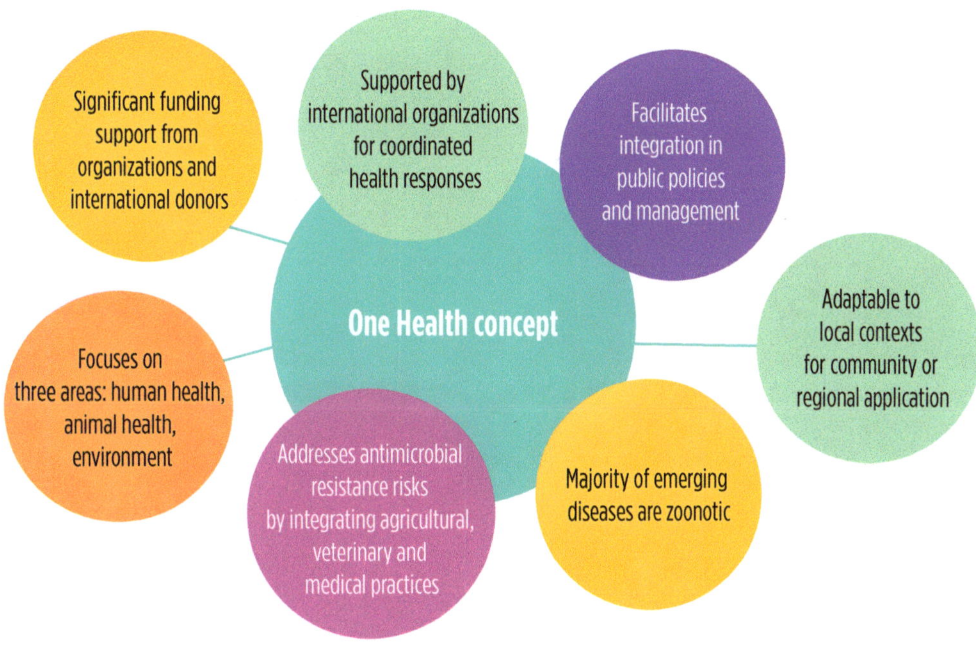

Figure 2. Why the One Health concept stands out compared to other approaches.

Zoonoses, agriculture and food security

This section explores the practical applications of One Health, mainly in research. It focuses on several zoonotic diseases while also addressing non-zoonotic animal diseases. But beyond diseases, it gives also special attention to agricultural systems, examining how farming practices, crop protection, soil health and aquaculture directly influence human and animal health outcomes. By integrating these diverse perspectives, this section highlights the essential role of sustainable agriculture in One Health strategies, particularly in preventing disease emergence, reducing antimicrobial resistance and ensuring food security. Together, these chapters demonstrate the power of One Health in addressing interconnected health, agriculture and environmental challenges.

Infection with SARS-CoV-2 in animals: a One Health challenge

Amélie Desvars-Larrive,
Liuhuaying Yang (illustrations)

Although the COVID-19 pandemic is sustained through human-to-human transmission, its causative agent, SARS-CoV-2, is a multi-host pathogen. Numerous domestic, captive, and free-ranging wild animals proved susceptible to natural or experimental infection with SARS-CoV-2, which presents a potentially wide-ranging, not yet fully elucidated, host spectrum (Figure 1). Natural infections of animals with SARS-CoV-2 have been confirmed in more than 40 countries (Figure 2) and in at least 31 animal species belonging to 18 taxonomic families (as of May 2023). As in humans, animal infection with SARS-CoV-2 is considered as an emerging disease and all Member States of the World Organisation for Animal Health (WOAH) are required to report cases of animal infection with SARS-CoV-2.

Animal infections generally originate from human-to-animal transmission events (colloquially called "spillback"), although natural transmission among conspecifics (e.g. in golden hamsters, white-tailed deer, and mink) or cross-species transmission (e.g. from mink to cats) have been described. Among SARS-CoV-2 animal hosts, dogs are very likely dead-end hosts, whereas white tailed-deer (*Odocoileus virginianus*) might become reservoir hosts in North America, in which co-circulation of different lineages could potentially result in the emergence of new virus variants.

SARS-CoV-2 represents the only known example of secondary spillover (i.e. cross-species transmission from a new animal host back to humans), with animal-to-human transmission reported in pet hamsters, farmed mink, cats, and white-tailed deer. This led to the culling of millions of commercial mink in multiple countries (2020–2021) and of some 2,000 hamsters in Hong Kong (January 2022). Finally, SARS-CoV-2 transmission to captive, rare, or endangered wild animal species, such as great apes, tigers, or hippopotami, demonstrates that it may also pose a threat to conservation efforts.

In a pandemic context, the accessibility of FAIR (Findable, Accessible, Interoperable, and Reusable) data across sectors and disciplines is a key element in developing robust One Health mathematical modelling and risk assessment frameworks. Timely and standardized reporting of both human and animal infections through a centralized platform is crucial for comprehending the complex, dynamic epidemiology of emerging zoonotic-origin pathogens at human-animal-environment interfaces and mitigating their impacts.

References

Nerpel A., Yang L., Sorger J., Käsbohrer A., Walzer C., Desvars-Larrive A. 2022. SARS-ANI: a global open access dataset of reported SARS-CoV-2 events in animals, *Scientific Data*, 9 (438). https://doi.org/10.1038/s41597-022-01543-8

World Organisation for Animal Health (WOAH). 2023. Terrestrial Animal Health Code. https://www.woah.org/en/what-we-do/standards/codes-and-manuals/terrestrial-code-online-access/

Naderi S., Chen P.E., Murall C.L., Poujol R., Kraemer S., Pickering B.S., Sagan S.M., Shapiro B.J. 2023. Zooanthroponotic transmission of SARS-CoV-2 and host-specific viral mutations revealed by genome-wide phylogenetic analysis, *eLife*, 12: e83685. https://doi.org/10.7554/eLife.83685

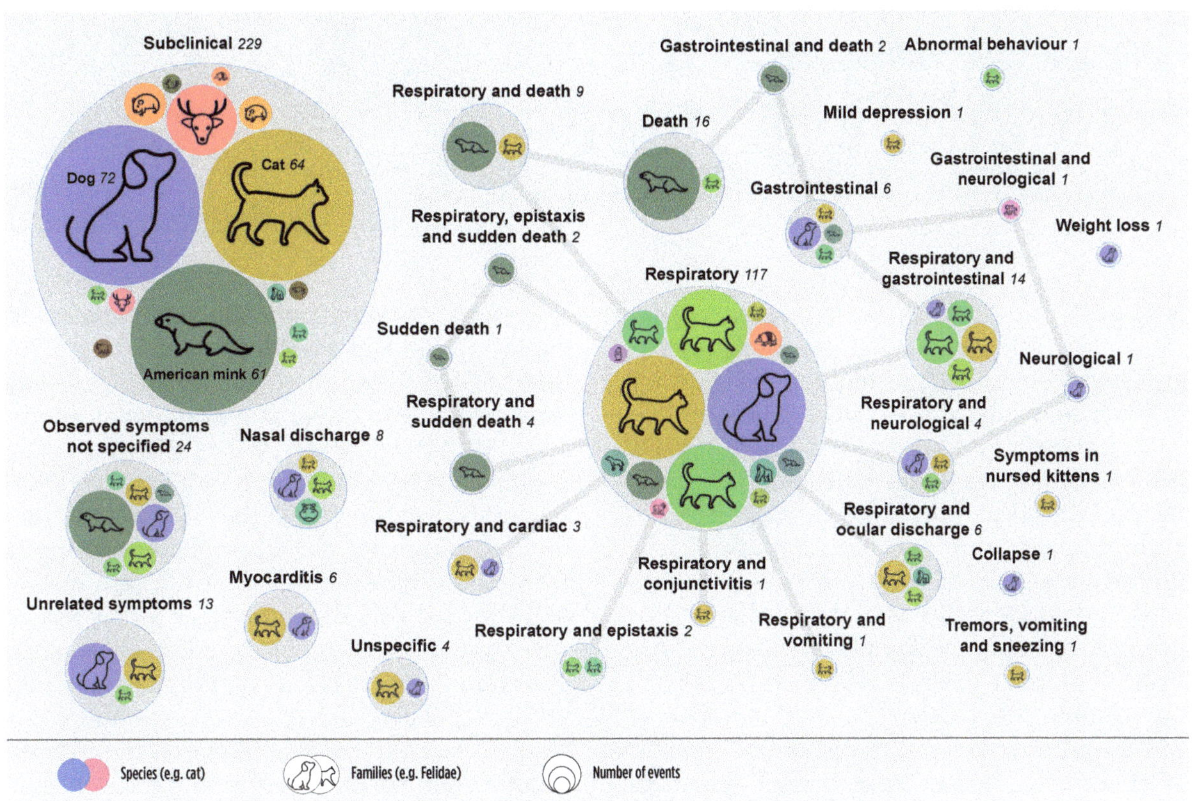

Figure 1. Reported clinical signs allegedly associated with natural SARS-CoV-2 infection in the different animal hosts worldwide.

Grey edges show when multiple symptoms were associated. The size of the circle is proportional to the number of events in which the clinical sign has been reported (we consider as an "event" to be when one single animal case or several epidemiologically related animal cases were identified by the presence of viral RNA and/or antibodies). The number on the right of each clinical sign corresponds to the total number of events mentioning this clinical sign. Colours of the circles correspond to animal species; animal symbols represent taxonomic families.Sources: World Animal Health Information System of the World Organisation for Animal Health (WOAH-WAHIS) and Program for Monitoring Emerging Diseases (ProMED-mail) of the International Society for Infectious Diseases. Structured data retrieved from Nerpel *et al.* 2022 on May 2023.

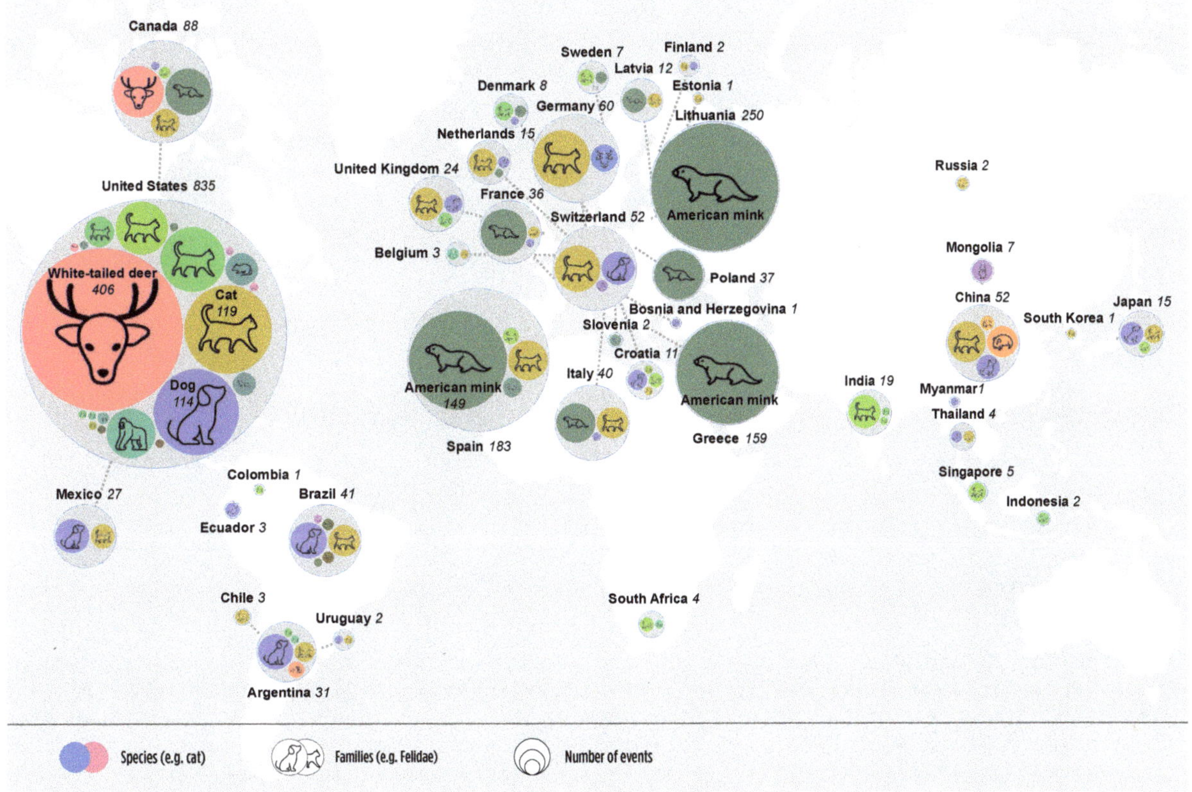

Figure 2. Distribution of the reported SARS-CoV-2 events (2,046 infections/exposures) in animals.

Mapping the areas at risk for transmission of a vector-borne disease (West Nile fever)

Annelise Tran, Benoit Durand, Véronique Chevalier

West Nile virus (WNV), responsible for West Nile fever (WNF), is a widespread arthropod-borne virus of the genus Flavivirus. The transmission cycle of WNV is an enzootic cycle between mosquitos and birds (Figure 1). Humans and horses are considered incidental hosts. While most human infections are asymptomatic and mild cases present only flu-like symptoms, more severe cases (encephalitis, meningoencephalitis or meningitis) can occur. In Europe, the circulation of WNV has been confirmed episodically since the 1960s, but during the particularly hot summer of 2010, a very large number of human cases of WNF was reported in areas previously free of the disease. Since then, there have been annual outbreaks, suggesting an endemic transmission cycle and thus a resurgent public health problem, including safety of the blood supply.

In order to identify the environmental determinants of WNF in Europe, a One Health approach has been proposed in order to consider all factors likely to impact the presence and abundance of mosquito vectors and avian hosts. Using a statistical approach, the relationship between WNV infection status (infected or uninfected) and several environmental factors was studied at the district level (Figure 2).

Results showed that the best model to explain the infected/uninfected status of human WNF cases by district includes July temperature and early June water index anomalies, WNV circulation in the previous year, presence of wetlands, type of avian migration route, and human population. All of these factors are positively and significantly correlated with the probability of infection. These results highlight the role of water areas in the risk of WNV transmission in Europe and the relevance of a water index derived from satellite imagery to detect water areas above seasonal averages in June, which may favour high mosquito densities during the summer months and virus transmission to humans in late summer and early fall.

Maps of the probability of WNV infection by district can be produced annually based on this model (Figure 2). Built using 2001–2011 epidemiological data, the model successfully predicted the occurrence of WNV human and horse' cases in the following years, demonstrating the predictive capacity of such an approach. Thus, it could be used to assess the risk of WNV infection in the future under different climate change scenarios.

References

Durand B., Tran A., Balança G., Chevalier V. 2017. Geographic variations of the bird-borne structural risk of West Nile virus circulation in Europe. *PloS One*, 12 (10): 15 p. https://doi.org/10.1371/journal.pone.0185962

Semenza J.C., Tran A., Espinosa L., Sudre B., Domanovic D., Paz S. 2016. Climate change projections of West Nile virus infections in Europe: Implications for blood safety practices. *Environmental Health*, 15 (28): 12 p. https://doi.org/10.1186/s12940-016-0105-4

Tran A., Sudre B., Paz S., Rossi M., Desbrosse A. ***et al.*** 2014. Environmental predictors of West Nile fever risk in Europe. *International Journal of Health Geographics*, 13 (26): 11 p. https://doi.org/10.1186/1476-072X-13-26

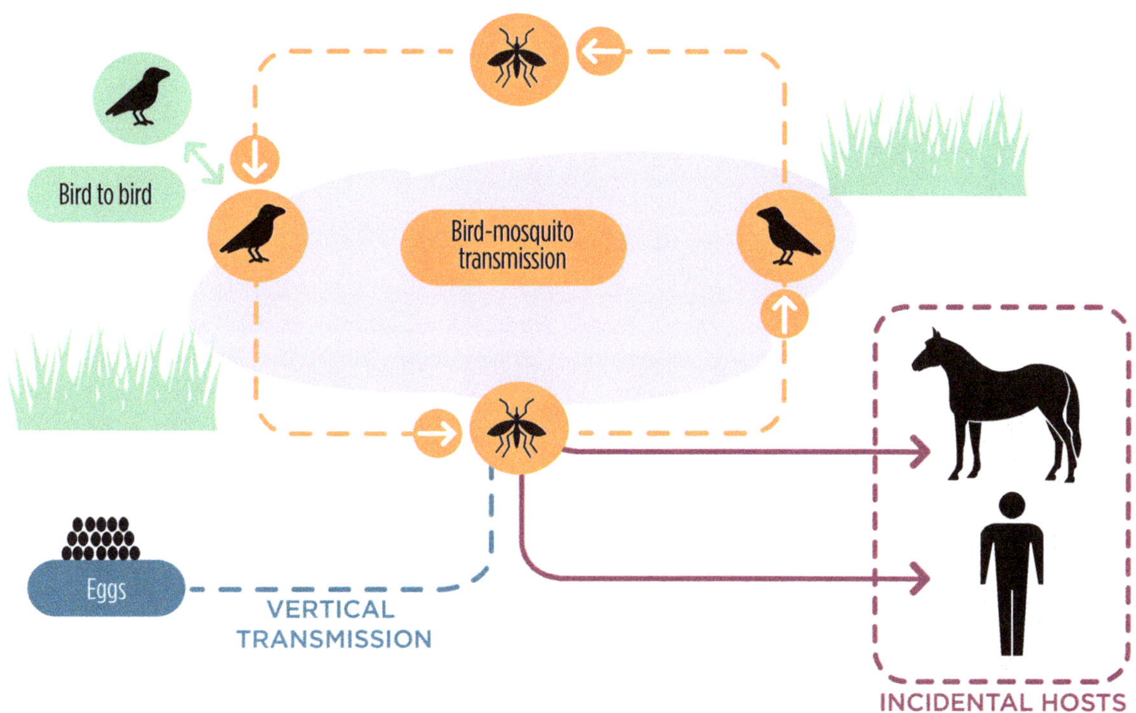

Figure 1. Schematic representation of the West Nile virus transmission cycle.

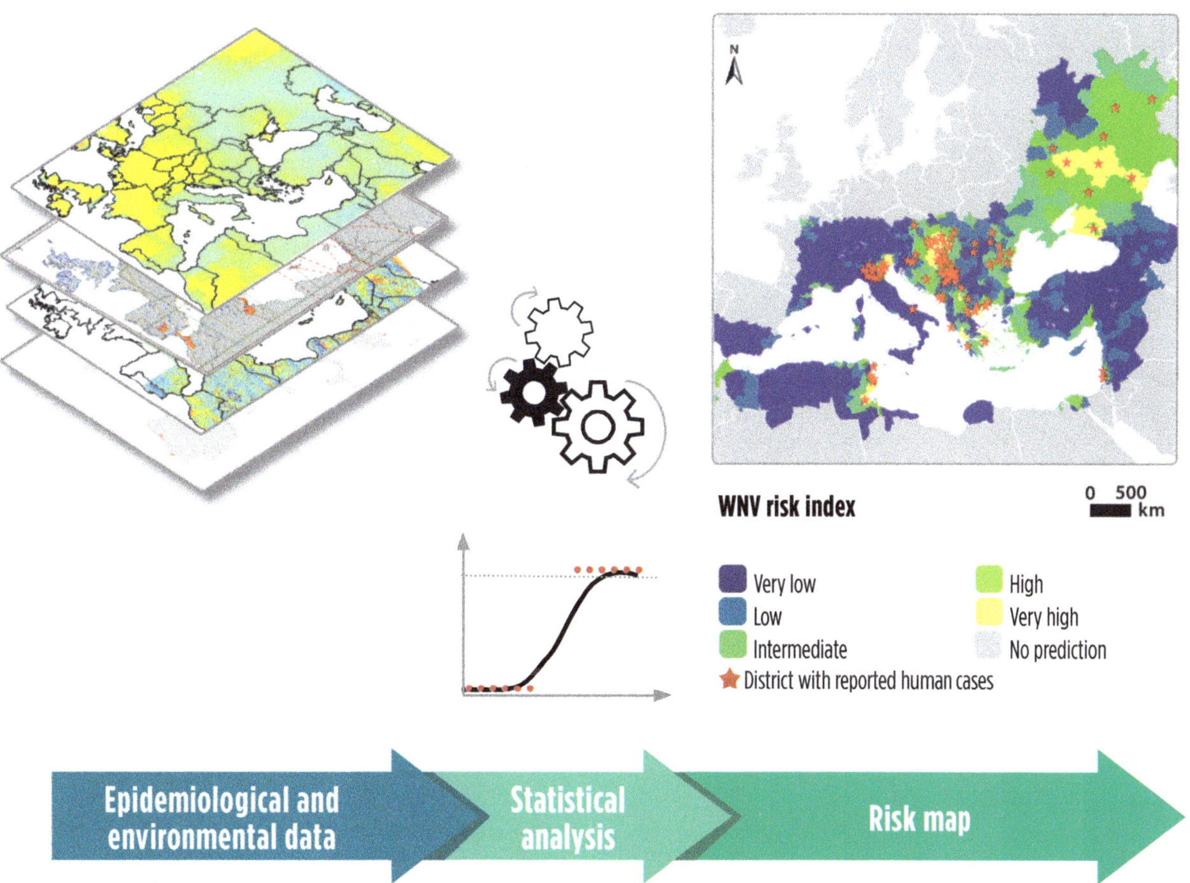

Figure 2. General framework for mapping West Nile virus (WNV) transmission risk areas in Europe based on environmental and meteorological indicators.

One Health game theory: eliminating dog rabies in Africa

Jakob Zinsstag, Alvar Bucher, Guenther Fink, Nakul Chitnis, Bassirou Bonfoh, Artemiy Dimov

Rabies, a neurological disease spread by dogs, is a neglected public and veterinary health issue in the Global South. Every year, rabies causes the deaths of approximately 59,000 people, 3.7 million years of life lost to disability, and USD 8.6 billion in economic damages. The WHO is working to eliminate dog-transmitted rabies deaths by 2030. Rabies deaths can be prevented by giving people who have been bitten with immediate post-exposure prophylaxis (PEP). In low- and middle-income countries, PEP is rare, observance is low and dog immunization is limited. Mass immunization can eliminate dog rabies if coverage is high enough. The main constraints of canine vaccination are access to vaccines and dog movements, often human-mediated, which allow the reintroduction of rabies at local and national levels. Indeed, active border protection may not always be possible to prevent reintroduction from the outside.

Due to the recurrent threat of rabies introduction, policy measures by specific countries depend on those of others. Strategic policy choices can be analysed mathematically using game theory (see Box). Using this type of framework, we evaluated the advantages and costs of policy measures for all actors and compared them to determine the most profitable ones, whether motivated by self-interest or collaboration. We used this method to analyse the benefits of cooperating for socially optimal policy equilibrium. Unvaccinated dogs and lack of PEP in people are common in all African countries. We showed that coordinated dog vaccination across countries and PEP would eradicate rabies in Africa, resulting in a USD 9.5 billion welfare gain (95% confidence interval, CI: 8.1–11.4 billion) (Figure 1). Mass dog vaccinations that are not coordinated between countries and insufficient human PEP reduce wellbeing and do not eliminate canine rabies. Even with a probable reintroduction of rabies from other countries, mass dog vaccination is the prevailing theoretical game strategy for many countries.

Thus, handling rabies control in this way in African countries could reduce mortality to almost zero and possibly eliminate rabies, improving social and environmental effects. Figure 2 suggests an approach based on different starting points. Vaccinating dogs and coordinating with Senegal, The Gambia and Mali would prevent the disease from returning to Mauritania via the Atlantic Ocean and Sahara Desert. Mali, Senegal and The Gambia would then coordinate major immunization operations with their southern and eastern neighbours. Platforms like the African Union Bureau for Animal Resources (AU-IBAR, https://au-ibar.org/) and ECOWAS (https://www.ecowas.int/) could coordinate such a programme. Strategic analysis in game theory employing a model of dog-human rabies transmission shows an incremental benefit of a One Health approach.

References

Bucher A., Dimov A., Fink G., Chitnis N., Bonfoh B., Zinsstag J. 2023. Benefit-cost analysis of coordinated strategies for control of rabies in Africa. *Nature Communications*, 14, 5370. https://doi.org/10.1038/s41467-023-41110-2

Mauti S., Léchenne M., Mbilo C., Nel L., Zinsstag J. 2019. Rabies. In: *Transboundary Animal Diseases in Sahelian Africa and Connected Regions* (Kardjadj M., Diallo A., Lancelot R., eds). Springer, p. 107–120.

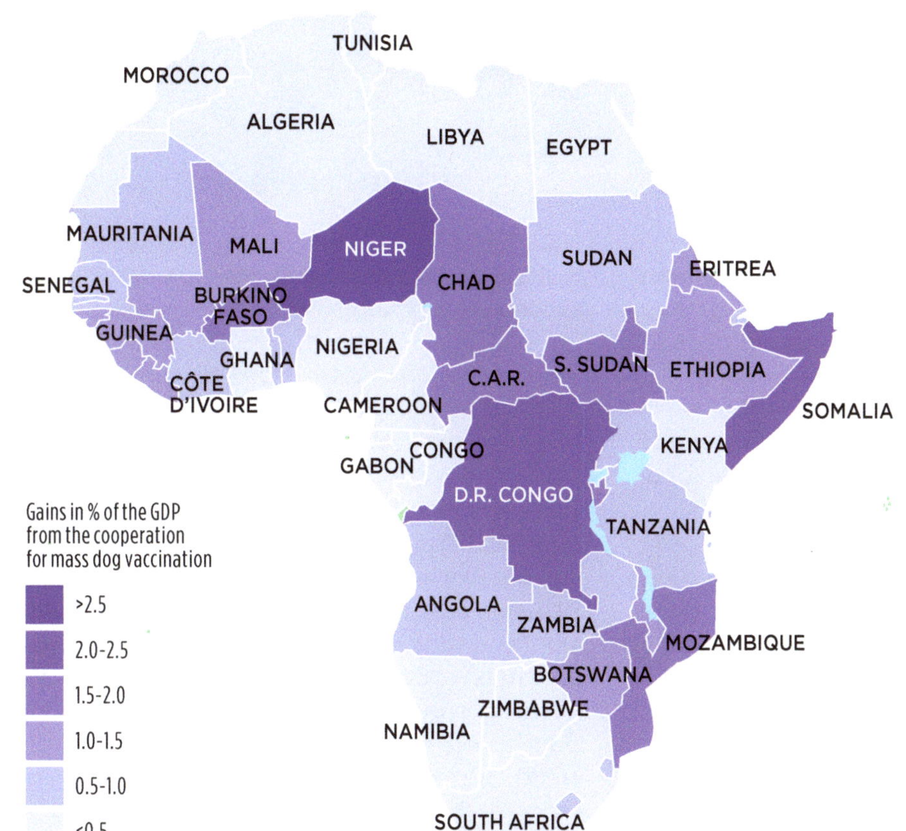

GAME THEORY

It is a branch of mathematics and economics that studies strategic decision-making among interacting actors/stakeholders, where each one's choices influence the outcomes for others. In epidemiology, it is used to model and analyse human behaviour in response to diseases, such as the decision to get vaccinated. It helps predict population responses and optimize public health strategies to control disease spread. For example, it supports understanding how individual choices can impact herd immunity.

Figure 1. Relative gain in billion USD per country as a percentage of the corresponding gross domestic product (GDP) between 2024–2054 (30 years). From: Bucher 2023.

Figure 2. Potential strategy for coordinated mass dog vaccination against rabies in Africa. From: Mauti 2019.

The way forward for rabies control and elimination in South Asia: the One Health approach

Kinzang Dukpa, Lesa Thompson, Pasang Tshering, Hirofumi Kugita, Marie-Marie Olive

Rabies is responsible for 60,000 human deaths worldwide each year, with 95% of those deaths concentrated in Africa and Asia. South Asia contributes about 45% of the global burden of human rabies. The disease is endemic in both humans and animals in seven of the eight South Asian Association for Regional Cooperation (SAARC) countries: Afghanistan, Bangladesh, Bhutan, India, Nepal, Pakistan and Sri Lanka (Figure 1). The Zero by 30 strategy is a global plan that seeks to end human deaths from dog-mediated rabies by 2030, drafted by the World Health Organization (WHO), World Organisation for Animal Health (WOAH), Food and Agriculture Organization of the United Nations (FAO) and Global Alliance for Rabies Control (GARC).

Since 99% of human rabies is transmitted by bites from rabid dogs, dog vaccination is key to stopping rabies transmission between dogs and from dogs to humans. Based on this knowledge, most SAARC countries have developed a national strategic plan on rabies, and some countries (e.g. Bangladesh, Bhutan and Sri Lanka) conduct nationwide mass dog vaccination campaigns combined with dog population management and human post-exposure prophylaxis.

In India's Goa State, a One Health approach was successful in controlling rabies with a 70% coverage in dog vaccination, culminating in the elimination of human rabies and a 92% reduction in monthly canine rabies cases (Figure 2). This case study shows that rabies is preventable with an integrated approach that includes effective policies focusing on community awareness, post-exposure prophylaxis and mass dog vaccination (Figure 3). India officially launched its National Action Plan for Dog Mediated Rabies Elimination (NAPRE) in 2023 under the National Rabies Control Program (NRCP), and the country is now in the process of developing a State-level Action Plan for Rabies Elimination (SAPRE). Bhutan could potentially be the first country in South Asia to achieve the global target of zero by 30 ahead of schedule, as it has recently completed nearly 100% sterilisation and vaccination of free-roaming dogs; this initiative is sustained through government funding (National Accelerated Dog Population Management and Rabies Control Program—NADPM&RCP).

References

WHO, FAO, OIE, Global Alliance for Rabies Control. 2018. Zero by 30: the global strategic plan to end human deaths from dog-mediated rabies by 2030. Geneva. https://www.woah.org/fileadmin/Home/eng/Media_Center/docs/Zero_by_30_FINAL_online_version.pdf

Lionel Harischandra P.A., Gunesekera A., Janakan N., Gongal G., Abela-Ridder B. 2016. Sri Lanka takes action towards a target of zero rabies death by 2020. *WHO South-East Asia Journal of Public Health*, 5(2), 113–116. https://doi.org/10.4103/2224-3151.206247

Gibson A.D., Yale G., Corfmat J., Appupillai M., Gigante C.M. *et al.* 2022. Elimination of human rabies in Goa, India through an integrated One Health approach. *Nature Communications*, 13, 2788. https://doi.org/10.1038/s41467-022-30371-y

Brunker K., Mollentze N.. 2018. Rabies virus. *Trends in Microbiology*, 26(10), 886–887. https://doi.org/10.1016/j.tim.2018.07.001

National Centre for Animal Health, Bhutan: https://ncah.gov.bt/nadpmrcp/

Thangaraj J.W.V., Navaneeth S.K., Shanmugasundaram D., Suganya E., Sudha R. *et al.* 2025. Estimates of the burden of human rabies deaths and animal bites in India, 2022–23: A community-based cross-sectional survey and probability decision-tree modelling study. *The Lancet Infectious Diseases*, 25(1), 126–134. https://doi.org/10.1016/S1473-3099(24)00490-0

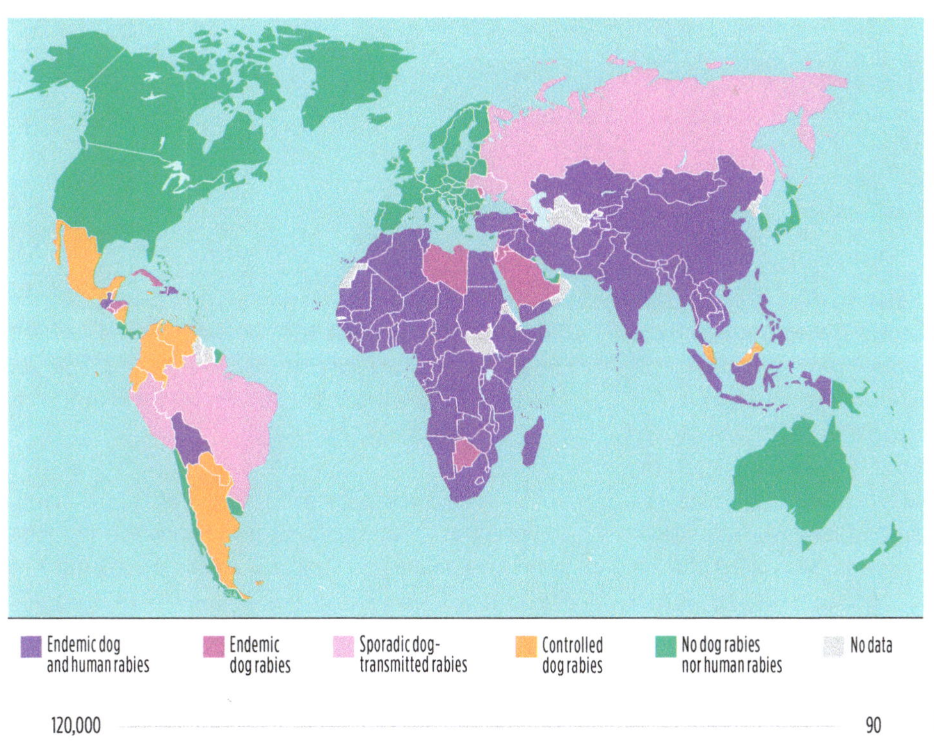

Figure 1. Global distribution of rabies cases. Source of data: WHO (https://www.who.int/data/gho/data/indicators/indicator-details/GHO/reported-number-of-human-rabies-deaths) and WAHIS-WOAH (https://wahis.woah.org/#/dashboards/qd-dashboard).

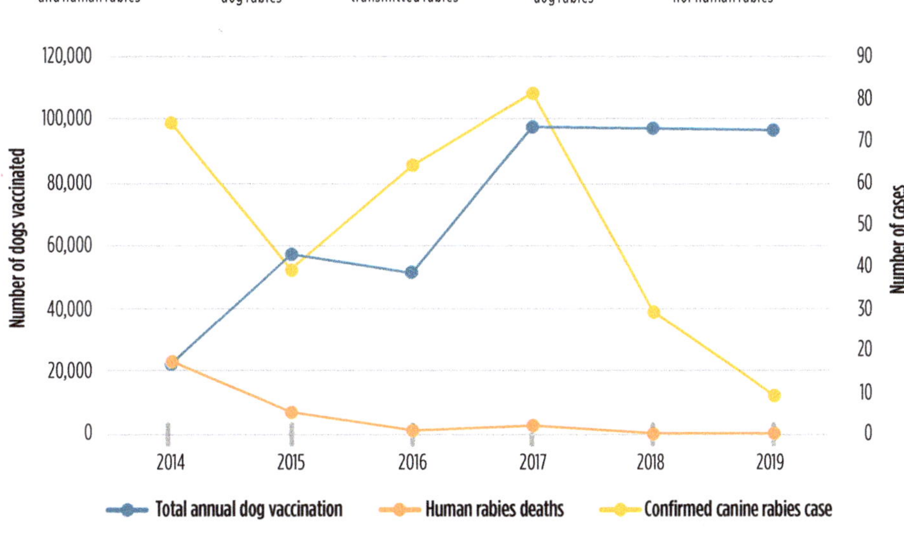

Figure 2. Successful implementation of the One Health approach focused on dog vaccination to control rabies in India. Data extracted from Gibson *et al.* 2022.

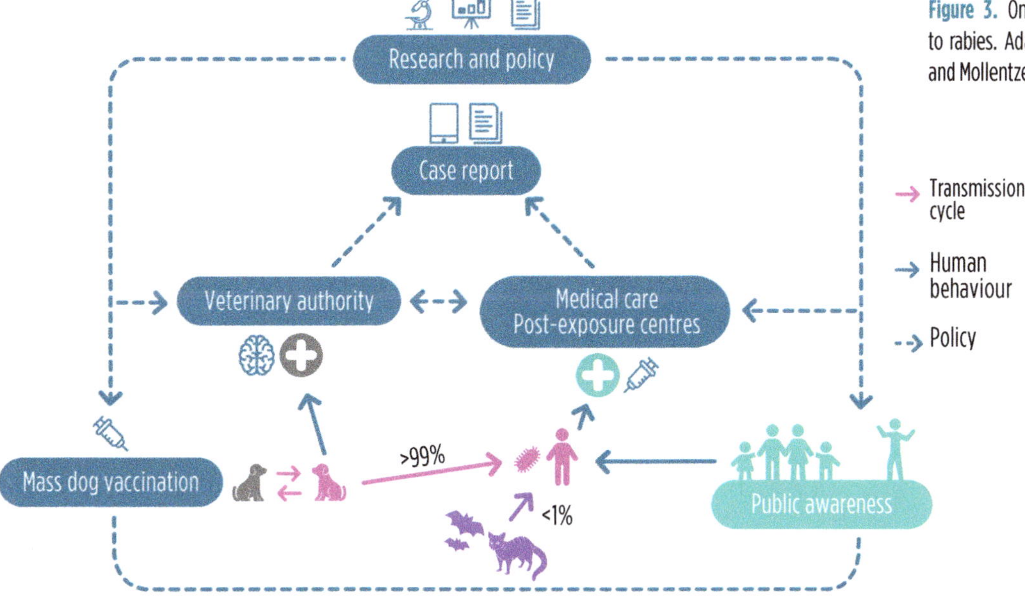

Figure 3. One Health approach to rabies. Adapted from Brunker and Mollentze 2018.

The One Health approach to manage Rift Valley fever transmission

Marie-Marie Olive,
Soa-Fy Andriamandimby,
Véronique Chevalier

Rift Valley fever virus (RVFV) is a zoonotic arthropod-borne virus that infects ruminants and humans throughout Africa, the South-West Indian Ocean islands, and the Arabian Peninsula. RVFV is spread by many mosquitoes, including the genera *Culex* and *Aedes* (Figure 1). Transovarian RVFV transmission has only been shown in *Aedes mcintoshi* in East Africa and may occur in other *Aedes* species. Inter-epizootic mechanisms are still poorly known. In dry mud, infected diapause eggs may survive inter-epizootic periods. Flooded eggs hatch as infected imagoes, starting the transmission cycle (Figure 1.A). Other mosquito genera spread the virus through ruminants, accelerating the cycle (Figure 1.B). Transhumance or trade of ruminants can transfer the virus to virus-free environments. Humans are mostly infected by bovine tissues or fluids after slaughter (Figure 1.C).

Climatic and environmental factors are known to drive RVFV outbreaks and boost mosquito emergence and multiplication. In East Africa, RVFV outbreaks are linked to heavy rainfalls closely related to the warm phase of the El Niño Southern Oscillation in the south-western Indian Ocean.

Although the amplification cycle between ruminants and mosquitoes must happen before human cases will occur, RVFV outbreaks first tend to be noticed when human cases appear. This underscores the need to reinforce One Health surveillance in risky areas, where mosquito abundance and animal and human syndromes are monitored along with environmental changes (Figure 2).

As an example of the One Health approach, an integrated analysis of environmental, cattle and human serological data was performed in Madagascar following the 2008–2009 epidemics in order to identify the at-risk area for Rift Valley fever (RVF) outbreaks in the country. Using a statistical approach, the relationship between both human and cattle RVFV serological status and several climate and landscape factors (a proxy for RVF vector abundance) and bovine density was studied at the commune level. The results suggest that the humid environment of the western, north-western and eastern coastal areas are suitable for RVFV transmission in both cattle and humans. A risk map was drawn up showing the most likely areas for RVFV transmission in Madagascar (Figure 3). This map predicted the areas affected by the 2021 RVF epidemic in Madagascar. When combined with anthropological studies to assess the social acceptability of measures taken, this study should help better target RVF prevention, surveillance and control efforts in Madagascar.

We believe that the burden of RVF can be reduced by adopting a One Health approach in broad collaboration between multidisciplinary research and health sectors.

References

Bird B.H., Nichol S.T. 2012. Breaking the chain: Rift Valley fever virus control via livestock vaccination. *Current Opinion in Virology*, 2, 315–323. https://doi.org/10.1016/j.coviro.2012.02.017

Linthicum K.J., Anyamba A., Tucker C.J., Kelley P.W., Myers M.F., Peters C.J. 1999. Climate and satellite indicators to forecast Rift Valley fever epidemics in Kenya. *Science*, 285, 397–400. https://doi.org/10.1126/science.285.5426.397

Lancelot R., Cêtre-Sossah C., Hassan O.A., Yahya B., Ould Elmamy B. *et al.* 2019. Rift Valley fever: One Health at play? *In Transboundary Animal Diseases in Sahelian Africa and Connected Regions* (Kardjadj M., Diallo A., Lancelot R., eds), pp.121–148. https://doi.org/10.1007/978-3-030-25385-1_8

Olive M.-M., Chevalier V., Grosbois V., Tran A., Andriamandimby S.-F. *et al.* 2016. Integrated analysis of environment, cattle and human serological data: risks and mechanisms of transmission of Rift Valley fever in Madagascar. *PLOS Neglected Tropical Diseases*, 10(8), e0004976. https://doi.org/10.1371/journal.pntd.0004827

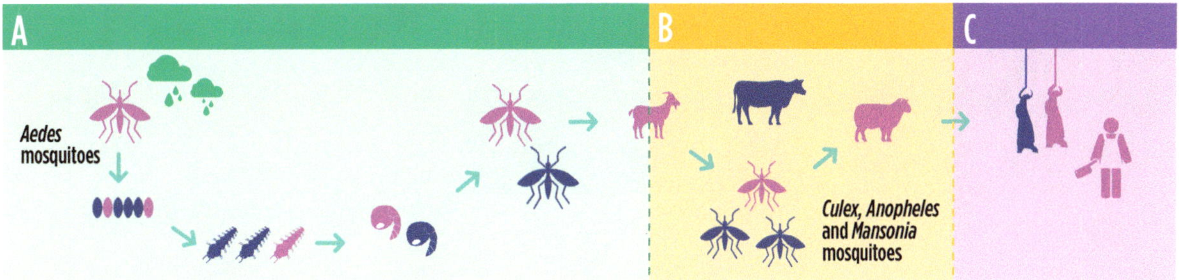

Figure 1. Schematic representation of the transmission cycle.

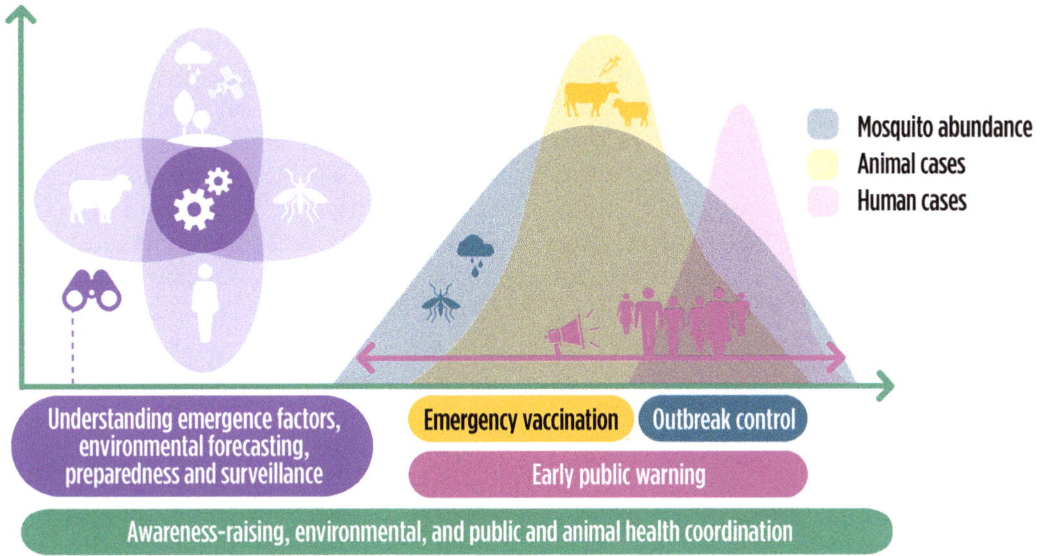

Figure 2. One Health approach to understand, forecast, prevent, ensure early detection of and control Rift Valley fever. Adapted from Lancelot *et al.* 2019; Bird and Nichols 2012.

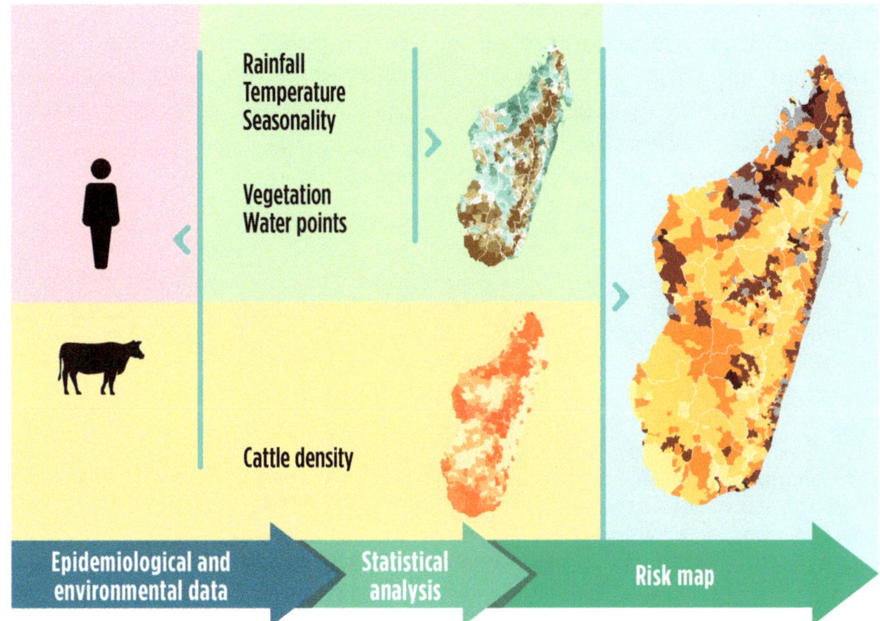

Figure 3. General framework for mapping Rift Valley fever virus transmission risk areas in Madagascar based on environmental indicators and cattle density.

Multi-scale spatial modelling to facilitate the implementation of One Health approaches

Annelise Tran, Hobiniaina Anthonio Rakotoarison, Larisa Lee-Cruz, Daouda Kassié

Epidemiological cycles can be considered complex systems. For infectious diseases involving several reservoir or vector species in particular, understanding transmission requires knowledge of the wild or domestic animals and arthropods involved, their interactions, and the impact of climatic conditions, landscape, human activities, etc. Modelling offers a methodological framework to reproduce and study a real system through abstraction. In the field of epidemiology, it has been used to simulate epidemic dynamics, the impact of control actions, and to map areas at risk of transmission. By considering the spatial dimension, we can implicitly consider the environment and spatial interactions, which thus favours One Health approaches.

Expert knowledge, biological surveys and participative approaches can be combined using spatial modelling techniques to tackle human and animal health issues. In Madagascar, a spatial multicriteria evaluation (MCE) approach was used to identify areas where pigs are at a higher risk of contamination with synthetic anabolic hormones, present in human contraceptives illegally used in pig farms. Data describing the drug supply chain in veterinary and human sectors were combined to produce risk maps at the country level (Figure 1), consistent with results from pig samples collected in slaughterhouses and with breeders' responses collected at local scale.

Similar approaches have been used to better understand the transmission cycle of the Ebola virus, which potentially involves several intermediate animal host species. A recent study at regional scale used a spatial MCE approach to combine different categories of data—climatic, environmental, anthropogenic, and the distribution of animal species potentially involved—to map the risk of Ebola emergence (Figure 2), which varies in space and time according to the seasons.

To facilitate the use of spatial models by non-specialists, tools like the QGIS "Full MCE for public health" plug-in[1] have been developed. Such tools allow public health actors to generate risk maps and integrate them into their action plans. These tools and methods can be used together to combine risk factors from different sectors and disciplines to help better integrate knowledge into health and health policy actions. They can also be useful in implementing systemic approaches such as the One Health approach.

1. www.geoinformations.developpement-durable.gouv.fr/qgis-r625.html

References

Lee-Cruz L., Lenormand T., Cappelle J., Caron A., De Nys H. *et al.* 2021. Mapping of Ebola virus spillover: Suitability and seasonal variability at the landscape scale. *PLoS Neglected Tropical Diseases*, 15(8), 29 p. https://doi.org/10.1371/journal.pntd.0009683

Rakotoarison H.A., Rasamimalala M., Rakotondramanga J.M., Ramiranirina B., Franchard T. *et al.* 2020. Remote sensing and multi-criteria evaluation for malaria risk mapping to support indoor residual spraying prioritization in the central highlands of Madagascar. *Remote Sensing*, 12(10), 1585. https://doi.org/10.3390/rs12101585

Ramahatafandry D.A. 2020. Cartographie des risques d'utilisation détournée du medroxyprogesterone acétate (MPA) dans les élevages de porcs à Madagascar. Institut Pasteur de Madagascar, 47 p.

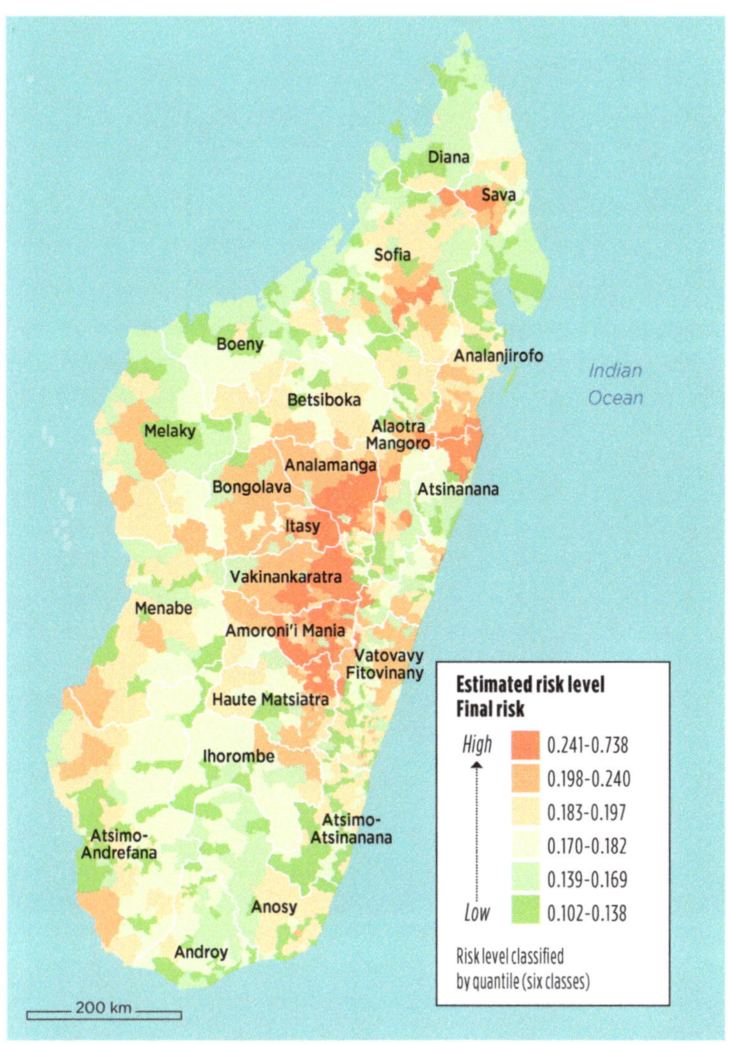

Figure 1. Risk map for residues of synthetic anabolic hormones (medroxyprogesterone acetate) in pigs in Madagascar, produced from expert knowledge, biological surveys and participative approaches.

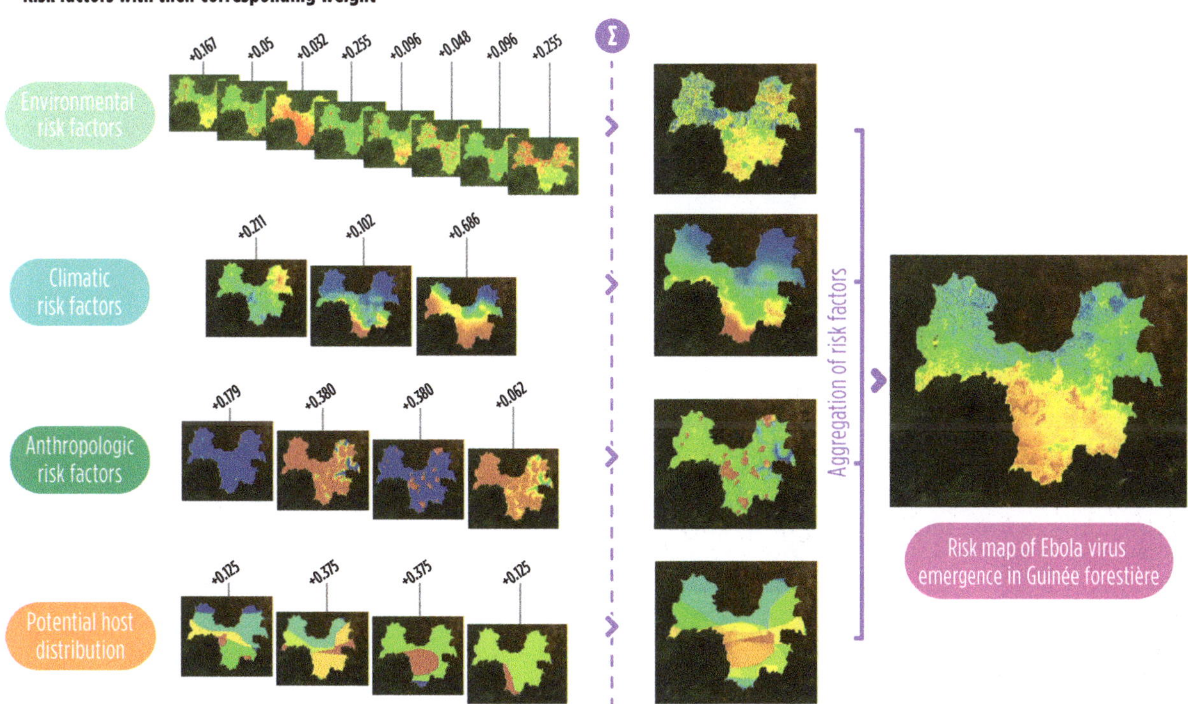

Figure 2. Illustration of the different stages of the spatial multi-criteria evaluation approach, using the example of Ebola virus emergence.

Tuberculosis unleashed: why One Health holds the key

François Roger

Tuberculosis (TB) is an ancient disease that has afflicted humankind for thousands of years. While the disease is primarily caused by the bacterium *Mycobacterium (M.) tuberculosis, M. bovis,* another member of the *M. tuberculosis* complex (MTBC), also plays a significant role, particularly in zoonotic TB. These two species share a common ancestor and have evolved to adapt to different hosts over time. Historically, *M. bovis* was a major cause of human TB, especially before the introduction of milk pasteurization and advances in veterinary practices.

Around 10 million people worldwide develop TB annually, mostly due to *M. tuberculosis. M. bovis* accounts for a much smaller percentage of human cases—typically fewer than 10%, and often below 2% in countries with stringent public health measures such as milk pasteurization and bovine TB control. The disease caused about 1.25 million human deaths in 2023 (WHO). Human infection with *M. bovis* typically occurs through direct contact with infected animals or indirectly via contaminated food. Although *M. tuberculosis* is primarily a human pathogen, it can also infect animals when humans and animals live in close proximity, such as with elephants in zoos, dogs, pigs and cattle (Figure 1). This bidirectional transmission underscores the critical need for integrated surveillance and control systems. Several countries have implemented surveillance programs for zoonotic TB due to *M. bovis* at varying levels of intensity (Figure 2).

However, recent studies have shown that zoonotic TB is even more complex than previously thought. Other MTBC members, including *M. orygis, M. caprae,* and *M. pinnipedii,* also contribute to zoonotic TB transmission. For instance, *M. orygis* is responsible for most zoonotic TB cases in South Asia, while *M. caprae* often spills over from infected goats and other livestock in Europe. *M. pinnipedii,* found in pre-Columbian human populations in South America, causes infections in people who come into contact with diseased seals and sea lions. These findings suggest that the current narrow definition of zoonotic TB—primarily as *M. bovis* infection—poorly reflects the full spectrum of TB of animal origin.

Managing TB within the One Health framework is extremely important, as it promotes interdisciplinary collaboration for disease surveillance and control. Furthermore, a deeper understanding of the broader ecological context in which MTBC subspecies circulate can support the development of more effective public health policies, ensuring a comprehensive approach to TB prevention and control.

References

de Macedo Couto R., Santana G.O., Ranzani O.T., Waldman E.A. 2022. One Health and surveillance of zoonotic tuberculosis in selected low-income, middle-income and high-income countries: A systematic review. *PLoS Neglected Tropical Diseases*, 16(6), e0010428. https://doi.org/10.1371/journal.pntd.0010428

Dean A.S., Zumla A., Abubakar I., Gilsenan A., Glaziou P. *et al.* 2018. A roadmap for zoonotic tuberculosis: A One Health approach to ending tuberculosis. *The Lancet Infectious Diseases*, 18(12), 137–138. https://doi.org/10.1016/S1473-3099(18)30013-6

Duffy S.C., Marais B., Kapur V., Behr M.A. 2024. Zoonotic tuberculosis in the 21st century. *The Lancet Infectious Diseases*, 24(4), 339–341. https://doi.org/10.1016/S1473-3099(24)00059-8

Etter E., Donado P., Jori F., Caron A., Goutard F., Roger F. 2006. Risk analysis and bovine tuberculosis, a re-emerging zoonosis. *Annals of the New York Academy of Sciences,* 1081(1), 61–73. https://doi.org/10.1196/annals.1373.006

Santos N. 2023. Overview of tuberculosis in animals. MSD Veterinary Manual. Retrieved from www.msdvetmanual.com/generalized-conditions/overview-of-tuberculosis-in-animals/overview-of-tuberculosis-in-animals

WILDLIFE
ENVIRONMENT
LIVESTOCK
HUMANS

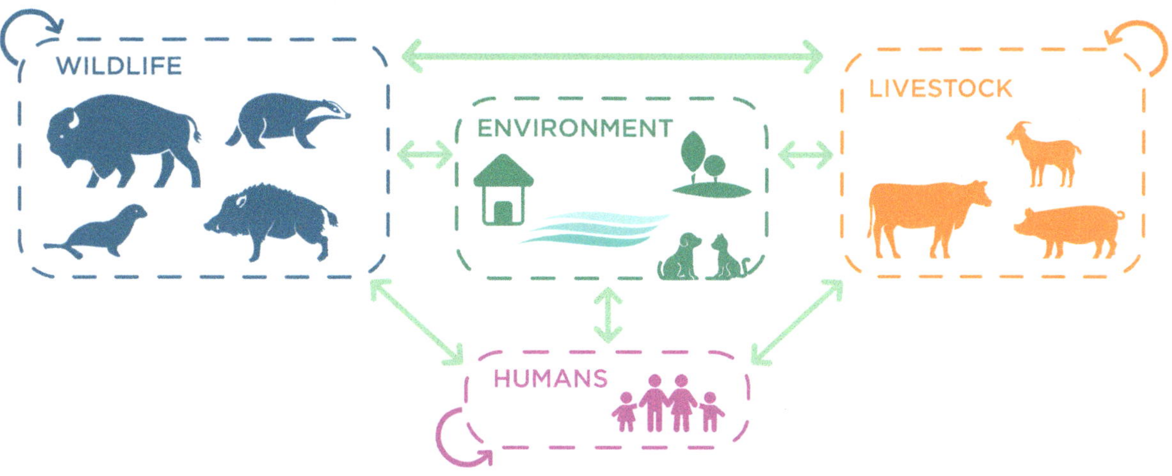

Figure 1. The spread of bovine tuberculosis between animals and humans illustrates the importance of the current One Health approach.

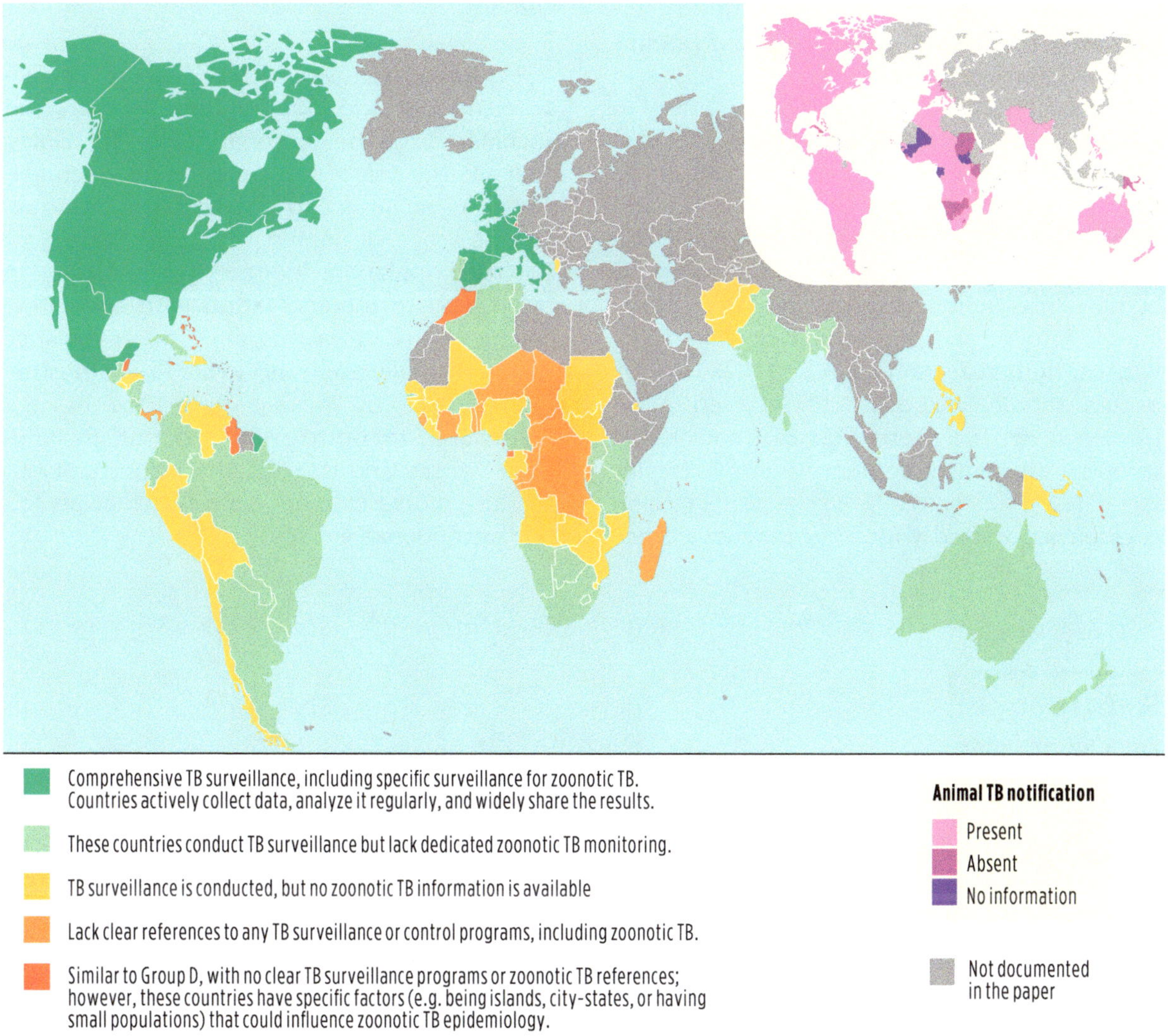

Comprehensive TB surveillance, including specific surveillance for zoonotic TB. Countries actively collect data, analyze it regularly, and widely share the results.
These countries conduct TB surveillance but lack dedicated zoonotic TB monitoring.
TB surveillance is conducted, but no zoonotic TB information is available
Lack clear references to any TB surveillance or control programs, including zoonotic TB.
Similar to Group D, with no clear TB surveillance programs or zoonotic TB references; however, these countries have specific factors (e.g. being islands, city-states, or having small populations) that could influence zoonotic TB epidemiology.
Animal TB notification
Present
Absent
No information
Not documented in the paper

Figure 2. Global map of human tuberculosis (TB) surveillance data related to cases caused by *Mycobacterium bovis* and animal TB. From de Macedo Couto *et al.* 2022.

Cross-species influenza threats: the critical role of One Health surveillance and control

Marisa Peyre,
Claire Hautefeuille

Influenza viruses are known to mutate easily and cross the species barrier, with phylogenetics studies showing how human and animal viruses are interconnected (Figure 1). For this reason, it is essential to adopt One Health approaches for surveillance and control of these viruses in animal and human populations.

The low pathogenic avian influenza H7N9 surveillance and control in Asia is a good illustration. H7N9 has low pathogenicity in poultry but has been responsible for many cases in humans. H7N9 was first detected in humans in February 2013 in Shanghai and Anhui, China. Between February 2013 and July 2018, more than 1,500 human cases were confirmed and more than 600 people died, mainly in China (Figure 2). Over this period, five waves of human infections occurred, typically starting in October and ending in June. Like H5N1 viruses, human transmission of H7N9 viruses occurs through exposure to infected poultry. However, the proportion of transmission through accidental contact is much higher for H7N9 than for H5N1, whose transmission is often associated with risky practices close to birds. Surveillance of live poultry markets combined with the closure of infected markets has reduced the number of human cases. The implementation of a vaccination campaign in domestic birds has been effective in reducing the prevalence of infection and the risk of transmission to humans. In the case of H7N9 management, the surveillance and the control of birds successfully reduced infections in humans.

Moreover, some countries like France recommend pig and poultry farm workers get vaccinated for the seasonal human flu. Doing so limits the risk of transmission between animals and humans and prevents the introduction of seasonal human virus into the pig or poultry population. Indeed, this introduction could lead to recombination with circulating animal influenza viruses and the emergence of a potentially zoonotic virus. Pigs are often considered to be "mixing-vessels" as they are susceptible to both avian and human influenza viruses and provide the opportunity to produce a reassortment from swine, avian and human influenza viruses.

References

EFSA, ECDC. 2024. *Avian influenza: One Health surveillance is key to prevent virus evolving.*

FAO. 2022. Avian Influenza A(H7N9) virus situation update. https://www.fao.org/animal-health/situation-updates/avian-influenza-A(H7N9)-virus/en

FAO. 2020. EMPRES-i—Global Animal Disease Information System. https://empres-i.apps.fao.org/general

Haut Conseil de la Santé Publique (HCSP). 2021. Avis relatif à la prévention de la transmission à l'homme des virus influenza porcins et aviaires. In: *Rapport de l'HCSP.* Haut Conseil de la Santé Publique. https://www.hcsp.fr/explore.cgi/avisrapportsdomaine?clefr=1142

Jiang W., Hou G., Li J., Peng C., Wang S. *et al.* 2019. Prevalence of H7N9 subtype avian influenza viruses in poultry in China, 2013-2018. *Transboundary and Emerging Diseases*, 66(4), 1758-1761. https://doi.org/10.1111/tbed.13183

Pardo-Roa C., Nelson M.I., Ariyama N., Aguayo C., Almonacid L.I. *et al.* 2025. Cross-species and mammal-to-mammal transmission of clade 2.3.4.4b H5N1 virus in Chile. *Nature Communications*, 16, 1234. https://doi.org/10.1038/s41467-025-57338-z

Swayne D.E., Ed. 2016. *Animal Influenza, 2nd edition.* Wiley.

Yu H., Wu J.T., Cowling B.J., Liao Q., Fang V.J. *et al.* 2014. Effect of closure of live poultry markets on poultry-to-person transmission of avian influenza A H7N9 virus: an ecological study. *The Lancet*, 383(9916), 541-548. https://doi.org/10.1016/S0140-6736(13)61904-2

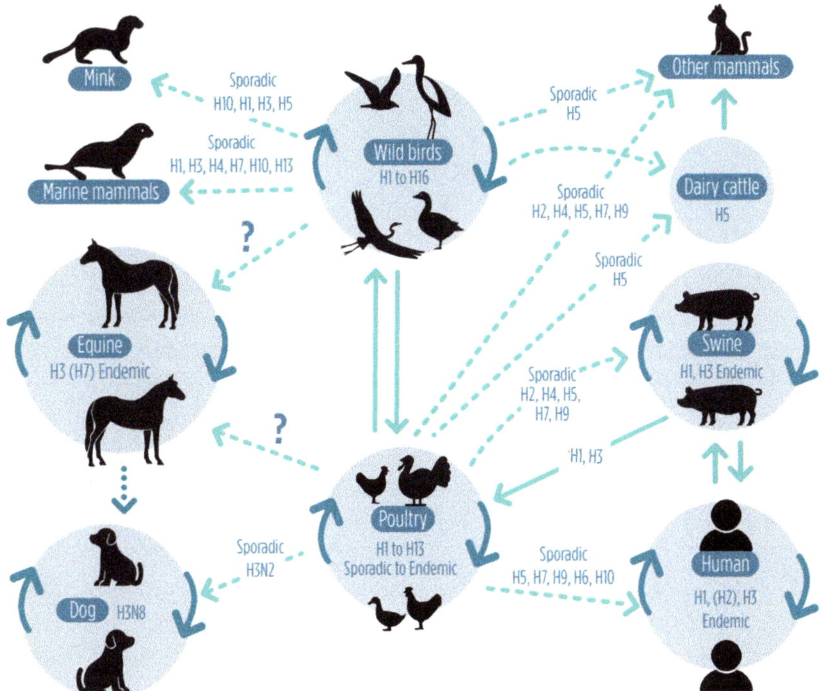

Figure 1. Influenza viruses transmission between species. H: hemagglutinin subtype. Adapted from Swayne 2016.

Figure 2. Farm outbreaks and human cases due to H7N9 viruses in 2017. HPAI: highly pathogenic avian influenza; LPAI: low pathogenic avian influenza. Source: FAO 2019.

One Health approach and African trypanosomiasis in Guinea

Mamadou Camara, Moïse Kagbadouno, Dimitri Justeau-Allaire, Bruno Bucheton, Jean-Mathieu Bart

African trypanosomiasis is a parasitic disease caused by a protozoan of the genus *Trypanosoma*, which is transmitted from one mammal to another through the bite of the tsetse fly. In 2025, the WHO officially declared that human African trypanosomiasis (HAT) had been eliminated as a public health problem in Guinea. This disease wreaked havoc during the 20th century, particularly in the Guinée forestière. Thanks to medical surveys and improved living conditions, such as the widespread installation of wells in villages, the last cases of HAT in the Guinée forestière date back more than 20 years. However, while Guinea's goal was to eliminate human transmission by 2030, the recent massive development of pig farming in the Guinée forestière raises the question of the potential role of these animals as a reservoir for the parasite. In addition to implementing an epidemiological surveillance strategy based on the diagnosis of suspected cases, the Guinean National Control Programme for Neglected Tropical Diseases, in collaboration with the Institut de recherche pour le développement (IRD, French National Research Institute for Sustainable Development), decided to update entomological and veterinary data according to the One Health approach (Figure 1).

Surveys are ongoing in the Guinée forestière to determine the trypanosomiasis transmission route from 2022 to 2024. A strong spatial distribution pattern was observed: tsetse flies were only captured in Gouécké area (northern zone), while none were found in the Yomou area (southern zone). Analysis of pig blood samples also revealed a specific immune imprint for trypanosomes in 27% of the pigs, with parasites found in 9.5% of them. Our data show a perfect spatial correlation: in the southern zone, where no tsetse flies were caught, pigs are trypanosome-free. In the northern zone, where tsetse flies were caught, a high proportion of pigs were trypanosome carriers. Molecular studies showed the presence of *T. congolense* (9%) and *Trypanozoon* (45%), including one pig harboring *T. b. gambiense*, the agent of HAT, which could present a risk of spillover from pigs to humans.

To understand the influence of the landscape configuration on this pattern, we are exploring the digital twin paradigm[1]. We plan to integrate heterogeneous spatial data from various sources (e.g. field inventories, remote sensing) and disciplines (e.g. epidemiology, ecology, social sciences) into an information system that will enable us to analyse the interrelationships between the various One Health dimensions. Subsequently, AI-assisted participatory planning will allow stakeholders to design and implement better-informed management actions (e.g. installation of tiny targets), whose effects will be measured and monitored through the digital twin.

1. The digital twin is a virtual representation of a real-world system or process, used to simulate, analyze, and optimize its behavior in real time.

References

Solano P., Courtin F., Kaba D., Camara M., Kagbadouno M. *et al.* 2023. Vers l'élimination de la maladie du sommeil [Towards elimination of human African trypanosomiasis]. *Medecine Tropicale et Santé Internationale*, 3(1). https://doi.org/10.48327/mtsi.v3i1.2023.317

Camara O., Kaboré J.W., Soumah A., Leno M., Bangoura M.S. *et al.* 2024. Conducting active screening for human African trypanosomiasis with rapid diagnostic tests: The Guinean experience (2016-2021). *PLoS Neglected Tropical Diseases*, 18(2), e0011985. https://doi.org/10.1371/journal.pntd.0011985

Traoré B.M., Koffi M., N'Djetchi M.K., Kaba D., Kaboré J. *et al.* 2021. Free-ranging pigs identified as a multi-reservoir of *Trypanosoma brucei* and *Trypanosoma congolense* in the Vavoua area, a historical sleeping sickness focus of Côte d'Ivoire. *PLoS Neglected Tropical Diseases*, 15(12), e0010036. https://doi.org/10.1371/journal.pntd.0010036

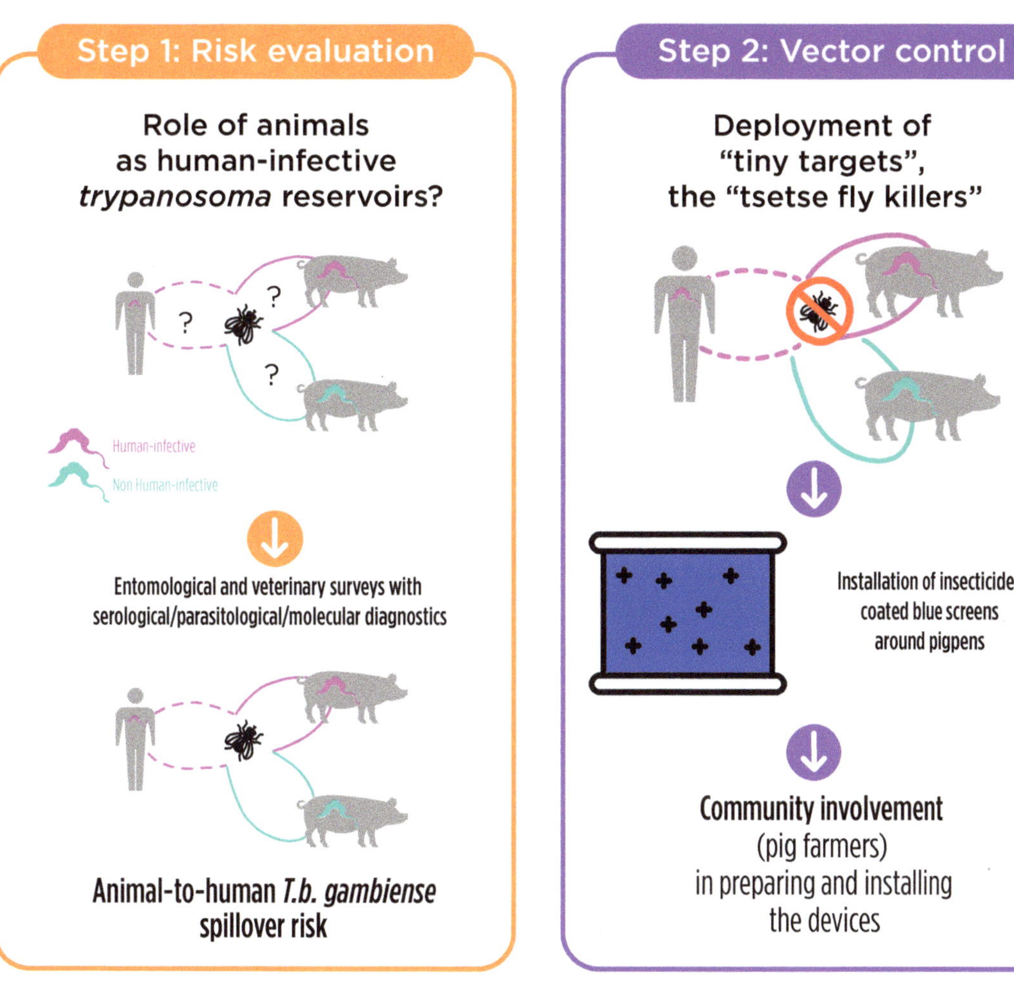

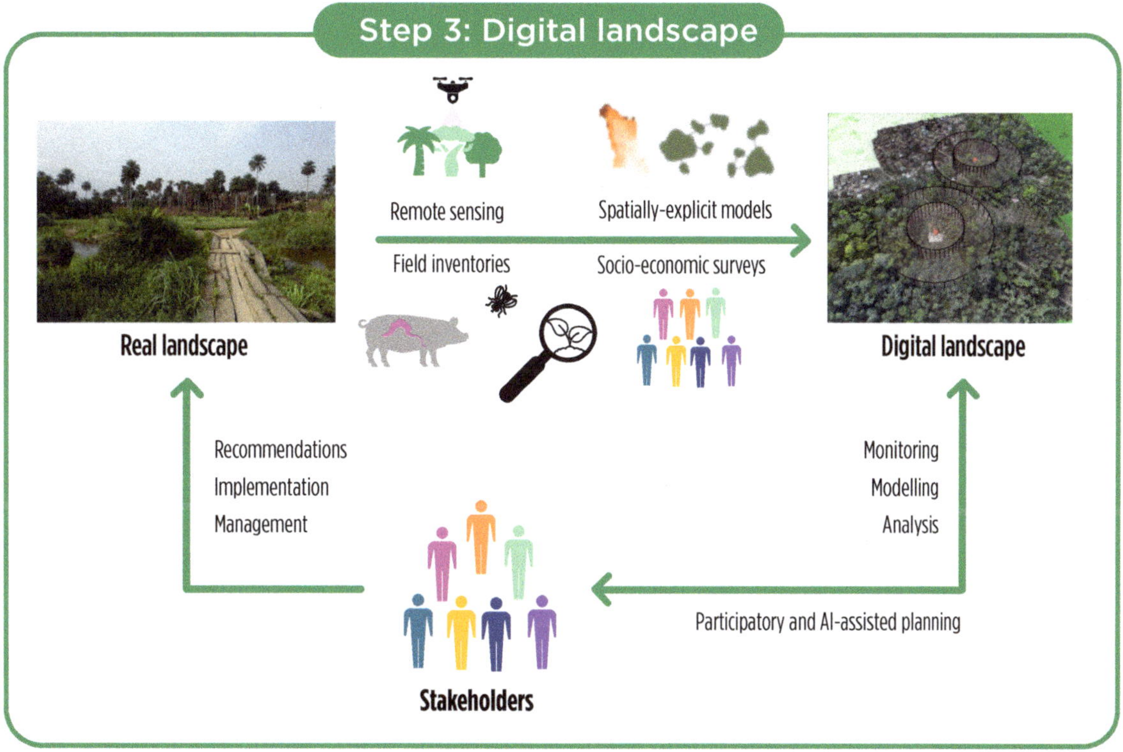

Figure 1. One Health strategies to unravel the role of animals as reservoirs of human African trypanosomiasis.

Combating non-zoonotic animal plagues: the power of One Health in protecting livelihoods

François Roger

African swine fever (ASF) and peste des petits ruminants (PPR) are among the most devastating animal diseases (Figure 1). These two animal plagues have microbiological, epidemiological and ecological characteristics that justify differentiated study but present comparable difficulties with regard to their control and which requires an integrated perspective. The One Health approach can offer a solution in the face of these diseases due to the large impact they have on food security, economic stability, geopolitics and public health.

ASF is a viral disease that mainly infects domestic and wild pigs. The ASF virus is extremely contagious, usually fatal and always causes major economic losses in areas where people are reliant on pigs for their protein (meat) supply and incomes. PPR is a highly infectious viral disease of small ruminants and other domestic and wild species. PPR causes high mortality among livestock, which reduces incomes of poor farmers.

The One Health approach offers a structured framework for dealing with diseases like these with major social and economic repercussions. Neither PPR nor ASF is zoonotic, but due to their consequential effects on animal health, they indirectly affect human well-being by impacting the incomes of family farmers. These diseases can lead to food supply problems in areas depending heavily on livestock for nutrition and livelihoods. Added to these issues are the ecological consequences—i.e. impacts on wildlife and ecosystem disruptions. Effective disease control will require the veterinary and agricultural sectors to work together with environmental management and rural communities (Figure 2).

Successful execution of a One Health approach in terms of surveillance of these types of diseases requires integrated surveillance systems for domesticated and wild animals. These systems must be coupled with strong biosecurity methods and vaccination efforts where possible, with greater community engagement at the local level. Indeed, by involving farming communities in the decision-making process, we can ensure that the measures adopted are both sustainable and appropriate from a cultural point of view. Finally, integrating animal health into more comprehensive public and environmental health strategies strengthens community resilience and could mitigate the economic impact of an animal plague epidemic.

References

Bakke H.J., Perez A.D., Miclat-Sonaco R., Perez A.M., Schambow R.A. 2025. Mental health impacts of African swine fever outbreaks on veterinarians in the Philippines. *Frontiers in Veterinary Science*, 12, 1519270. https://doi.org/10.3389/fvets.2025.1519270

Lane J.K., Kelly T., Bird B., Chenais E., Roug A. *et al.* 2025. A One Health approach to reducing livestock disease prevalence in developing countries: Advances, challenges, and prospects. *Annual Review of Animal Biosciences*, 13, 277–302. https://doi.org/10.1146/annurev-animal-111523-102133

Otu A., Onwusaka O., Meseko C., Effa E., Ebenso B. *et al.* 2024. Learning from One-Health approaches to explore links between farming practices, animal, human and ecosystem health in Nigeria. *Frontiers in Nutrition*, 11, 1216484. https://doi.org/10.3389/fnut.2024.1216484

Roger F.L., Fournié G., Binot A., Wieland B., Kock R.A. *et al.* 2021. Peste des petits ruminants (PPR): Generating evidence to support eradication efforts. *Frontiers in Veterinary Science*, 7, 636509. https://doi.org/10.3389/fvets.2020.636509

Roth J.A., Galyon J. 2024. Food security: The ultimate one-health challenge. *One Health*, 100864. https://doi.org/10.1016/j.onehlt.2024.100864

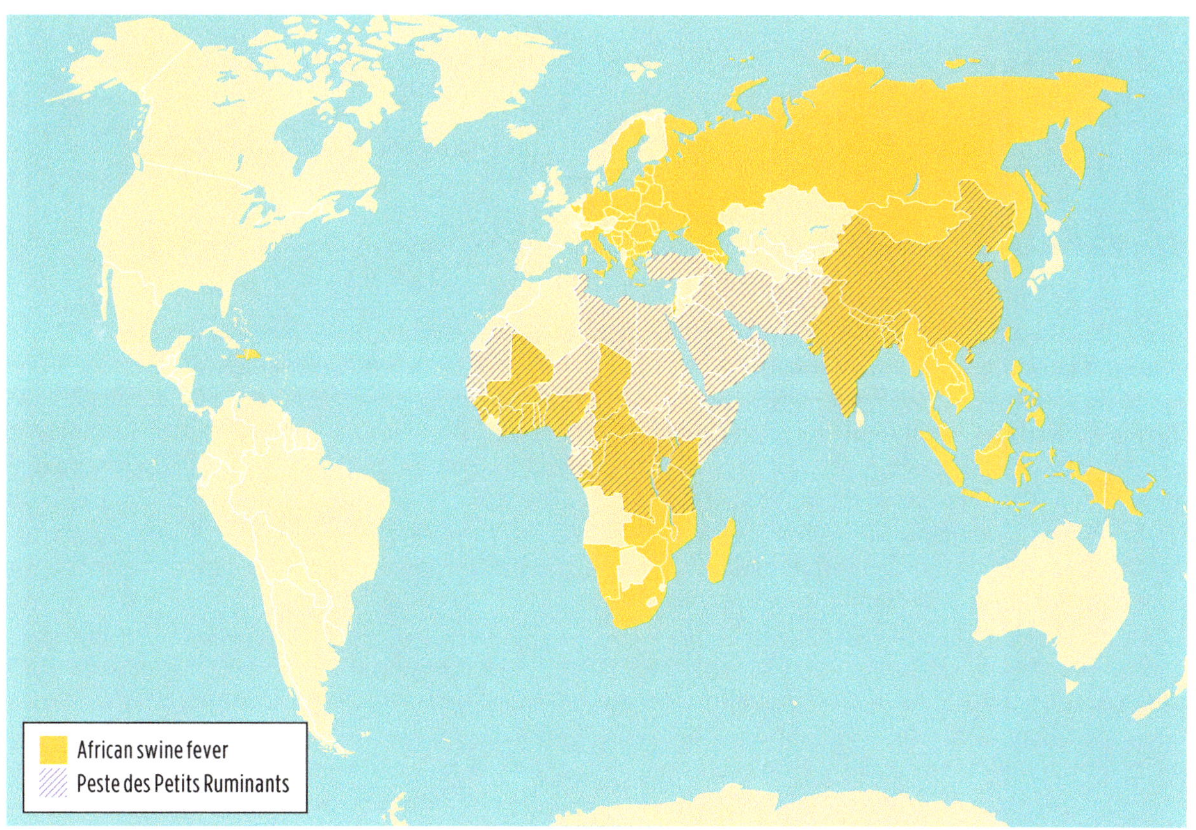

Figure 1. African swine fever (ASF) and peste des petits ruminants (PPR) distribution in 2024. From: World Animal Health Information System (WAHIS) – WOAH 2024.

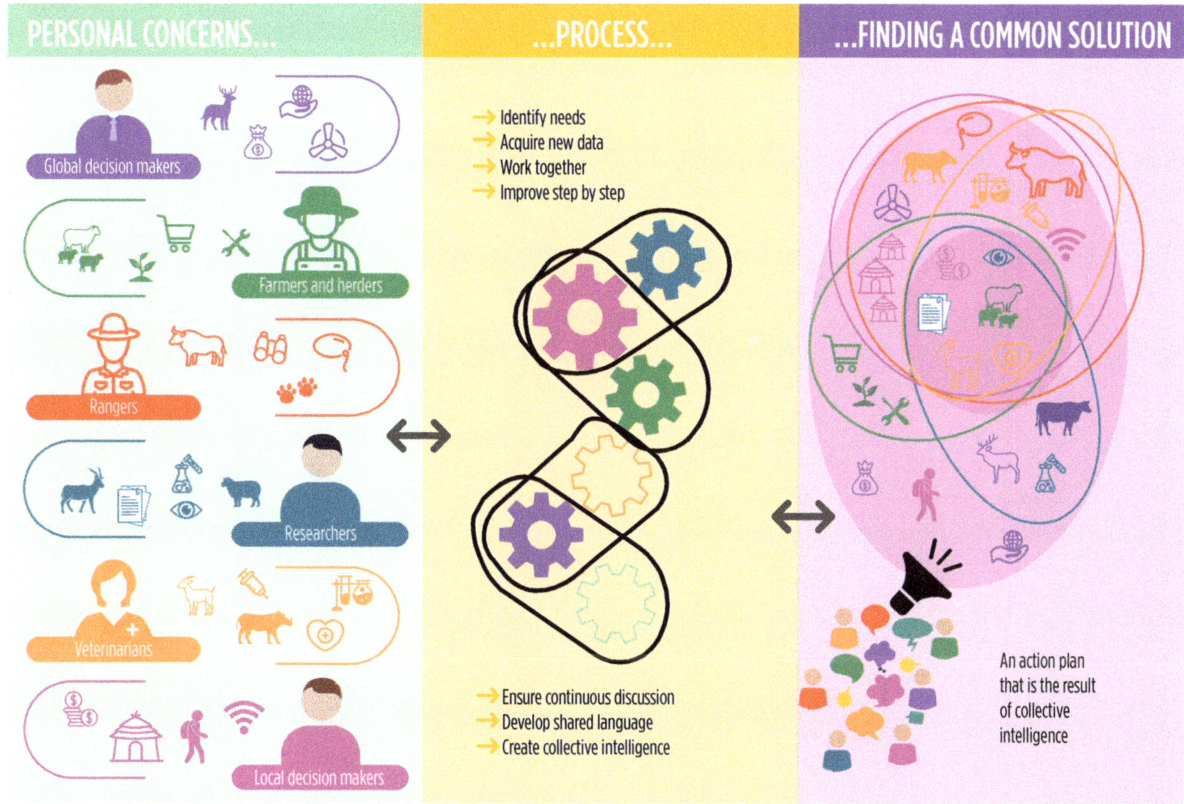

Figure 2. A collaborative process for controlling PPR, involving stakeholders such as global and local decision-makers, farmers, herders, rangers, researchers, and veterinarians, each with specific concerns like animal health and environmental sustainability. Central interlocking gears represent a four-step process: identifying needs, collecting data, collaborating, and continuous improvement. This process depends on ongoing dialogue, developing a common language, and fostering collective intelligence. On the right, stakeholders' concerns merge into a shared solution, depicted by overlapping circles. The co-developed action plan emerges through a funnel, integrating input from the diverse group. From: Roger *et al.* 2021.

How One Health can transform food systems

Michel Duru

Public policymakers face challenges in creating healthier, more sustainable food systems because environmental, agricultural, food and health issues are often disconnected from each other. One Health has proven to be a useful approach in dealing with antibiotic resistance and zoonoses. An approach to health that considers four areas—humans, agroecosystems, livestock and the Earth system (Figure 1, central area)—can better account for the links between agriculture, food and non-communicable diseases. Earth-system health is characterized by planetary boundary indicators.

When developing public policies to achieve the Sustainable Development Goals related to food, three dimensions need to be considered (Figure 1, peripheral area): the degree of complexity of life, the spatiotemporal scales at which flows occur, and the political, social and economic factors on a local to international level. Biodiversity of organisms and their habitats is a result of games and power relations between stakeholders with different objectives, health perspectives, political agendas and media access. This can be true for both food system and health stakeholders, and depends on whether their investments are more curative than preventive.

The European food system (Figure 2), which is based on industrial agriculture and a typical Western diet (high intake of animal-based products and ultra-processed foods), can illustrate the approach discussed here. This food system creates significant environmental impacts, both locally and globally, and increases the risk of non-communicable diseases. It is also extremely dependent on imports (some of which are linked to deforestation) that can raise the risk of zoonoses.

Taking the French food system as an example, an additional 9 million hectares of agricultural land i.e. 34% of the country's utilized agricultural area (UAA)—would be required to produce that food (Figure 3). Two-thirds of that UAA (arable lands and grasslands) is used for livestock farming. Redirecting some of the arable land towards agroecology and shifting to a more plant-based diet would increase soil and landscape biodiversity due to decrease synthetic fertilizers and pesticides used, and improve human health. Moreover, producing renewable energies based on anaerobic digestion of animal waste and adopting intercropping practices could reduce the production of first-generation agrofuels.

The current system is efficient at producing high volumes of cheap food, but the hidden costs have been estimated at 8% of the GDP in developed countries. The holistic vision enabled by the One Health approach makes it possible to craft a convincing narrative to establish comprehensive public policies to prevent rebound effects and improve buy-in from target audiences.

References

Duru M. 2021. Le Covid-19 et le dérèglement climatique appellent à refonder notre système alimentaire. *Pour*, 239(1), 109-118. https://doi.org/10.3917/pour.239.0109

Duru M., Therond O. 2024a. Usage des terres et agriculture durable. *Regards*, R116. https://sfecologie.org/regard/r116-nov-2023-duru-et-therond-agriculture-et-usage-des-terres/

Duru M., Therond O. 2024b. *One Health* (Une seule santé) pour concevoir des alternatives crédibles aux défaillances des systèmes alimentaires. *Cahiers Agricultures*, 33, 18. https://doi.org/10.1051/cagri/2024016

FAO. 2023. Hidden costs of global agrifood systems worth at least $10 trillion. https://www.fao.org/newsroom/detail/hidden-costs-of-global-agrifood-systems-worth-at-least--10-trillion/en

Kahn LH. 2021. Developing a one health approach by using a multi-dimensional matrix. *One Health*, 13, 100289. http://doi.org/10.1016/j.onehlt.2021.100289

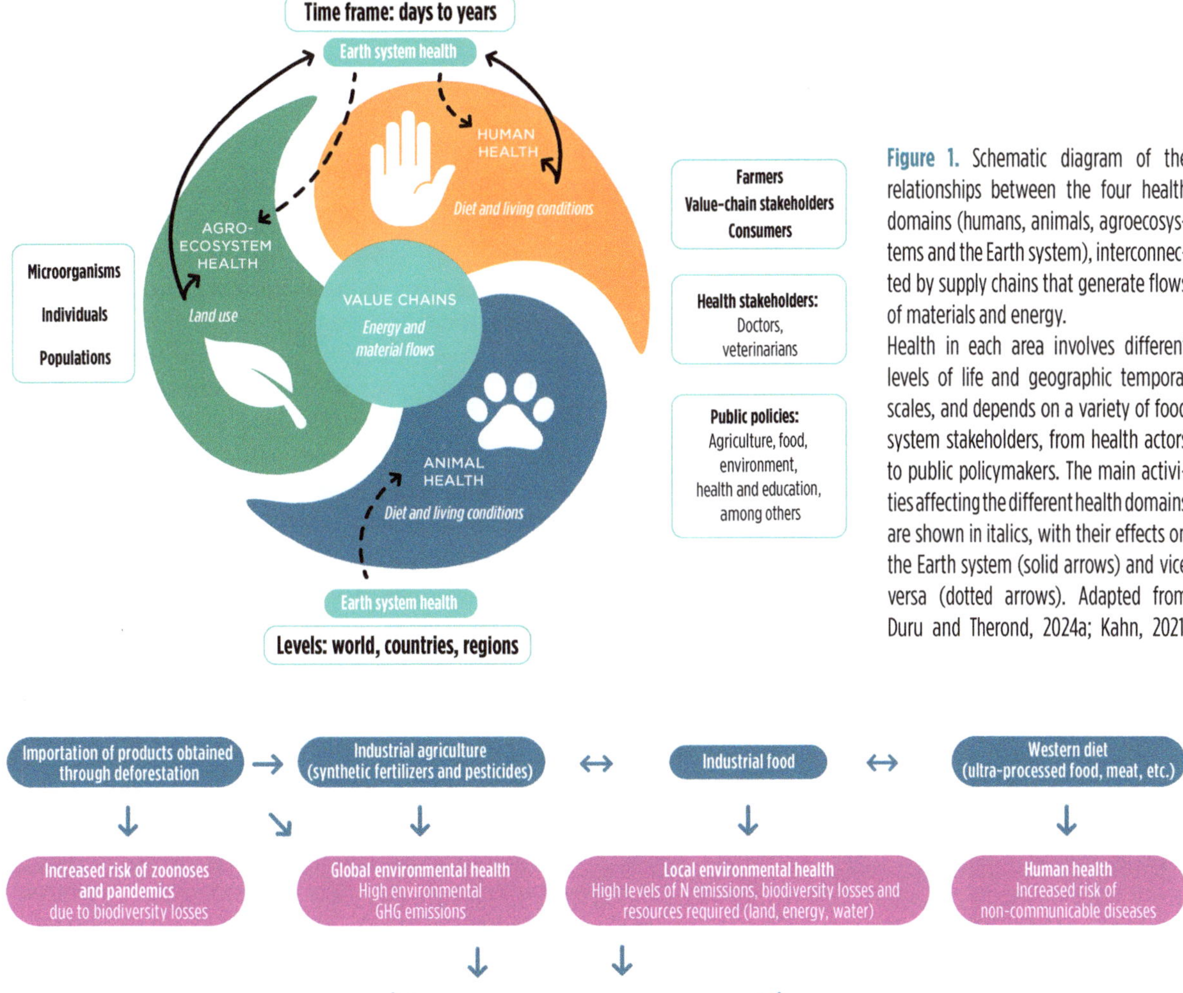

Figure 1. Schematic diagram of the relationships between the four health domains (humans, animals, agroecosystems and the Earth system), interconnected by supply chains that generate flows of materials and energy.
Health in each area involves different levels of life and geographic temporal scales, and depends on a variety of food system stakeholders, from health actors to public policymakers. The main activities affecting the different health domains are shown in italics, with their effects on the Earth system (solid arrows) and vice versa (dotted arrows). Adapted from Duru and Therond, 2024a; Kahn, 2021.

Figure 2. Schematic diagram of the food system, from field to plate, and its impacts on and costs to environmental and human health. Adapted from Duru 2021.

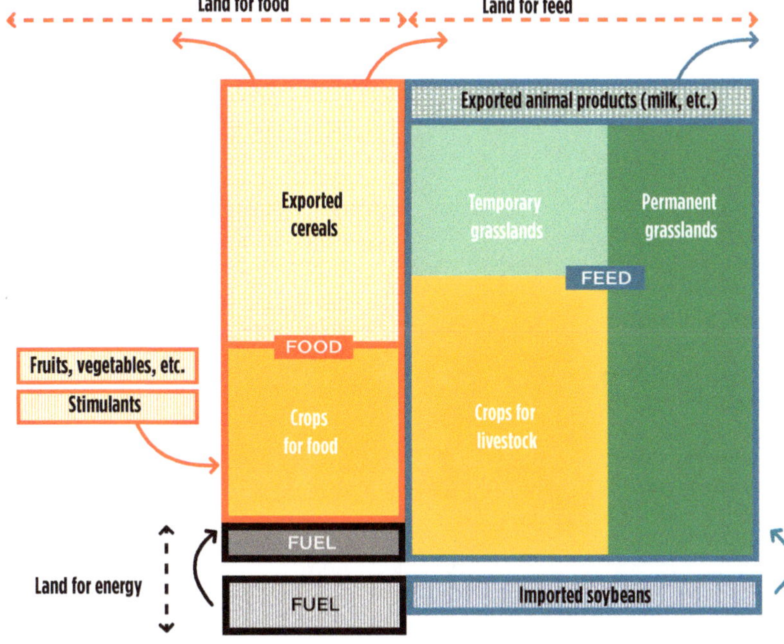

Figure 3. Schematic diagram of the main uses of agricultural land, as well as net imports and exports.
Areas used for our food supply in plant products (food: red outlines), in animal products (feed areas used for domestic livestock farming: blue outlines) and for energy production (fuel: black outlines). The production areas located in France are in the large central rectangle, while the "imported areas" are in the three outer rectangles. Adapted from Duru and Therond 2024b.

Crop production practices and infectious hazards: a One Health call to action

Alain Ratnadass,
Samira Sarter

Pathogenic microbes posing problems for human health can be transmitted through various cropping practices. Bacterial contaminations and infections are the predominant source of human infections. Studies conducted primarily in the United States, and to a lesser extent in China and South Africa (Figure 1), showed that coliforms (mainly via antibiotic resistance) were the leading cause of bacterial health issues. The most problematic parasites were protozoan *Cryptosporidium* spp. and *Giardia* spp. along with helminthic *Ascaris* spp. and *Trichuris* spp., with most studies (6 out of 38) conducted in Vietnam. Concerning viruses, enteric noroviruses and rotaviruses accounted for the largest share of literature references, with a majority of studies (5 out of 28) from the United States.

Crop production practices with objectives other than crop protection (e.g., irrigation, fertilization) negatively impact human infectious risks when based on the principles of agroecology and organic farming, while producing mixed impacts when based on conventional agriculture principles (Figure 2). Land-use change (i.e., mainly deforestation) can lead to very different zoonotic pathogen transmission and disease epidemic outcomes depending on the quality of the resulting matrix, which can act as a pathogen barrier (with a diversified agroforest) or a pathogen incubator (with a monospecific plant stand).

Integrating crop plant health within the One Health concept is thus relevant and valuable regarding human infectious disease risk reduction, regardless of benefits to other components. These actions are not restricted to crop protection practices, which improve crop plant health by alleviating biotic stresses but may negatively impact both natural ecosystem and agroecosystem health. They also include other cropping practices, such as land-use change (with deforestation degrading water resources, particularly reservoir water quality), crop irrigation (which can improve or harm water resources and soil health, depending on contamination levels), and fertilization (which affects plant nutrition and soil health, generally in a positive way).

While soil health is vital for crop plant health in agroecosystems, natural ecosystem health depends on wild fauna health, which is strongly affected by land-use changes and interactions with farming systems. Human and domestic animal health are impacted by both natural and agroecosystem health (including potential direct contamination by food and feed) (Figure 3).

References

Perfecto I., Chaves L.F., Fitch G.M., Hajian-Forooshani Z., Iuliano B. *et al.* 2023. Looking beyond land-use and land-cover change: Zoonoses emerge in the agricultural matrix. *One Earth*, 6(9), 1131–1142. https://doi.org/10.1016/j.oneear.2023.08.010

Ratnadass A., Deberdt P., Martin T., Sester M., Deguine J.-P. 2023. Impacts of crop protection practices on human infectious diseases: Agroecology as the preferred strategy to integrate crop plant health within the extended "One Health" framework. In: *One Health: Human, Animal and Environment Triad* (Vithanage M., Prasad M.N.V., eds.). Chichester: John Wiley and Sons, pp. 287–308. https://doi.org/10.1002/9781119867333.ch21

Ratnadass A., Sarter S. 2023. How agricultural practices affect the risk of human contamination by infectious pathogens: The need for a "One Health" perspective. *CABI Reviews*, 2023, 30 p. https://doi.org/10.1079/cabireviews.2023.0003

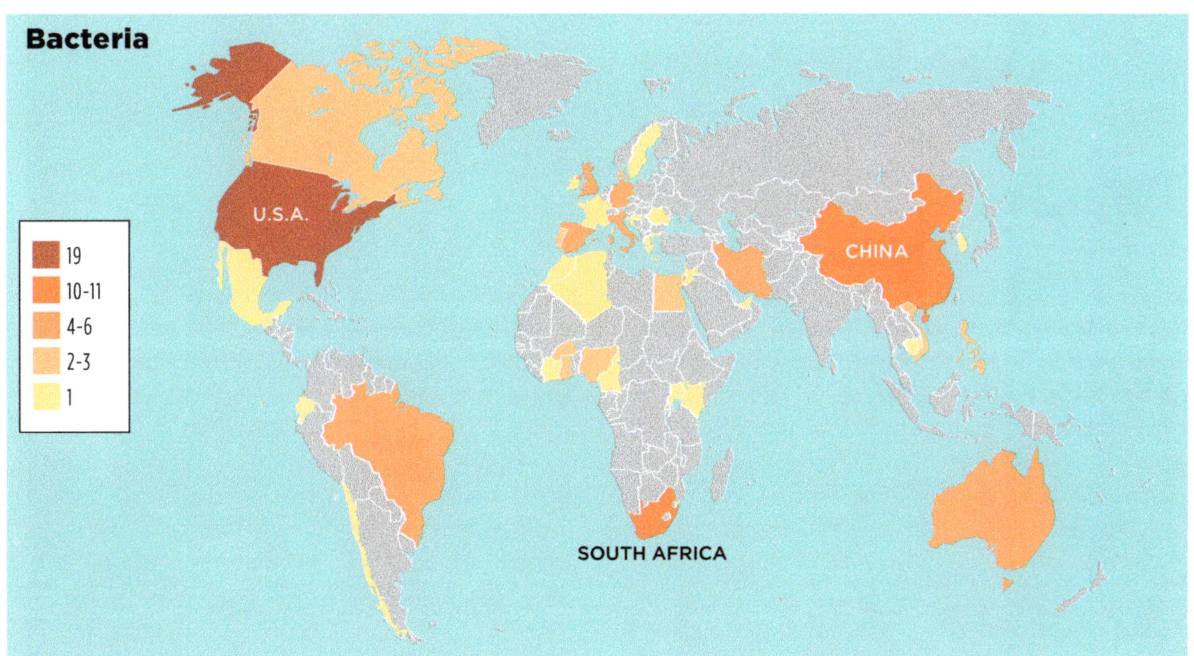

Figure 1. Geographical distribution of the number of studies on interactions between crop production practices and human bacterial infection hazards as reported in Ratnadass and Sarter (2023).

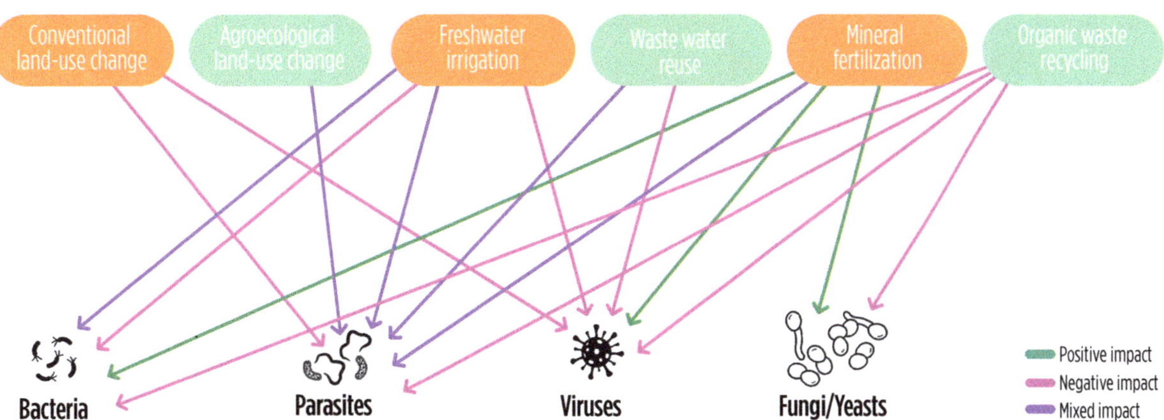

Figure 2. Impacts of crop production practices either conventional (orange boxes) or agroecological/organic (green boxes) on biological hazards to human health.

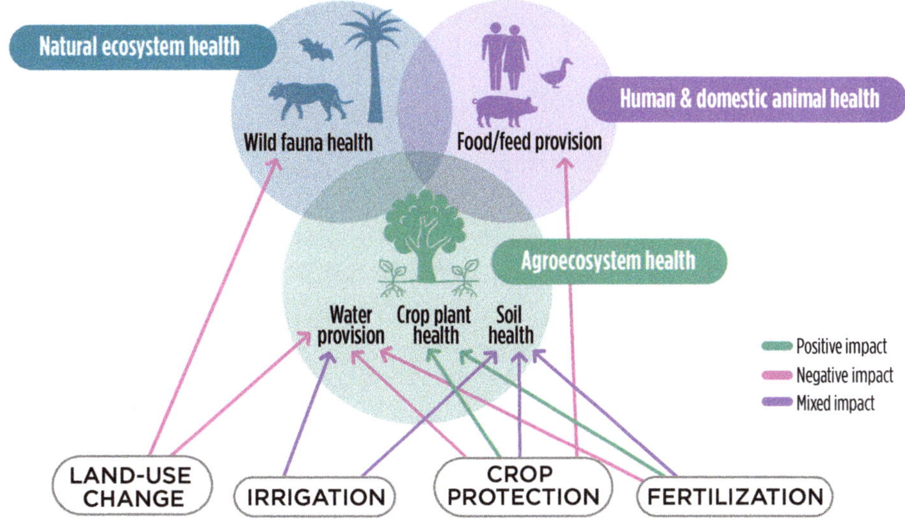

Figure 3. A new "One Health" conceptual diagram bringing crop plant health within an "Agroecosystem health" sphere, at par with "Natural ecosystem" and "Human & domestic animal" health spheres.

Agroecological crop protection: a special link between agroecology and One Health

Jean-Philippe Deguine,
Alain Ratnadass

In the scientific literature, human health and animal health are the main two pillars of the One Health concept (Figure 1). The third pillar, ecosystem health, is less often taken into account. Plant health is rarely mentioned and cultivated plant health even less so, despite a very recent spike, mainly from French teams (Figure 2).

Yet there are clear interactions between (crop) plant health and the other pillars. Agroecological crop protection (ACP) is the application of agroecological principles to crop protection (Figure 3). With its three dimensions (scientific, agronomic and social) and its underpinnings (prioritizing pest and disease prevention, promoting biodiversity and fostering soil health), ACP positively impacts the health of crops, soils, agroecosystems and food systems. ACP also has a positive impact on the health of ecosystems in general, contributing to the reduction of biodiversity erosion, to climate change adaptation and mitigation, to human and animal health, and to animal welfare.

In addition, recent studies have shown that ACP reduces the risk of viral zoonotic diseases and infectious fungal diseases (mycoses) and has variable effects on other human infectious diseases (Figure 4). Thus, ACP represents a special link between agroecology, which aims to provide healthy agriculture and food systems, and One Health, which aims for the overall health of living communities.

As in Figure 4, the "plant health" sphere should be systematically added to the other three spheres that are generally the only ones found in illustrations of One Health in the literature. Similarly, the scientific work related to plant health must also be taken into account in the description and development of the One Health concept (as reflected in Figure 1 in the early 2020s).

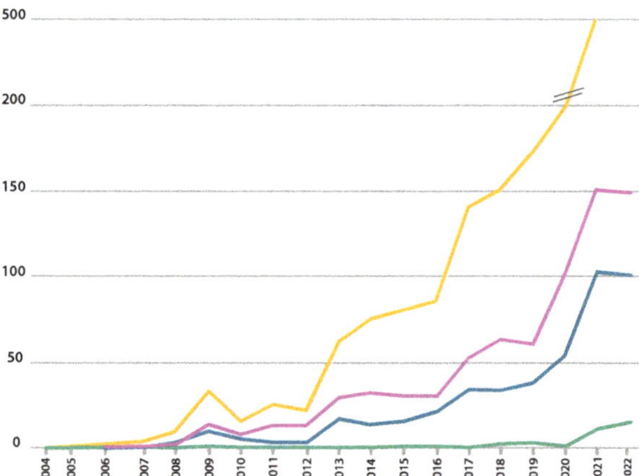

Figure 1. Evolution of the number of publications linking One Health and each of the four types of health (human, animal, ecosystem and plant). Searches on the WoS Core collection (https://www.webofscience.com/wos/woscc/ba-sic-search) for the terms shown in parenthesis, from 2004 (date when the One Health concept appeared) to 2022.

References

Ratnadass A., Deberdt P. 2021. Pratiques de protection des cultures en agroécosystèmes tropicaux et risques de maladies humaines et animales d'origine bactérienne. *Cahiers Agricultures*, 30, 42. https://doi.org/10.1051/cagri/2021028

Ratnadass A., Deguine J.-P. 2021. Crop protection practices and viral zoonotic risks within a One Health framework. *Science of the Total Environment*, 774, 145172. https://doi.org/10.1016/j.scitotenv.2021.145172

Ratnadass A., Sester M. 2023. Crop protection practices and risks associated with human fungal infectious diseases: a One Health perspective. *Cahiers Agricultures*, 32, 7. https://doi.org/10.1051/cagri/2022036

Figure 2. Global interest in the search term "plant health" based on Google Trends data (2004–2022). Data source: Google Trends (https://trends.google.com), accessed in 2023.

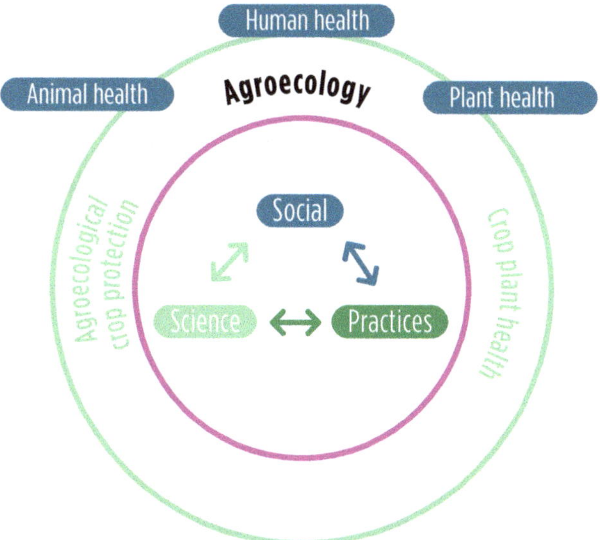

Figure 3. Agroecological crop protection (ACP) is positioned at the intersection of agroecological principles and the One Health concept.

This figure illustrates how ACP, by integrating three key dimensions—scientific, agronomic and social—promotes the health of cultivated plants while simultaneously supporting and enhancing agroecosystem health. It also demonstrates ACP's direct and indirect positive impacts on human and animal health, biodiversity conservation and resilience to climate change. By adopting practices aimed at disease and pest prevention, fostering functional biodiversity and maintaining soil health, ACP serves as a crucial link between agricultural sustainability and the overall health of living communities. It thus embodies a holistic approach, recognizing and actively promoting interactions among plant, animal, human and environmental health.

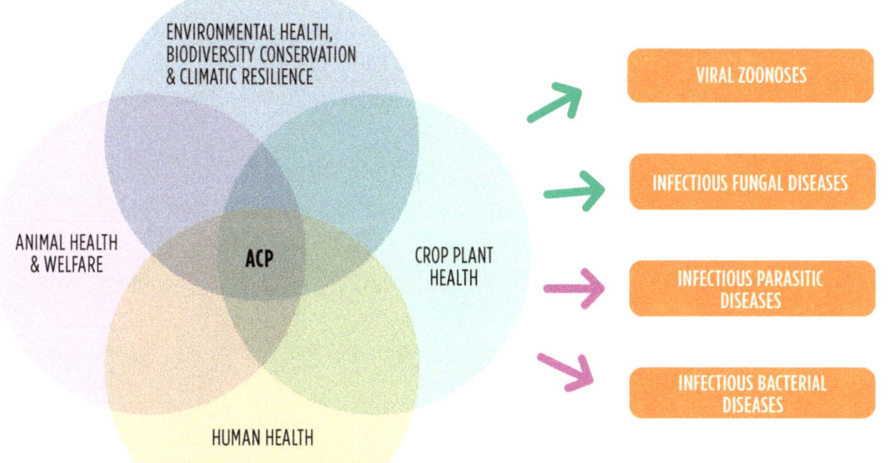

Figure 4. The four components of One Health and the effect of agroecological crop protection (ACP) on human diseases, namely viral zoonoses, infectious fungal diseases, infectious parasitic diseases and infectious bacterial diseases. Green arrows: reduction of disease risk; purple arrows: mixed impact on disease risk.

Soil health must be included in One Health

Laetitia Herrmann,
Pranaba Nanda Bhattacharyya,
Didier Lesueur

Soil health is defined as the continued capacity of soil to function as a vital living ecosystem that sustains plants, animals and humans, and connects agricultural and soil science to policy, stakeholder needs and sustainable supply chain management. Initiatives aimed at investigating how human health is influenced by animal and plant health and the environment should automatically include soil health.

Soil, the most biodiverse habitat on Earth, supports human health through essential ecosystem services like nutrient cycling and pollutant remediation. Soil is a natural source of antimicrobial resistance. Moreover, overuse of antimicrobials in human and animal health has increased antimicrobial resistance (AMR) in soils, with negative impacts on human health. It is becoming critical to have effective antimicrobial drugs for human health, and more research on soil microbiomes could be highly beneficial.

With a growing human population, sustainable crop practices must be prioritized to maintain environmental balance. Agroecological practices include reduced chemical inputs and the use of beneficial microorganisms to optimize yields and reduce pathogen attacks. Adopting these practices significantly reduces greenhouse gas emissions while improving soil biodiversity and soil health in general. While intensive management can optimize crop yields in the short term, promoting sustainable agricultural practices like agroecology that prioritize soil biodiversity will have a lasting impact on soil and human health and make a major contribution to the concept of One Health.

There is growing interest in soil health and adopting environmentally-friendly practices in the agricultural sector. However, it takes time to implement new strategies, and actions are needed to promote these practices for sustainable agriculture. In response to this need, CIRAD, the Agricultural Genetics Institute, the Alliance of Bioversity & CIAT and Deakin University in Melbourne, Australia, created the Common Microbial Biotechnology Platform (CMBP) network in January 2019[1] (Figure 1). The CMBP network, with 70 partners across 20 countries, fosters collaboration on agroecology and soil health in the Asia-Pacific region (Figure 2).

Our research activities have a positive impact on the soil health of smallholder farms, leading to increased and more sustainable livelihoods for farmers. Several scientific publications were published to reach the international scientific community. The network is open to new partners, ideas and projects, and we look forward to moving ahead with new opportunities with the common goal of achieving sustainable agriculture in Southeast Asia.

1. www.cmbp-network.org

References

Banerjee S, van der Heijden M.G.A. 2023. Soil microbiomes and one health. *Nature Reviews Microbiology,* 21, 6–20. https://doi.org/10.1038/s41579-022-00779-w

Herrmann L., Lesueur D. 2023. Restoring soil health and sustaining crop yields with agroecological practices, the two main goals of the CMBP Asia-Pacific network: a real opportunity for the horticultural production in Asia. *Proceedings of AHC2023 The 4th Asian Horticultural Congress,* 28-31August 2023, Tokyo, Japan, p. 54.

Lehmann J., Bossio D.A., Kögel-Knabner I., Rillig M.C. 2020. The concept and future prospects of soil health. *Nature Reviews Earth & Environment,* 1, 544–553. https://doi.org/10.1038/s43017-020-0080-8

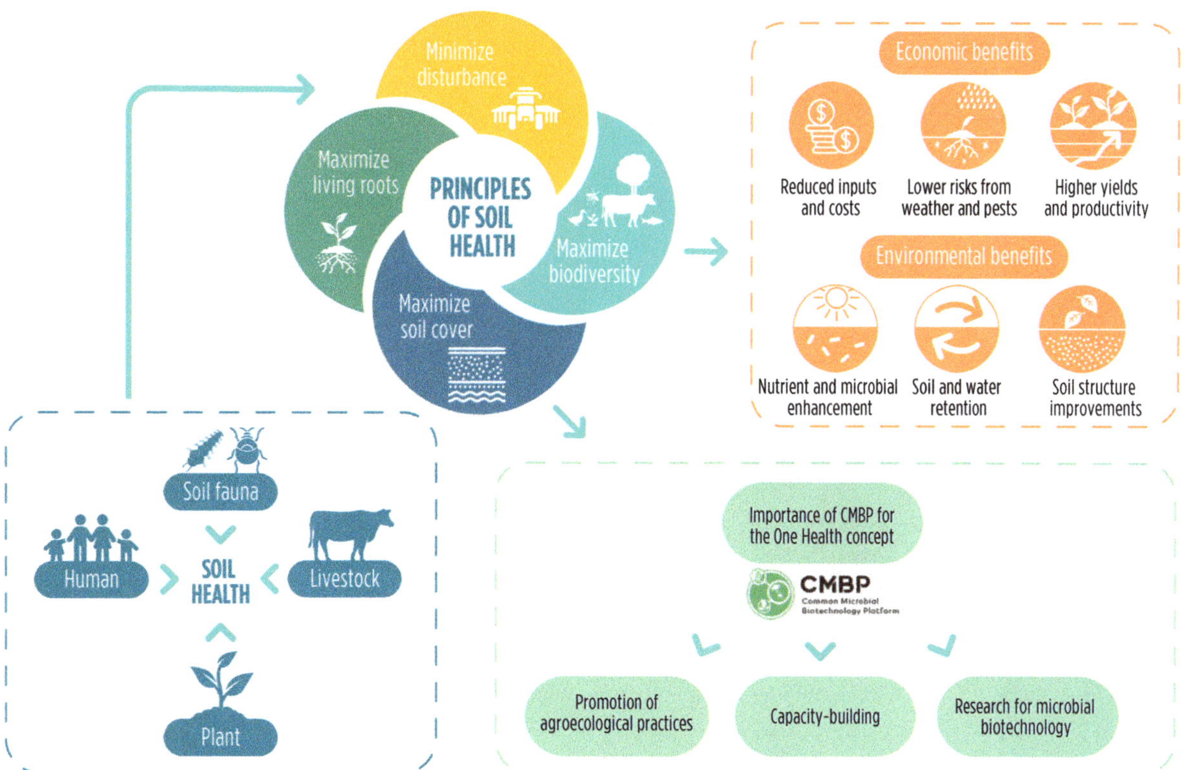

Figure 1. Principles of soil health and its relevance to One Health.

Soil health is crucial for sustainable agriculture and must be considered within the One Health initiative. The strategic significance of the Asia-Pacific network CMBP is to promote agroecological practices that enhance soil health.

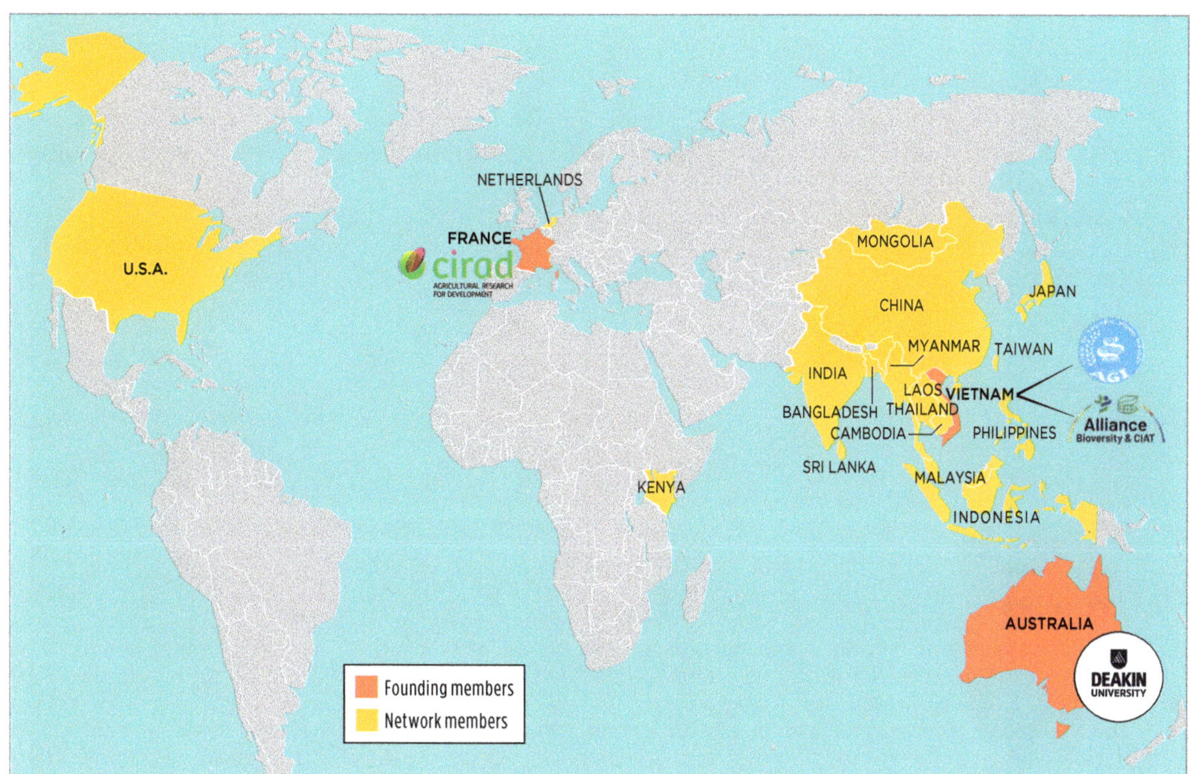

Figure 2. Map displaying the countries from which members joined the CMBP network. Source: CMBP network (www.cmbp-network.org).

One Health in aquaculture: antibiotic use in an era of global warming

Kazi Ahmed Kabir,
Partho Pratim Debnath,
Channarong Rodkhum,
Tran Minh Phu, Samira Sarter

One Health in aquaculture is more a conceptual idea than part of an operationalization framework. One Health is typically viewed through the narrow lens of zoonosis. But because zoonosis is less reported in aquaculture, the use of antibiotics, antimicrobial resistance (AMR) and the worsening effect of AMR due to global warming in this sector is often considered a lesser priority.

Where aquaculture is concerned, antibiotic use is highest in southern and southeast Asia, where 85% of global fish is produced. Antibiotic use per tonne of fish production is highest in India, followed by Indonesia and China (Figure 1). The most frequently reported antimicrobial-resistant bacteria in aquaculture is *Vibrio* spp. (25%), which can cause severe infections in fish and humans and pose a serious One Health threat. Given the threats to the aquaculture sector and the associated human health risks, major aquaculture-producing countries have begun including aquaculture in their national AMR action plans. Despite commitments to regulate antibiotic use in aquaculture, the use of these products is expected to continue to grow until 2030 (Figure 1) in China (the world's largest producer), with a slight decline anticipated among other major producers.

Antimicrobials in aquaculture are administered mainly through therapeutic and prophylactic approaches. While efforts are under way to better regulate therapeutic antibiotic use by 2030, the future use prophylactic antimicrobials remains an open question. AMR can increase over time and will likely worsen with global warming; interactions with other chemicals (e.g. disinfectants) used in aquaculture can also be detrimental. Moreover, the risks from aquaculture are numerous and the level of unpredictability is higher since aquaculture systems are often connected (directly or indirectly) with open aquatic environments that are reservoirs of antimicrobial-resistant bacteria. These bacteria can come from a variety of sources, including sewage waste, hospital effluents and agricultural run-off from livestock and crops. A cumulative hazard perspective is needed to address these issues. Dealing with such complex challenges will require establishing a global coordinated initiative, ensuring better representation of aquaculture in One Health platforms and developing tools to operationalize One Health in aquaculture (Figure 2).

References

Caputo A., Bondad-Reantaso M.G., Karunasagar I., Hao B., Gaunt P. *et al.* 2023. Antimicrobial resistance in aquaculture: A global analysis of literature and national action plans. *Reviews in Aquaculture,* 15, 568–578. https://doi.org/10.1111/raq.12741

Reverter M., Sarter S., Caruso D., Avarre J.-C., Combe M. *et al.* 2020. Aquaculture at the crossroads of global warming and antimicrobial resistance. *Nature Communications*, 11, 1–8. https://doi.org/10.1038/s41467-020-15735-6

Schar D., Klein E.Y., Laxminarayan R., Gilbert M., Van Boeckel T.P. 2020. Global trends in antimicrobial use in aquaculture. *Scientific Reports*, 10, 1–9. https://doi.org/10.1038/s41598-020-78849-3

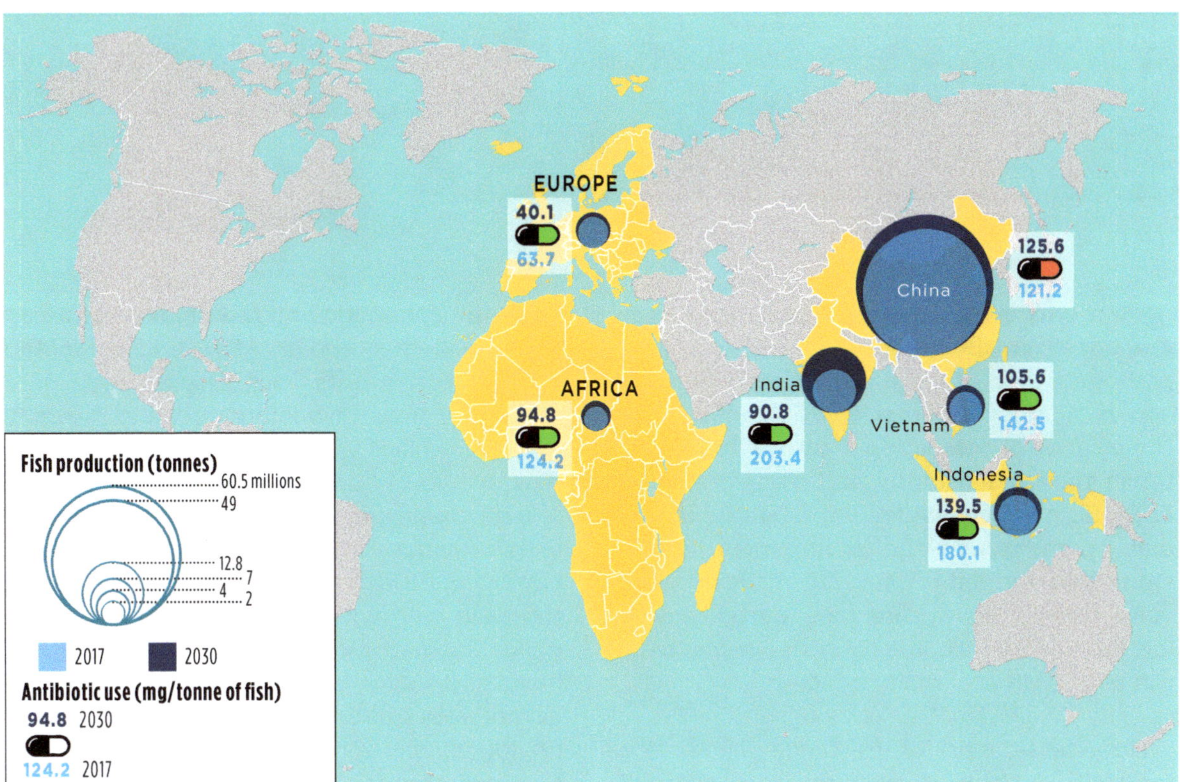

Figure 1. Projected antimicrobial use in aquaculture by country (2017 vs 2030).
This figure presents estimated antibiotic use in aquaculture across major producing countries, comparing 2017 levels with projections for 2030. The data highlight both total usage (in metric tonnes) and intensity of use (mg per tonne of fish produced). These trends raise concerns about the continued expansion of antimicrobial inputs, particularly in countries where regulatory frameworks and surveillance systems remain limited. Adapted from Schar *et al.* 2020.

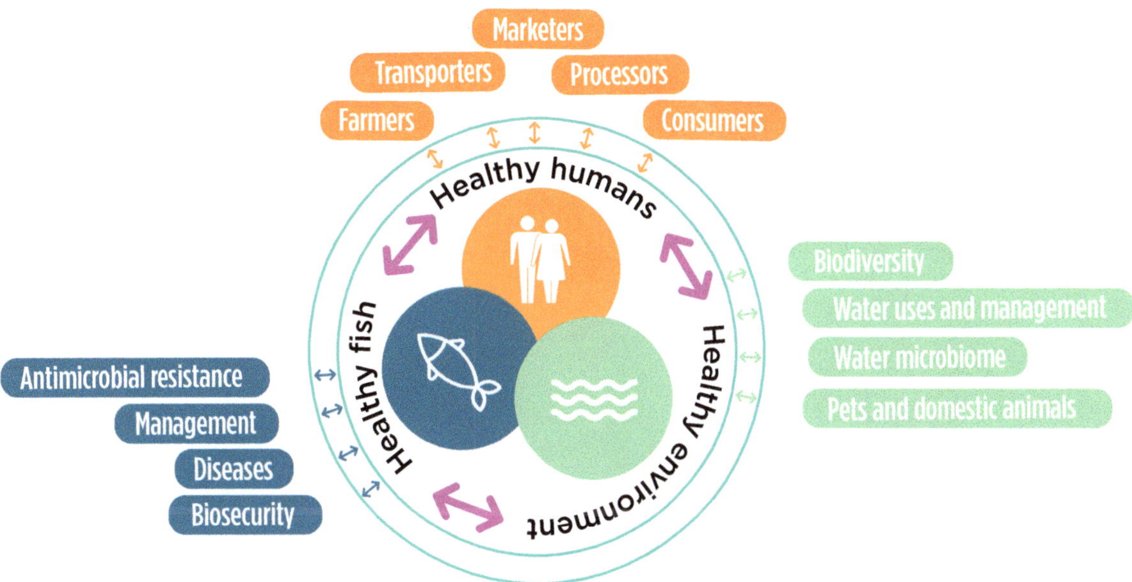

Figure 2. One Health framework for managing antimicrobial resistance risks in aquaculture.
This diagram illustrates the complex web of interactions in aquaculture systems that contribute to antimicrobial resistance (AMR). It highlights the interconnected roles of fish farmers, animal health, water quality, biodiversity, consumers, and the environment. The flow of antimicrobials and resistant bacteria across compartments—including through feed, water, and effluents—shows how AMR risks extend beyond farm boundaries, affecting human health and ecosystems. The figure advocates for integrated risk management approaches under a One Health framework to safeguard aquatic production, environmental sustainability and public health.

One Health and the spread of antibiotic resistance

Adrien Rieux, Noellie Gay,
Éric Cardinale, Étienne Loire

Escherichia coli that produces extended-spectrum β-lactamase (ESBL-Ec) is a pathogen that poses a threat to healthcare and is a crucial marker for epidemiological surveillance of antimicrobial resistance (AMR). While the epidemiology and transmission of ESBL-Ec has been widely investigated in animals, humans and the environment, global studies performed on several of these reservoirs simultaneously are rare, particularly in low- and middle-income countries in which the threat of AMR is of major concern.

To assess the risk of resistance genes being transmitted between bacteria from different sources, we collected 1,454 bacterial isolates from various human and animal hosts (humans, pigs, chicken, ducks, cattle, etc.) and the environment (drinking and irrigation water) in rural Madagascar. The isolates underwent whole genome sequencing and were analysed using cutting-edge phylogenomic methods to characterize population genetic structure and infer presumed transmission events.

ESBL-Ec were detected in almost 40% of the samples taken (Figure 1), regardless of the source. By sequencing and analysing the core genomes of 510 strains, we reconstructed their phylogenetic relationships, identified the genes that confer AMR, and estimated transmission events. Our results revealed a prominent level of bacterial genetic diversity within a limited geographical zone, including new, previously unknown groups of ESBL-Ec. The most unexpected finding, in clear contrast with previous observations made in Global North countries (in this case, England), was the ease with which bacterial and resistance genes seemed to circulate between different hosts and source compartments. Transmission events were so frequent that our reconstructed phylogeny showed a total lack of association between genomic diversity and compartment structure (Figure 2).

This study, one of the first performed on such a large scale in a country in the Global South, illustrates the need for an integrated approach encompassing human, animal and environmental health to combat a phenomenon as complex as AMR. In-depth knowledge of the various mechanisms that lead to the emergence and dissemination of AMR on a global level, in both the Global South and Global North, is vital for building appropriate, effective surveillance and control strategies.

References

Gay N., Rabenandrasana M.A.N., Panandiniaina H.P., Rakotoninidrina M.F., Ramahatafandry I.T. *et al.* 2023. One Health compartment analysis of ESBL-producing Escherichia coli reveals multiple transmission events in a rural area of Madagascar. *Journal of Antimicrobial Chemotherapy*, 78(8), 1848–1858. https://doi.org/10.1093/jac/dkad125

Ludden C., Raven K.E., Jamrozy D., Gouliouris T., Blane B. *et al.* 2019. One Health genomic surveillance of *Escherichia coli* demonstrates distinct lineages and mobile genetic elements in isolates from humans versus livestock. *mBio,* 10(1), e02693-18. https://doi.org/10.1128/mBio.02693-18

Rousham E.K., Unicomb L., Islam M.A. 2018. Human, animal and environmental contributors to antibiotic resistance in low-resource settings: Integrating behavioural, epidemiological and One Health approaches. *Proceedings of the Royal Society B: Biological Sciences*, 285, 1–9. https://doi.org/10.1098/rspb.2018.0332

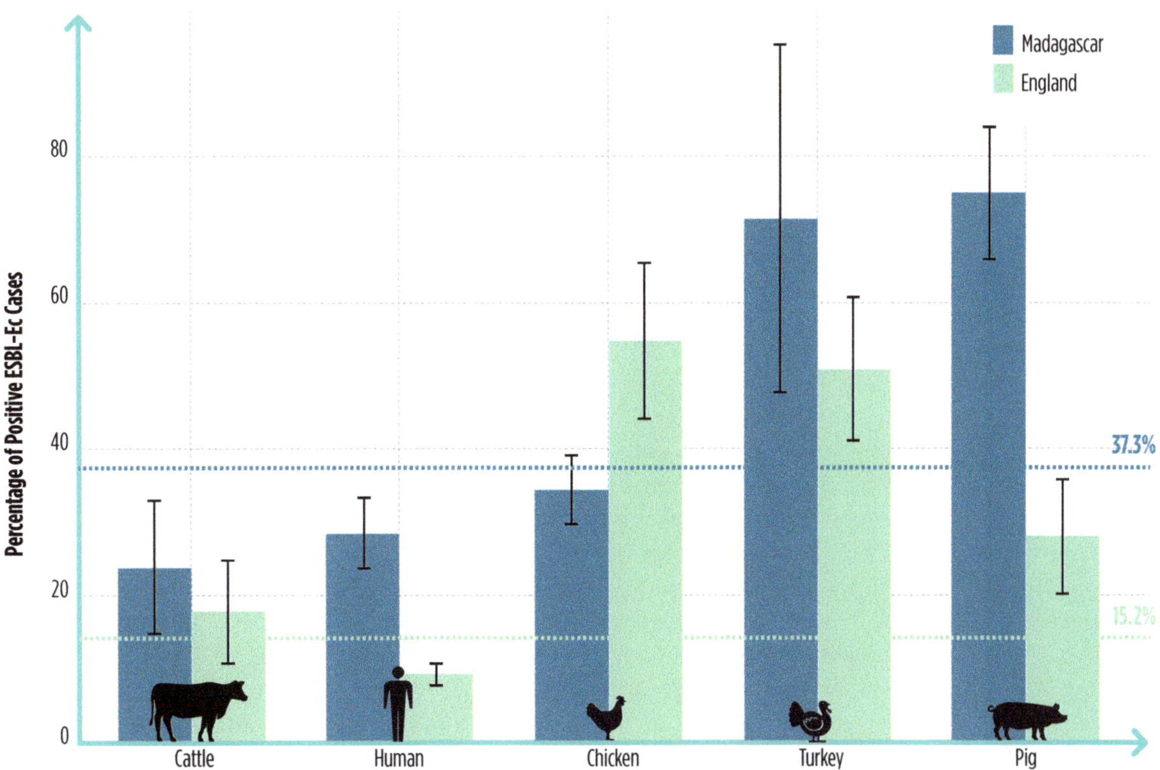

Figure 1. Significant differences were observed in humans, chickens and pigs (no significant difference in cattle or turkeys) may be due to variations in healthcare, veterinary and agricultural practices, antibiotic usage, or environmental factors between Madagascar and England. New epidemiological studies should be conducted to confirm or refute these hypotheses about the different ESBL-Ec profiles between Madagascar and England.

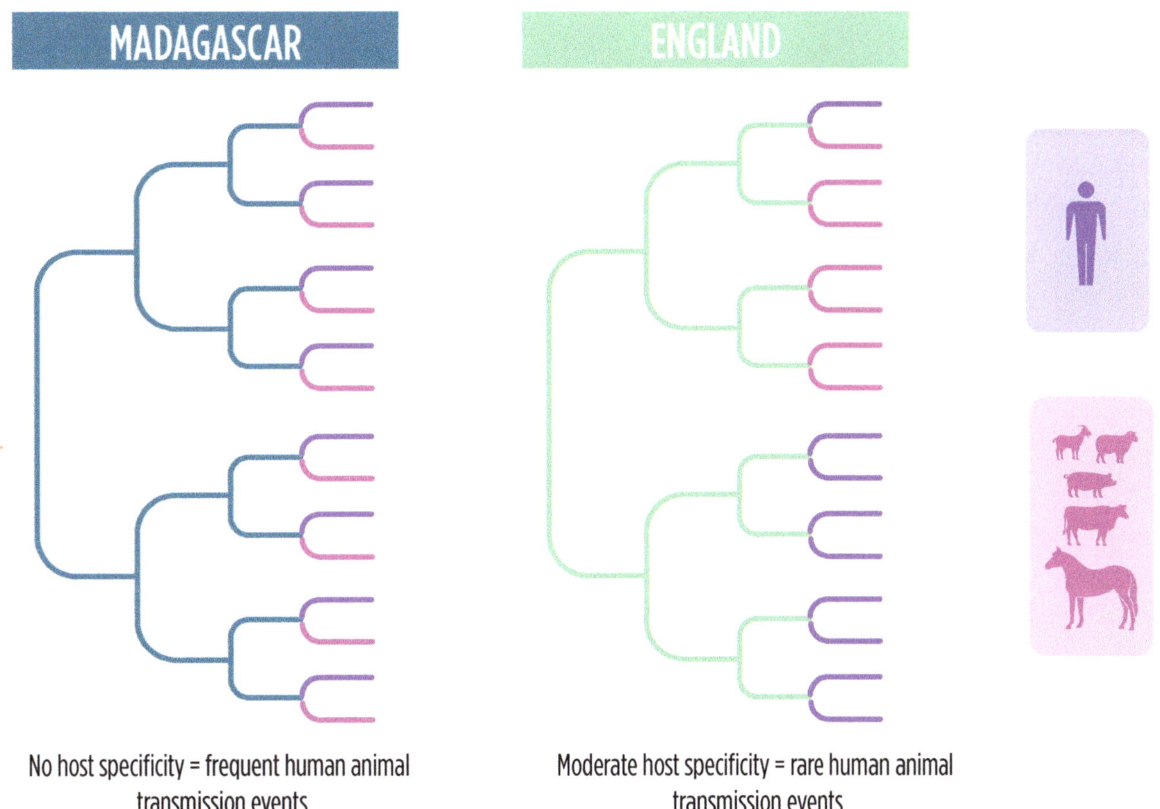

Figure 2. Schematization of global phylogeny obtained from ESBL-Ec strains sampled in Madagascar (left panel) and England (right panel). Hosts are indicated in violet (humans) and pink (animals).

One Health surveillance of antimicrobial resistance

Cécile Aenishaenslin,
Jane Parmley

Antimicrobial resistance (AMR) is a major threat to human and animal health. Although the development of resistant pathogens in human populations is primarily driven by antimicrobial consumption in human populations, the use of antimicrobials in animals to prevent or treat infections also contributes to the development of resistant microorganisms in animal populations with potential for human exposure through direct contact, the food chain or environmental pathways. In addition, antimicrobial residues and resistant microorganisms are released into ecosystems through waste from hospitals, livestock and aquaculture farms, and manufacturing units, while upstream determinants of antimicrobial use are multiple and cross-sectoral, which add complexity to the management of this emerging problem (Figure 1).

Consequently, surveillance systems for AMR should integrate the data about resistance in microorganisms circulating in humans, in animals and in the environments they occupy, in coherence with One Health. One Health surveillance systems built on collaboration and strong networks are essential to help advance knowledge on AMR epidemiology with a system perspective and to inform the development and implementation of effective interventions and policies (Figure 2).

One example of such a system is the Canadian Integrated Program for Antimicrobial Resistance Surveillance (CIPARS), which has been in operation since 2002. CIPARS is a national programme coordinated by the federal government of Canada and is dedicated to the collection, integration and analysis of data about antimicrobial use (AMU) and AMR in selected bacteria from humans, animals and animal-derived food sources in Canada (Figure 3). CIPARS provides an integrated approach to monitor and communicate trends in AMU and AMR in humans and animals, facilitates assessment of the public health impact of antimicrobials used in humans and agricultural sectors, and allows accurate comparisons with data from other countries that use similar surveillance systems.

References

Aenishaenslin C., Häsler B., Ravel A., Parmley J., Stärk K., Buckeridge D. 2019. Evidence needed for antimicrobial resistance surveillance systems. *Bulletin of the World Health Organization*, 97(4), 283-289. https://doi.org/10.2471/BLT.18.218917

Bordier M., Uea-Anuwong T., Binot A., Hendrikx P., Goutard F.L. 2020. Characteristics of One Health surveillance systems: A systematic literature review. *Preventive Veterinary Medicine*, 181, 104560. https://doi.org/10.1016/j.prevetmed.2018.10.005

CIPARS website: https://www.canada.ca/en/public-health/services/surveillance/canadian-integrated-program-antimicrobial-resistance-surveillance-cipars.html

Lambraki I.A., Cousins M., Graells T., Léger A., Henriksson P., Harbarth S., Troell M. *et al.*, 2022. Factors influencing antimicrobial resistance in the European food system and potential leverage points for intervention: A participatory, One Health study. *PLOS ONE* 17(2), e0263914. https://doi.org/10.1371/journal.pone.0263914

Robinson T.P., Bu D.P., Carrique-Mas J., Fèvre E.M., Gilbert M. *et al.* 2016. Antibiotic resistance is the quintessential One Health issue. *Transactions of the Royal Society of Tropical Medicine & Hygiene*, 110(7), 377–380. https://doi.org/10.1093/trstmh/trw048

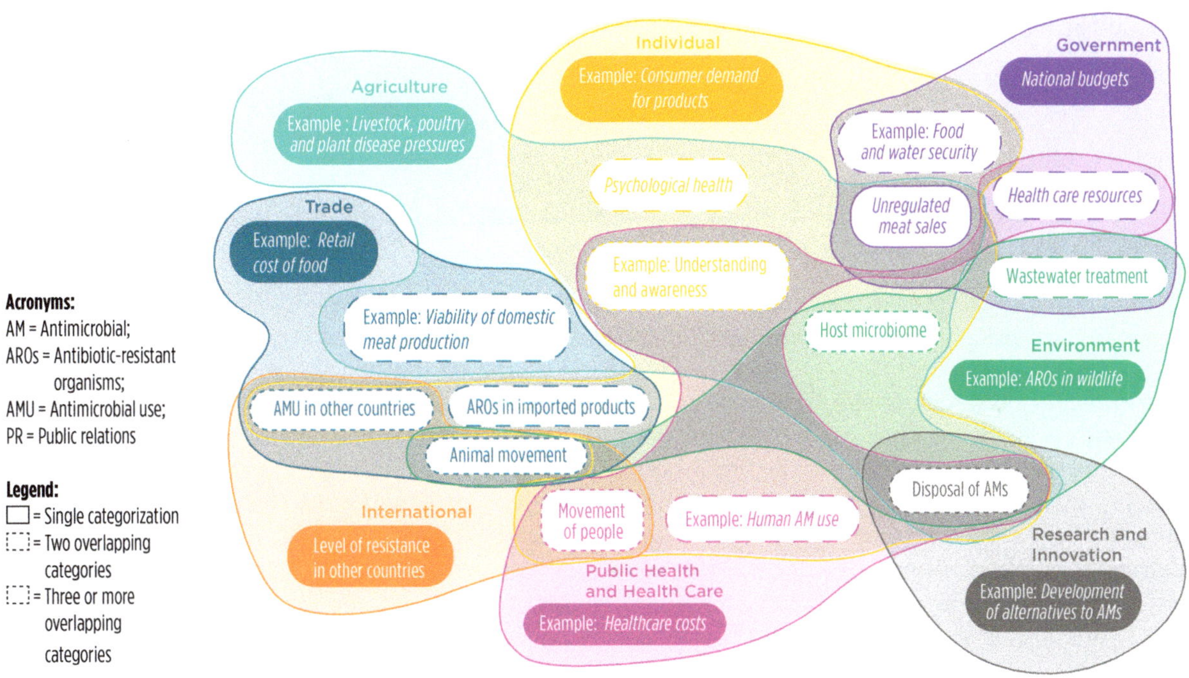

Figure 1. Pathways of antimicrobial resistance across sectors. Adapted from Lambraki *et al.* 2022.

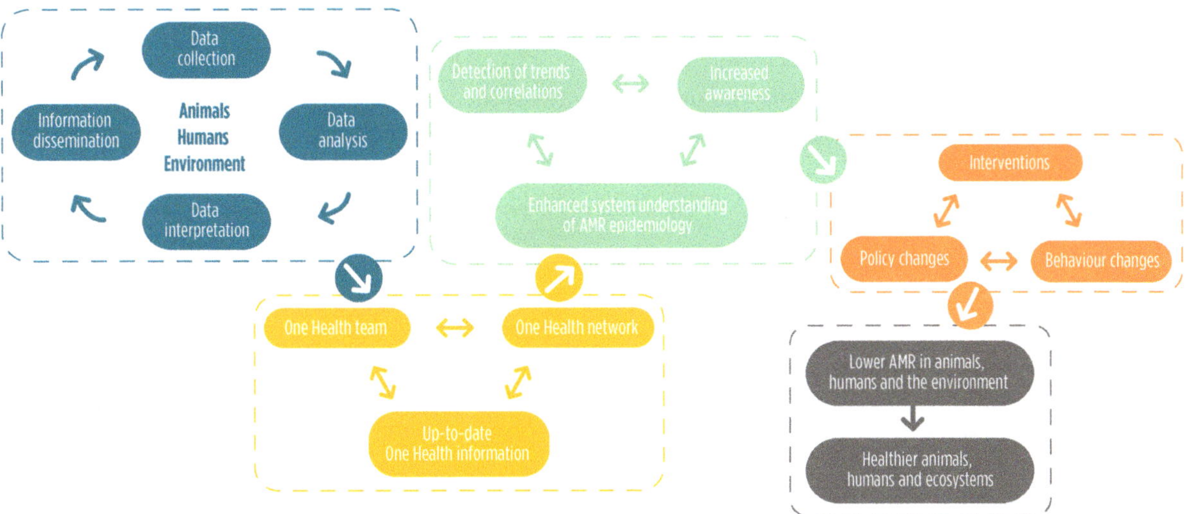

Figure 2. Expected outcomes of One Health surveillance of antimicrobial resistance (AMR).

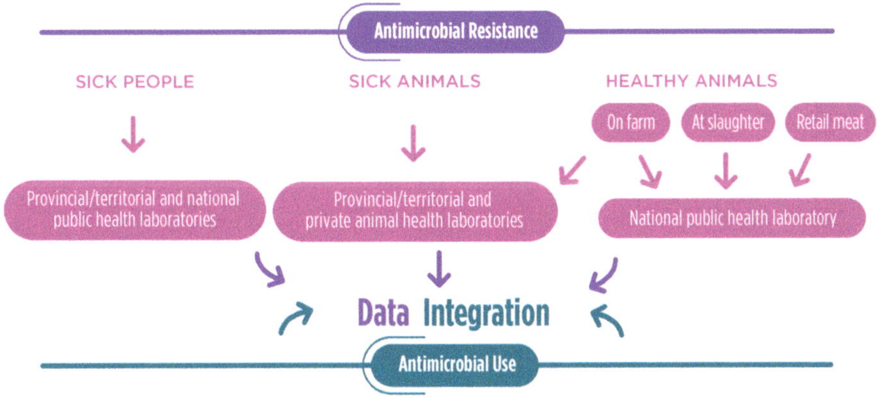

Figure 3. Structure of the Canadian Integrated Program for Antimicrobial Resistance Surveillance (CIPARS). Adapted from: https://www.canada.ca/en/public-health/services/surveillance/canadian-integrated-program-antimicrobial-resistance-surveillance-cipars/about-cipars.html

One Health for food safety

François Roger,
Marie-Marie Olive

Food safety is a fundamental pillar of public health. It aims to prevent foodborne diseases and ensure the safety of food throughout the supply chain. The globalization of supply chains, agricultural intensification, zoonoses and antimicrobial resistance have made this issue increasingly complex (Figure 1). Food safety encompasses measures to prevent risks associated with pathogens (e.g. *Salmonella* spp., *E. coli, Listeria* spp.), chemical residues (pesticides, antimicrobials), heavy metals and natural toxins (Figure 2). These risks often arise during food production, processing or consumption, and are exacerbated by inadequate sanitary infrastructure and unsuitable agricultural practices. For instance, the use of contaminated water for irrigation or food processing can introduce pathogens into the food chain (Figure 1).

Reducing microbial risks is essential for improving food safety. For example, adopting safer agricultural practices, such as proper composting and the rational use of antibiotics, helps minimize the spread of resistant bacteria and harmful pathogens. Advanced technologies like genomic and metagenomic sequencing provide unprecedented capabilities to monitor and detect emerging pathogenic strains, enabling rapid responses to potential threats. Additionally, building educational capacity is crucial; raising awareness among producers and consumers alongside training professionals through One Health-integrated educational frameworks plays a pivotal role in promoting the adoption of safer practices across the food system.

The One Health approach makes it possible to address these combined risks by promoting collaboration among veterinarians, agronomists, epidemiologists, sociologists, economists and public health experts. This collaborative framework can support sustainable resource management and reduce contamination. In some less advanced countries, socioeconomic inequalities exacerbate food safety challenges. Limited access to infrastructure, inadequate hygiene practices and a lack of education increase the prevalence of foodborne diseases. The One Health approach incorporates these social dimensions to propose tailored solutions, emphasizing awareness, robust regulatory frameworks and responsible practices among farmers and consumers. Integrating One Health principles into food safety encourages resilient, equitable and sustainable food systems while protecting public health.

References

Garcia S.N., Osburn B.I., Jay-Russell M.T. 2020. One Health for food safety, food security, and sustainable food production. *Frontiers in Sustainable Food Systems*, 4(1), 1-9. https://doi.org/10.3389/fsufs.2020.00001

Ezzati M., Lopez A.D., Rodgers A., Vander Hoorn S., Murray C.J., the Comparative Risk Assessment Collaborating Group. 2002. Selected major risk factors and global and regional burden of disease. *The Lancet*, 360(9343), 1347-1360. https://doi.org/10.1016/S0140-6736(02)11403-6

Mather A.E., Gilmour M.W., Reid S.W.J., French N.P. 2024. Foodborne bacterial pathogens: Genome-based approaches for enduring and emerging threats in a complex and changing world. *Nature Reviews Microbiology*, 22(9), 543-555. https://doi.org/10.1038/s41579-024-01051-z

Rodríguez-Melcón C., Alonso-Calleja C., Capita R. 2024. The One Health approach in food safety: Challenges and opportunities. *Food Frontiers*, 5(5), 1837-1865. https://doi.org/10.1002/fft2.458

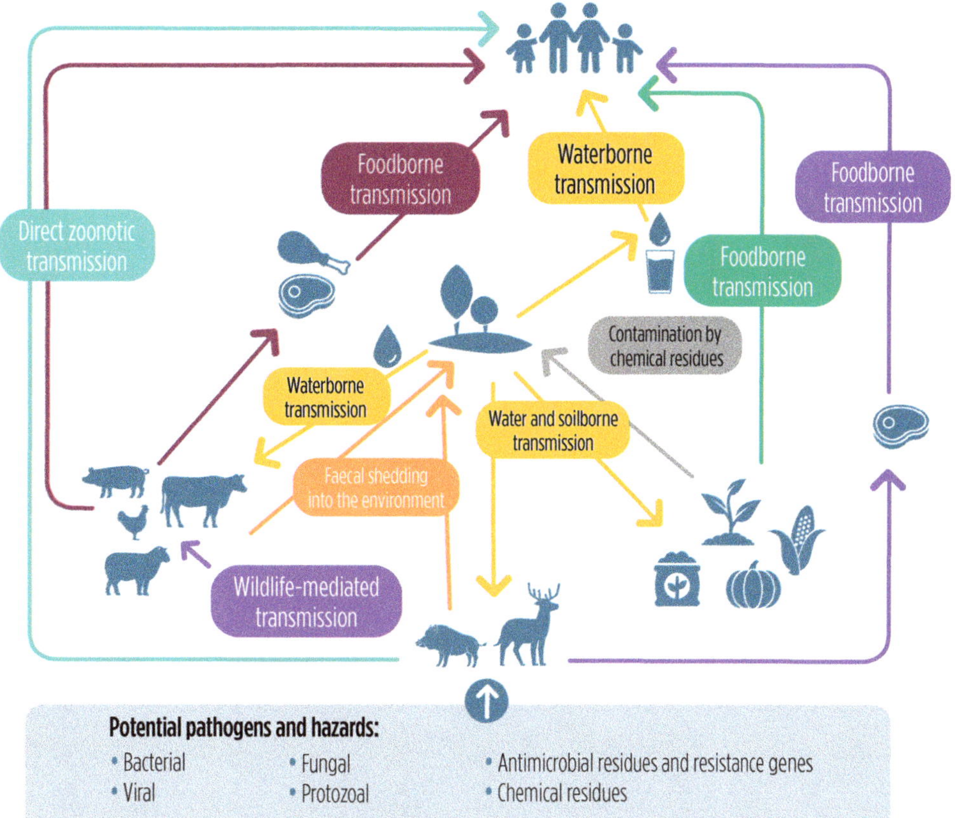

Figure 1. Sources of food contamination from vegetable crops, livestock farming, wildlife and the environment. Based on Mather *et al.* 2024.

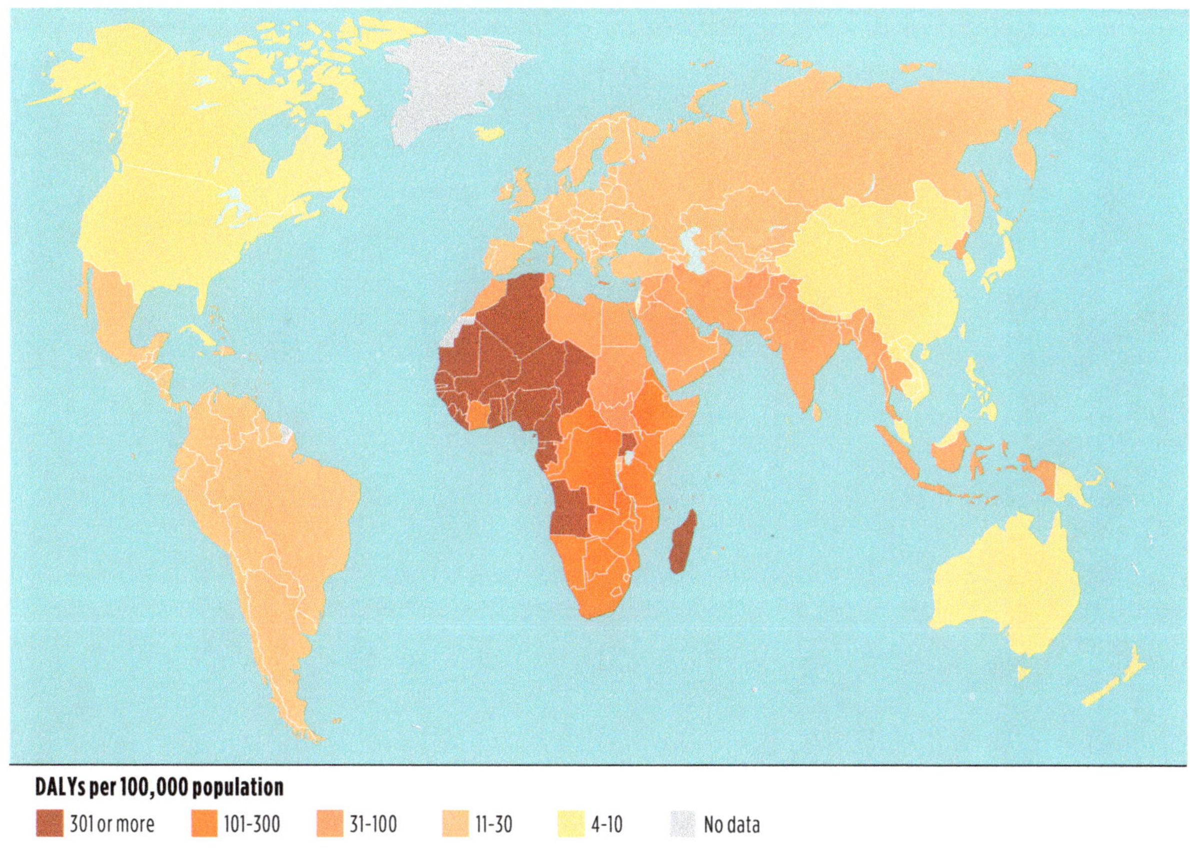

Figure 2. Map showing median foodborne DALYs per 100,000 population due to non-typhoidal *Salmonella enterica* in 2010. Source: WHO 2024. Data based on estimates from Ezzati *et al.* 2002. Map produced by WHO GIS Centre for Health, DNA/DDI.

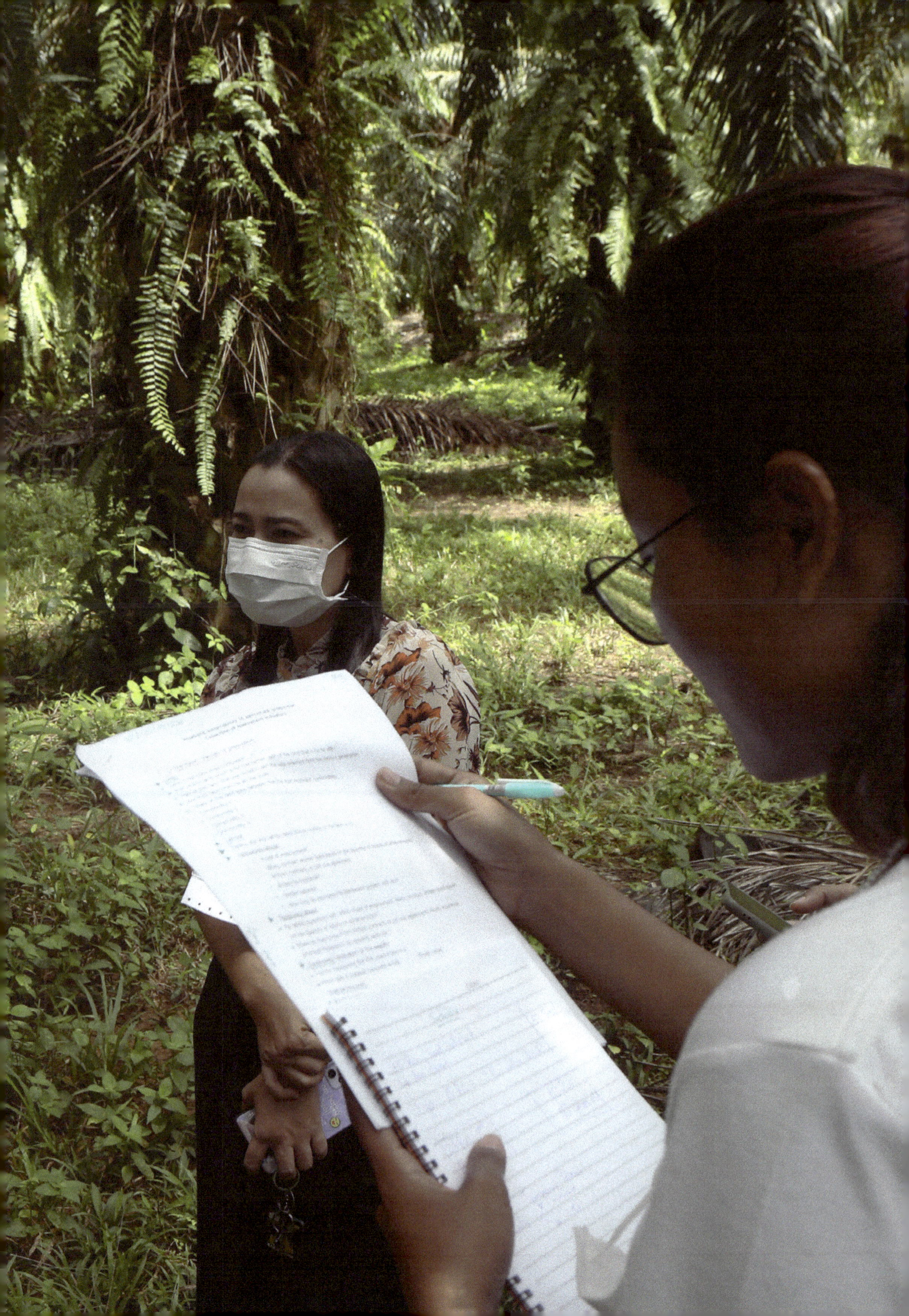

3

Education, networks and governance: One Health in action

This section focuses on the operationalization of One Health through education, capacity-building, governance structures, integrated projects and collaborative networks. It examines the role of academic programmes, professional training and institutional capacity-building initiatives like COHESA in Africa and the One Health Partnership in Vietnam to show how One Health principles are put into action across diverse contexts. This comprehensive approach demonstrates the integration of education, governance and networks to build sustainable One Health systems that address local, regional and global challenges.

Embracing complexity early on: incorporating One Health into bachelor's programmes

Sergio Guerrero-Sanchez, Chun Ting Angus Lam, Anne Conan, Dirk Pfeiffer

One Health courses have increasingly been incorporated in recent years into bachelor's programmes in such areas as the veterinary, medical and biological sciences. These courses highlight the interconnectedness of humans, animals and ecosystems, offering students broader perspectives that extend beyond their specific fields. They focus on interdisciplinary collaboration and cultivate vital skills in systems thinking, critical thinking and reflexivity.

Many bachelor's degree programmes still follow narrow reductionist models, especially in the natural sciences, which leads to isolated thinking among students. However, universities like UC Davis and Guelph are breaking these boundaries with integrated One Health courses that combine animal and human health, environmental sciences and social sciences.

Such inter-institutional programmes highlight the interconnectedness of life in our ecosystem, promote interdisciplinary collaboration, and encourage critical thinking to address global challenges. In the past decade, it has become essential to integrate social and ecological components into the One Health framework.

Addressing complex issues like climate change, biodiversity loss and pandemics requires embracing diverse perspectives. Interdisciplinary dialogues on animal and human health, equity, and conservation can challenge students' anthropocentric views and expand their ways of thinking.

To adequately prepare students, educators must develop a comprehensive One Health curricula aligned with bachelor-level programmes in veterinary, ecology, public health and social sciences. Continuous updates to core competencies and integration of diverse programmes are essential. Collaboration with stakeholders (farmers, government and non-government agencies) whose values align with the One Health approach will further enrich students' understanding of real-world challenges across socioeconomic contexts.

An integrated, inter-institutional One Health course fosters interdisciplinary dialogue between students and lecturers. Unlike traditional courses that focus on specialization, introducing an interdisciplinary framework early in students' academic journeys encourages a holistic mindset. This approach enhances collaboration among diverse disciplines (Figure 1) and better prepares students for effective problem-solving and critical thinking in their future careers.

References

Wilcox B.A., Aguirre A.A., De Paula N., Siriaroonrat B., Echaubard P. 2019. Operationalizing One Health employing social-ecological systems theory: lessons from the Greater Mekong sub-region. *Frontiers in Public Health*, 7, 85. https://doi.org/10.3389/fpubh.2019.00085

Togami E., Gardy J.L., Hansen G.R., Poste G.H., Rizzo D.M. ***et al.*** 2018. Core competencies in One Health education: What are we missing? *National Academy of Medicine Perspectives,* Discussion paper. https://nam.edu/perspectives/core-competencies-in-one-health-education-what-are-we-missing/

Clark S.G., Wallace R.L. 2015. Integration and interdisciplinarity: Concepts, frameworks, and education. *Policy Sciences*, 48, 233–255. https://doi.org/10.1007/s11077-015-9210-4

Patel N.S., Puah S., Kok X.-F. K. 2024. Shaping future-ready graduates with mindset shifts: studying the impact of integrating critical and design thinking in design innovation education. *Frontiers in Education*, 9, 1358431. https://doi.org/10.3389/feduc.2024.1358431

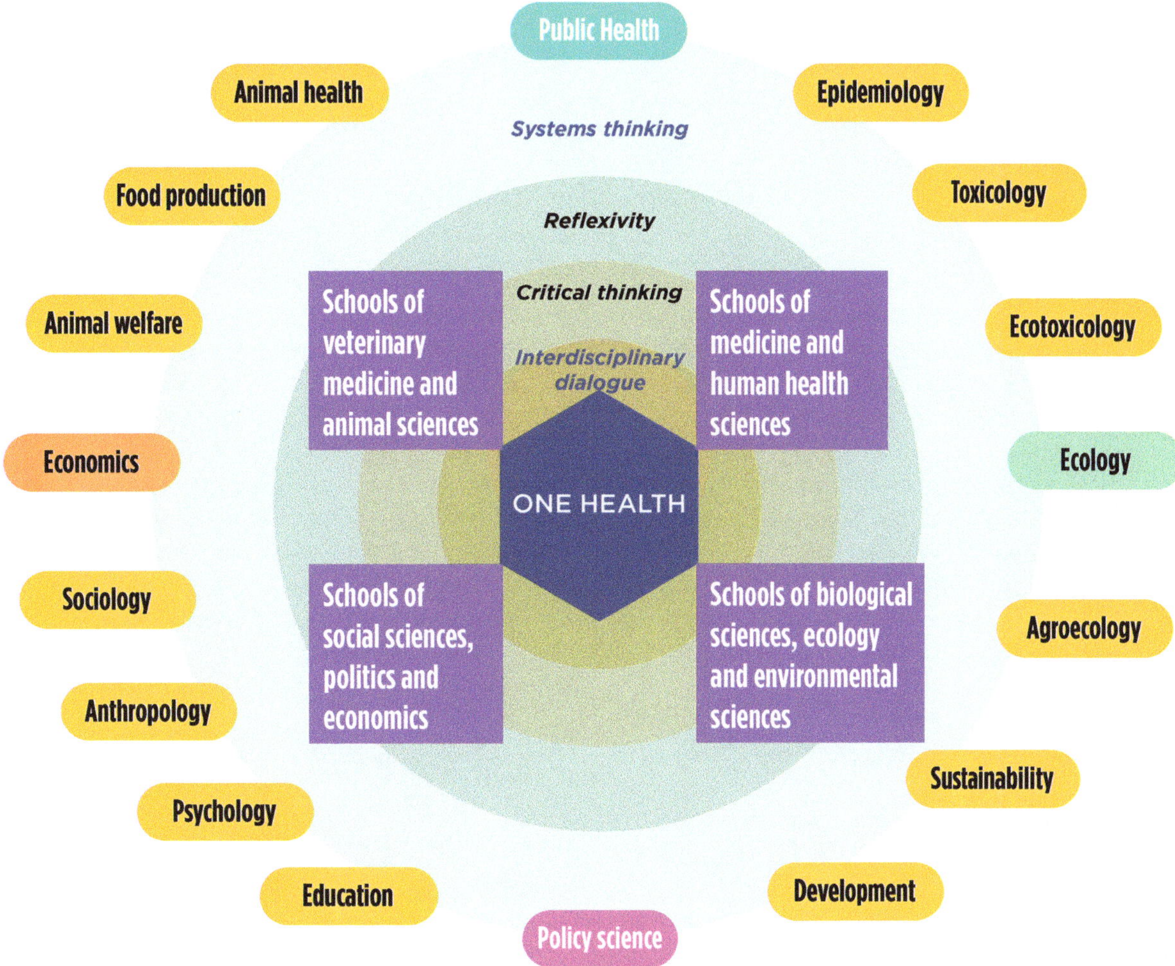

Figure 1. The complexity behind One Health.
The interdisciplinary nature of the One Health approach, along with the complexity of our reality, necessitates a learning process that begins with constructive dialogue amo ng various experts. This dialogue fosters a systems-thinking perspective achieved through critical thinking and reflexivity. While we can identify four main axes—policy, public health, ecology and economics—there are many topics related to the One Health approach that multiple disciplines can address. To encourage interdisciplinary dialogue and avoid imposing limitations on competencies and responsibilities, connectors are intentionally absent from this diagram.

INTERNATIONAL STUDENT ONE HEALTH ALLIANCE (ISOHA)

ISOHA is a global network founded in 2017 to connect and empower students and early-career professionals engaged in human, animal and environmental health. It brings together One Health clubs and student associations from universities and research institutions worldwide, offering a dynamic platform for leadership, collaboration and education. ISOHA supports its members through webinars, mentorship programs, advocacy initiatives and student-led publications. It also organizes international One Health student conferences and encourages youth engagement in global health diplomacy and research.
https://isohaonehealth.wordpress.com

Collaborative approach to building the future One Health workforce in South-East Asia

Mark Jaypee Gonzales, Yong Chyna Suit-B, André Furco, Tikiri Wijayathilaka, Kinley Choden, Ronello Abilla

The One Health approach, which recognizes the interconnectedness of human, animal and environmental health, has been adopted in South-East Asia for decades. However, to maintain momentum, producing a new generation of One Health professionals is critical. To support this objective, the World Organisation for Animal Health (WOAH), through its Sub-Regional Representation for South-East Asia, is developing a comprehensive training catalogue.

The South-East Asia One Health Training Catalogue (OHTC), which provides an easy-to-update platform for professionals, governments and organizations (Figure 1), was launched in February 2024. It collates One Health-related training courses across the region (Figure 2). This tool was designed to ensure accessibility, avoid duplication of efforts and help identify gaps in One Health training.

Although the project is in its early stages, the ongoing analysis highlights a wide range of training options already available in South-East Asia, including degree and non-degree programs, short courses and specialized One Health modules (Figure 3). Collaboration is sought between the WOAH Sub-Regional Representation for South-East Asia, South-East Asia One Health University (SEAOHUN), and the Tripartite Zoonoses Guide (TZG) team of the Quadripartite for ensuring that all training efforts in the near future are aligned with the core principles of One Health. The ultimate goal is to strengthen the One Health workforce and streamline capacity-building efforts.

The repository is expected to be hosted on an accessible platform like the Association of Southeast Asian Nations (ASEAN) One Health Network or the Joint One Health Learning Initiative or SEAOHUN. By fostering partnerships across sectors, this initiative will ensure a well-structured framework to train the next generation of One Health leaders, helping prevent zoonotic diseases and enhance health at the human-animal-environment interface.

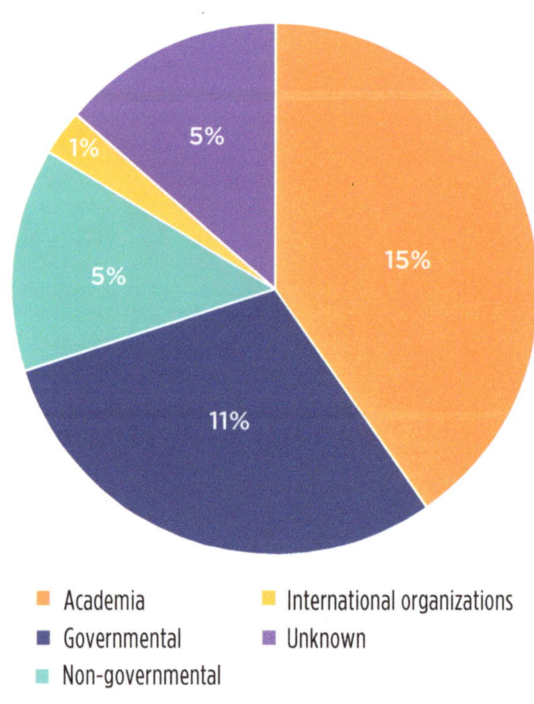

Figure 1. Hosting institutions of the One Health-related training courses.

References

Sullivan A., Ogunseitan O., Epstein J., Kuruchittham V., Nangami M. *et al.* 2023. International stakeholder perspectives on One Health training and empowerment: a needs assessment for a One Health Workforce Academy. *One Health Outlook*, 5, 8. https://doi.org/10.1186/s42522-023-00083-4

Togami E., Barton Behravesh C., Dutcher T.V., Hansen G.R., King L.J. *et al.* 2023. Characterizing the One Health workforce to promote interdisciplinary, multisectoral approaches in global health problem-solving. *PLoS ONE*, 18(5), e0285705. https://doi.org/10.1371/journal.pone.0285705

Figure 2. Geographical scope of One Health-related training courses in South-East Asia.

Figure 3. Type of One Health-related training courses.

Visualization of One Health education in South-East Asia

Duyen Truong Thi,
Flavie Goutard,
Bao Truong Dinh

South-East Asia is experiencing rapid environmental changes due to economic development, globalization and trade intensification. The region has been identified as a hotspot for endemic zoonoses and emerging infectious diseases (EIDs), with vulnerabilities arising from intensive agricultural practices. One Health (and the need for education and training on the topic) has taken on growing importance as these regions prepare for EIDs and other One Health challenges.

Zoonoses management requires better integration of animal health science, public health, social science, agriculture and livestock, engineering, and ecological and environmental sciences. A recent international agreement on the One Health approach encourages and legitimizes collaboration among animal, public and environmental health institutions to combat diseases that threaten animal and human health. International organizations from the quadripartite, including the World Organisation for Animal Health (WOAH), the Food and Agriculture Organization (FAO), the World Health Organization (WHO), and the United Nations Environment Programme (UNEP), fully support the approach. There is a need for a paradigm shift that includes human medicine, veterinary medicine, public health and environmental data with experts from various disciplines (e.g. social sciences, engineering, economics, education and public policy) working together. To bridge gaps between disciplines and sectors and facilitate effective collaboration and communication among both traditional and non-traditional One Health actors, various soft skills such as communication, leadership and cultural competence will be key.

One Health education (Figure 1) fosters interdisciplinary collaboration and gives professionals a platform where they can share insights, perspectives and methodologies to deepen their overall understanding of health challenges. Individuals trained in One Health principles are better able to engage in meaningful discussions across disciplines and serve as a vital link between scientific knowledge and societal impact. They are equipped to effectively communicate with a wide range of stakeholders, from policymakers to farmers and the general public, and to translate scientific findings into actionable measures. The end result is more relevant scientific research that can better inform policies and practices that reflect the diverse needs of communities to make a tangible impact on global health and well-being (Figure 2).

References

Mishra C., Samelius G., Khanyari M., Srinivas P.N., Low M. *et al.* 2022. Increasing risks for emerging infectious diseases within a rapidly changing High Asia. *Ambio*, 51(3), 494–507. https://doi.org/10.1007/s13280-021-01599-7
Mettenleiter T.C., Markotter W., Charron D.F., Adisasmito W.B., Almuhairi S. *et al.* 2023. The One Health High-Level Expert Panel (OHHLEP). *One Health Outlook*, 5(1), 1-8. https://doi.org/10.1186/s42522-023-00085-2
Laing G., Duffy E., Anderson N., Antoine-Moussiaux N., Aragrande M. *et al.* 2023. Advancing One Health: Updated core competencies. *CABI One Health*. https://doi.org/10.1079/cabionehealth.2023.0002

Figure 1. One Health-related Master's courses in South-East Asia (18 courses).

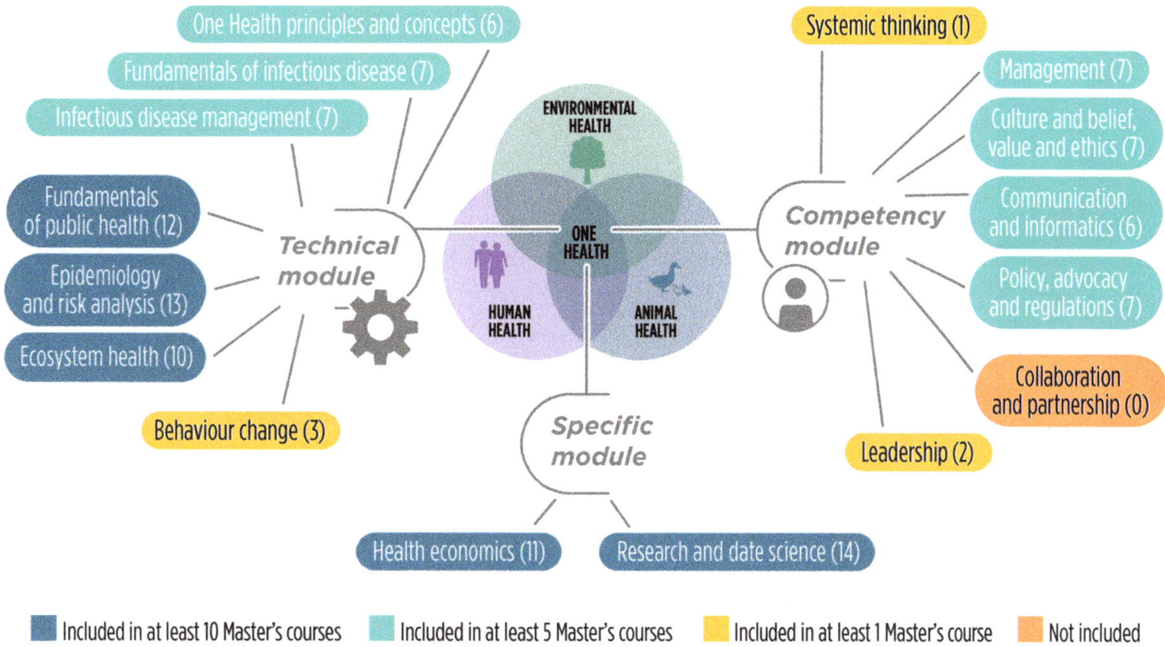

Figure 2. Southeast Asia One Health University Network's One Health module and its inclusion in Master's courses (16 courses).
A group of South-East Asian universities and colleges have formed the Southeast Asia One Health University Network (SEAOHUN) to strengthen the region's One Health workforce. SEAOHUN's mission is to combat zoonotic diseases and other associated health issues by fostering better multidisciplinary cooperation among medical, veterinary and environmental experts. To promote the One Health approach, SEAOHUN is actively involved in education, research and community outreach. This approach is collaborative, multisectoral and transdisciplinary.

One Health Research, Education and Outreach Centre in Africa (OHRECA)

Bernard Bett

The One Health Research, Education and Outreach Centre (OHRECA) was established by the International Livestock Research Institute (ILRI) in 2020 with supports from the German Federal Ministry for Economic Cooperation and Development (BMZ) to address One Health challenges in Sub-Saharan Africa. Its vision is a continent with healthier people, animals and the environment supported by strong health delivery systems and well-coordinated public-private institutions. Activities are conducted in Kenya, Uganda, Ethiopia, Tanzania, Malawi, Burkina Faso, Senegal, Mali, Burundi, Côte d'Ivoire and Nigeria. OHRECA follows two impact pathways: (1) applied research to generate new evidence on the benefits of One Health approaches, and (2) One Health training for communities, graduate fellows and policymakers in multiple countries (Figure 1).

The Centre's research focuses on four main One Health challenges: neglected zoonotic diseases (NZD), emerging infectious diseases (EID), antimicrobial resistance (AMR) and food safety.

For NZD, OHRECA developed evidence-based control measures for pork tapeworm (*Taenia solium*) in northern Uganda through multisectoral partnerships, identifying porcine cysticercosis hotspots and implementing rabies vaccination campaigns in Kenya. The EID team developed risk maps for re-emerging vector-borne diseases like Crimean-Congo haemorrhagic fever, Rift Valley fever (RVF), and Q fever. Their findings were integrated into surveillance manuals and contingency plans. They also supported foundational work on sequencing and phylogenetic analysis of RVF virus isolates in Kenya, Rwanda and Burundi.

The AMR team played an active role in international boards and working groups, translating research into policy influence. They have also contributed to environmental studies on the impact of antibiotic residues in livestock manure on greenhouse gas emissions.

Meanwhile, food safety scientists engaged in high-level One Health and policy initiatives with the United Nations Committee on World Food Security, African Union (AU), East African Community (EAC), Inter-University Council for East Africa (IUCEA), FAO, WHO and Codex Alimentarius.

OHRECA has trained 1,224 individuals, including surveillance officers, policymakers and market actors, in several countries (Figure 2).

References

Mishra C., Samelius G., Khanyari M., Srinivas P.N., Low M., Esson C., Venkatachalam S., Johansson Ö. 2022. Increasing risks for emerging infectious diseases within a rapidly changing High Asia. *Ambio*, 51(3), 494–507. https://doi.org/10.1007/s13280-021-01599-7

Mettenleiter T.C., Markotter W., Charron D.F., Adisasmito W.B., Almuhairi S., Behravesh C.B., Bilivogui P., Bukachi S.A., Casas N., Becerra N.C., Chaudhary A., *et al.*, 2023. The One Health High-Level Expert Panel (OHHLEP). *One Health Outlook*, 5(1), 1–8. https://doi.org/10.1186/s42522-023-00085-2

Laing G., Duffy E., Anderson N., Antoine-Moussiaux N., Aragrande M., Luiz Beber C., Berezowski J., Boriani E., Canali M., Pedro Carmo L., Chantziaras I., *et al.*, 2023. Advancing One Health: Updated core competencies. *CABI One Health*. https://doi.org/10.1079/cabionehealth.2023.0002

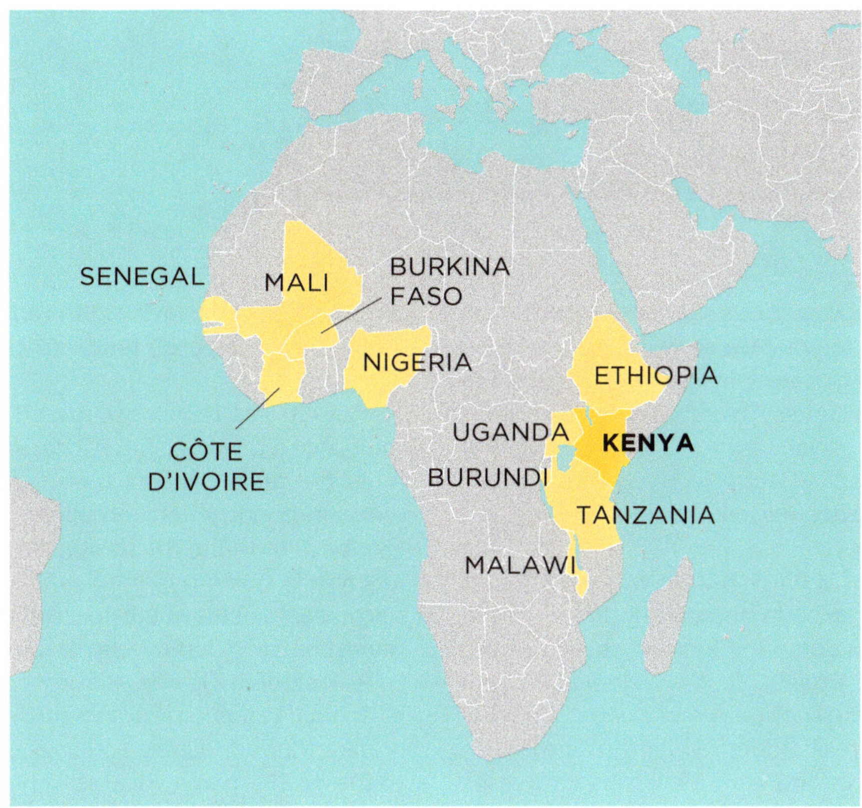

Figure 1. OHRECA's target countries.

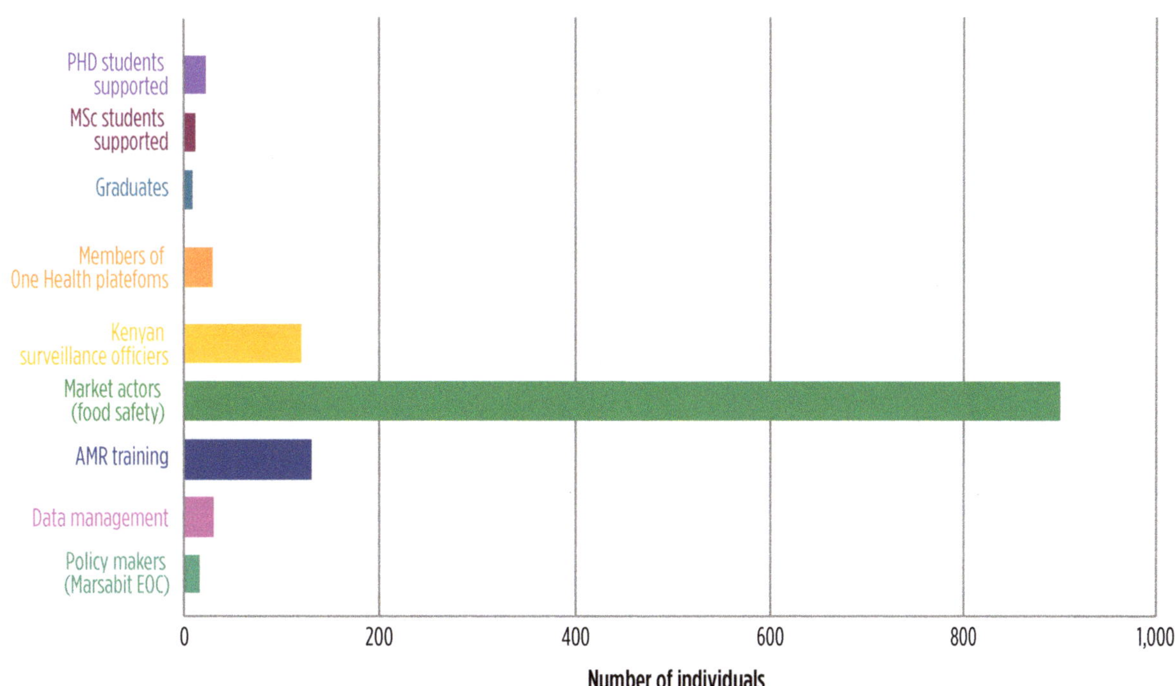

Figure 2. The impact of OHRECA training.

OHRECA has trained 1,224 individuals across various sectors and supported 22 PhD students and 11 MSc students, with eight obtaining their degrees so far. Short-term One Health training has involved 29 members of One Health platforms (Mali, Burkina Faso, Côte d'Ivoire, Senegal). Additionally, 120 Kenyan surveillance officers have been trained on zoonotic disease syndromic surveillance; 900 market actors on food safety, 130 on antimicrobial resistance (AMR) and 30 on data management; and 15 policymakers from Marsabit County's Emergency Operations Centre (EOC) on data management and analysis.

Innovative tools for One Health surveillance: ALERT, a collaborative serious game

Marie-Marie Olive, Sophie Muset, Mathias Talla Mba, Marisa Peyre

The use of serious games—which are designed to educate as well as entertain—has grown significantly in the health sectors in recent last decades. These games offer valuable tools to enhance interest in training and engage players more effectively to support learning and encourage behaviour change.

The ALERT board game was developed by a team of experts[1] in conjunction with the serious games company Bioviva as part of the EBO-SURSY2 Project. It is an innovative tool designed to foster engagement and collaboration among stakeholders involved in One Health surveillance systems (Figure 1) and addresses the need for community-based approaches to early disease detection. Engaging local stakeholders in health surveillance is critical to ensure the timely identification of emerging diseases. However, existing methods often struggle to fully integrate local actors, including those from different sectors (such as environmental, public, animal health) and citizens.

ALERT promotes good practices in One Health event-based surveillance systems through an interactive and collaborative format. The game helps raise awareness on the roles and responsibilities that each actor and sector plays in ensuring effective surveillance. By simulating real-life health events, players learn how to build a surveillance chain, from the local to the central level, and identify situations that may constitute health alerts. Players also learn how to foster communication and cooperation among the different stakeholders (human, animal, etc.) featured by the game.

ALERT was developed based on a rigorous, multiphase process. After formalizing its objectives, the team tested a prototype of the game with in-country stakeholders, incorporating feedback to refine the design. Facilitators play a key role in guiding gameplay and ensuring that participants achieve the learning objectives. The game has been deployed in several African countries, including Guinea, Cameroon and Senegal, with local stakeholders and students (Figure 2). A pilot study in Guinea suggested that, thanks to the ALERT game, local stakeholders acquired or strengthened their knowledge in One Health surveillance systems, highlighting the need to involve communities and reinforce multisectoral collaboration within zoonotic disease surveillance systems (Figure 3).

The ALERT project team will continue to deploy the game in various contexts and evaluate its impact to ensure its long-term success. Thanks to its flexibility in the situations generated, this game makes it possible to consider diverse local cultural, economic and social perspectives, including those of rural populations, when building surveillance systems, which would contribute to a better and long-term adoption of surveillance practices.

1. The team included experts from the French Agricultural Research Centre for International Development (CIRAD), World Organisation for Animal Health (WOAH), French National Research Institute for Sustainable Development (IRD) and Institut Pasteur.

2. https://rr-africa.woah.org/en/projects/ebo-sursy-en/

References

Wattanasoontorn V., Boada I., García R., Sbert M. 2013. Serious games for health. *Entertainment Computing*, 4, 231-247. https://doi.org/10.1016/j.entcom.2013.09.002

Luz S., Masoodian M., Rangel Cesario R., Cesario M. 2016. Using a serious game to promote community-based awareness and prevention of neglected tropical diseases. *Entertainment Computing*, 15, 43-55. https://doi.org/10.1016/j.entcom.2015.11.001

Olive M.-M., Tolno S.A., Ndong Bass I., Barry M.A., Talla Mba M. *et al.* 2024. ALERT: a collaborative board game to sustainably engage stakeholders in the One Health surveillance in Africa. Poster, 8th World One Health Congress, Cape Town, 20-23 September 2024.

Figure 1. The ALERT serious game.

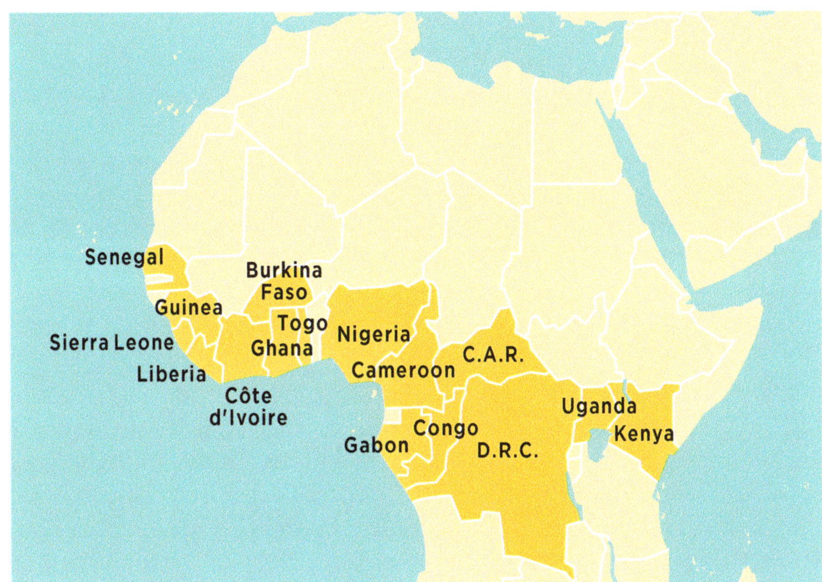

Figure 2. Countries where the ALERT serious game has been disseminated.

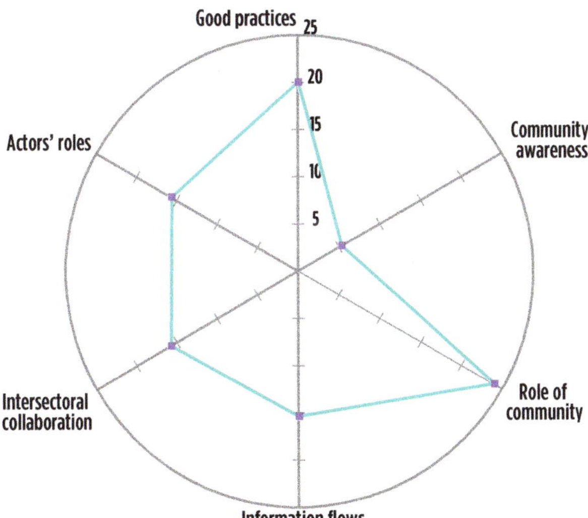

Figure 3. Result of proportional piling of topics that the game highlights among local stakeholders who participate in the ALERT serious game in Guinea.

One Health approach in Guinea

Marie-Marie Olive, Alpha Kabinet Keita, Claire Costis, Martine Peeters, Hélène De Nys, Abdoulaye Touré[1]

Guinea has faced repeated and concurrent outbreaks of infectious diseases since the 2014–2016 West African Ebola virus disease outbreak that originated there. In 2021 alone, Guinea experienced single virus disease outbreaks of Ebola and Marburg, several Lassa fever outbreaks and the second wave of COVID-19 (Figure 1).

In July 2017, a joint decree from the Guinean ministries of Health, of Livestock and Animal Production, and of Environment, Water and Forests set up a national One Health platform that was later decentralized to community level. Although the platform needs additional support, especially at local level, Guinea now has resources and mechanisms for coordinated health reporting and proven intervention mechanisms. National research centres have also been established and include the CERFIG (2017) and Institut Pasteur de Guinée (2018).

Various multidisciplinary research projects were launched in Guinea during the Ebola epidemic. In 2015, the post-EBOGUI research project drew on a pluridisciplinary approach to describe and analyse the clinical, immunovirological, psychological and socioanthropological consequences of Ebola after patients were released from treatment centres. Additional One Health projects, such as RESERVOIR, EBOHEALTH, EBO-SURSY and BCOMING, were set up to understand animal reservoirs and evaluate emergence risks and to develop early detection and response approaches at the human-wildlife interfaces. Meanwhile, projects such as DOPERAUS and AfriCam-PREACTS are working to further operationalize and decentralize the One Health platform (Figure 2). Most of these projects are carried out with local stakeholders (including communities) on the front lines of zoonotic disease emergence. Some projects (GRET) are also operationalizing the One Health concept using a territorial approach. Decentralized One Health platforms go beyond managing epidemic outbreaks, and also support communities with integrating to enhance their capacity to help communities integrate One Health issues based on their concerns. One Health platforms promote agricultural, land-use and natural-resource-use changes and play an important role in local dialogue.

Guinea has developed a One Health approach over the last decade that extends from the central to the local level (Figure 3). It has been resilient to Ebola epidemics and has provided insights from its experiences, leading to major epidemiological advances in our understanding of Ebola. Researchers and decision makers are working together, and projects are underway to create a local One Health approach that involves local stakeholders (including citizens) in actions to manage infectious disease outbreaks.

1. With the contribution of Saa André Tolno, Dobo Onivogui, Souana Goumou, Abdoul Karim Soumah.

References

Guenin M.-J., De Nys H.M., Peyre M., Loire E., Thongyuan S. *et al.* 2022. A participatory epidemiological and One Health approach to explore the community's capacity to detect emerging zoonoses and surveillance network opportunities in the forest region of Guinea. *PLoS Neglected Tropical Diseases*, 16(7), e0010462. https://doi.org/10.1371/journal.pntd.0010462

Keita A.K., Koundouno F.R., Faye M., Düx A., Hinzmann J. *et al.* 2021. Resurgence of Ebola virus in 2021 in Guinea suggests a new paradigm for outbreaks. *Nature*, 597, 539–543. https://doi.org/10.1038/s41586-021-03901-9

Maltais S., Brière S., Yaya S. 2022. Comment la résilience post-Ebola en Guinée contribue à la gestion de la COVID-19 ? *Santé Publique*, 34, 557–567. https://doi.org/10.3917/spub.224.0557

https://gret.org/en/projet/promoting-a-global-approach-to-health-at-territorial-level-in-forest-guinea/

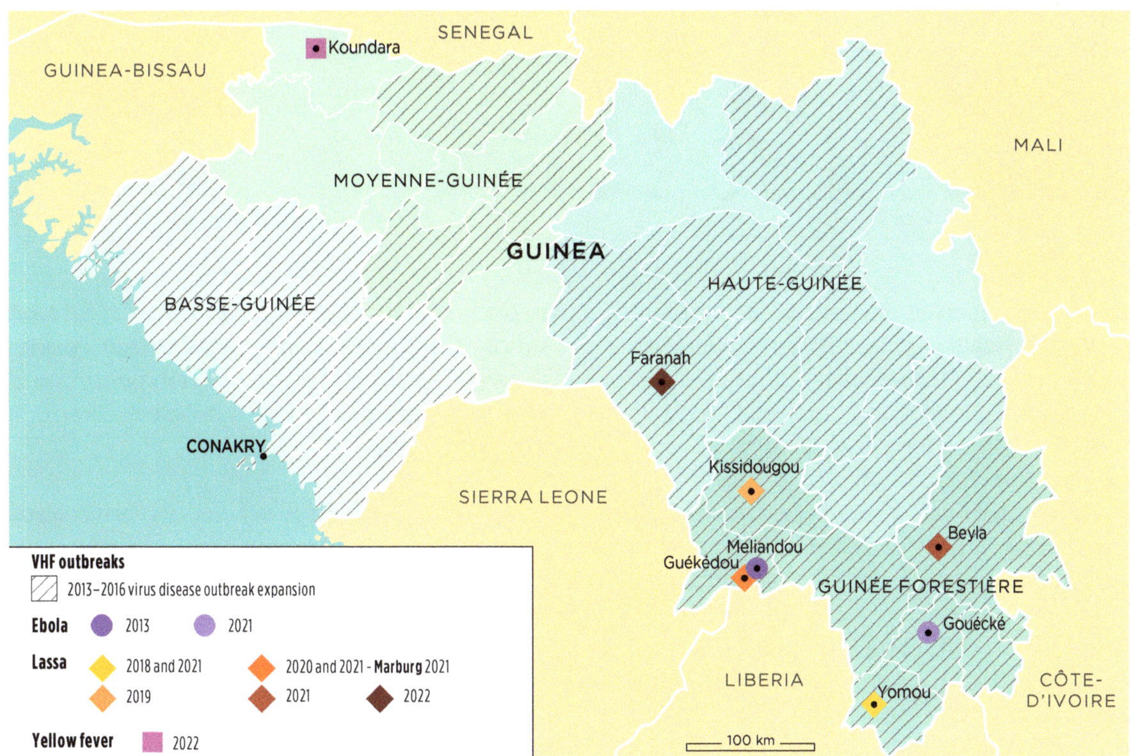

Figure 1. Viral haemorrhagic fever (VHF) outbreaks occurred in Guinea from 2013 to 2022.

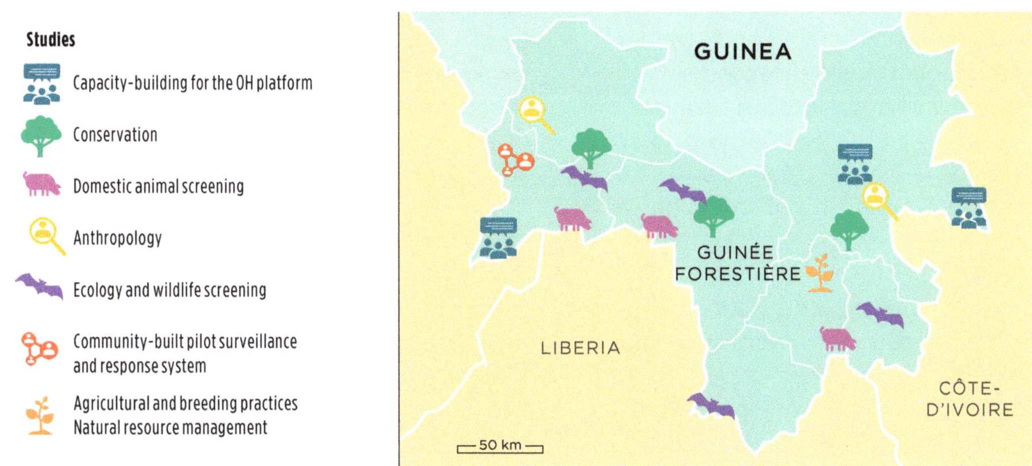

Figure 2. One Health research field studies in Guinée forestière.

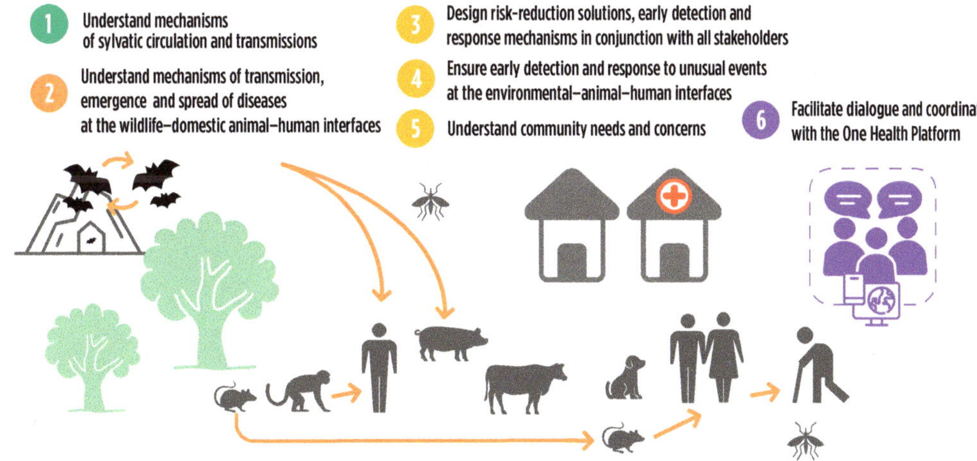

Figure 3. The One Health approach in Guinea.

CGIAR Initiative on One Health

Hung Nguyen-Viet,
Vivian Hoffmann

The official objective of the CGIAR Initiative on One Health (2022-2024) was "to demonstrate how One Health principles and tools integrated into food systems [could] help reduce and contain zoonotic disease outbreaks, improve food and water safety and reduce anti-microbial resistance, benefiting human, animal and environmental health" in low- and middle-income country settings.

The research initiative was operationalized through five thematic work packages (Figure 1). Each work package was designed to generate contextually relevant evidence and solutions, develop capacities for their use, and engage in policy and practice change. The initiative focused on testing and evaluating One Health innovations on disease surveillance, food safety and antimicrobial resistance intervention in seven countries: Vietnam, India, Bangladesh, Ethiopia, Kenya, Uganda and Côte d'Ivoire (Figure 2). Key results of the initiative include:

- Development of integrated surveillance at the wildlife–livestock–human interface, including genomic surveillance of pathogens to track viral evolution and predict outbreaks. Climate-responsive disease risk mapping support tools were created to target surveillance and resource allocation and enhance prevention efforts in vulnerable regions with scarce resources.
- Evaluation of innovations to improve slaughterhouse hygiene (Kenya) and food safety in pork markets (Vietnam) and in red meat shops (Ethiopia). Impact evaluation results showed positive changes in behaviours and attitudes, but more work is needed to reduce bacterial contamination.
- Crucial insights into drivers of antimicrobial misuse in livestock and aquaculture, along with recommendations based on this evidence.
- Better understanding of the spatial and temporal distribution of waterborne zoonotic pathogens and antimicrobial resistance in selected watersheds in Ethiopia and India to inform planning for pollution control and health risk mitigation. Business models for resource recovery from livestock waste in these countries were analysed and co-developed to prevent water pollution from livestock waste.
- Cooperation with local authorities to operationalize the One Health concept on the ground. One Health research sites were established and supported in Vietnam and Kenya, which serve as One Health practice sites that bring together actors from diverse disciplines and communities to work on topics prioritized by communities.

Significant progress was achieved in knowledge discovery, as evidenced by the publication of 105 peer-reviewed articles to date and other knowledge outputs, outcomes and impacts being reported.

References

CGIAR Initiative on One Health: https://www.cgiar.org/initiative/one-health/
CGIAR Initiative on One Health outputs: https://hdl.handle.net/10568/114641
Lam S., Hoffmann V., Bett B,. Fèvre E.M., Moodley A. *et al.* 2024. Navigating One Health in research-for-development: Reflections on the design and implementation of the CGIAR Initiative on One Health. *One Health*, 8,100710. https://doi.org/10.1016/j.onehlt.2024.100710

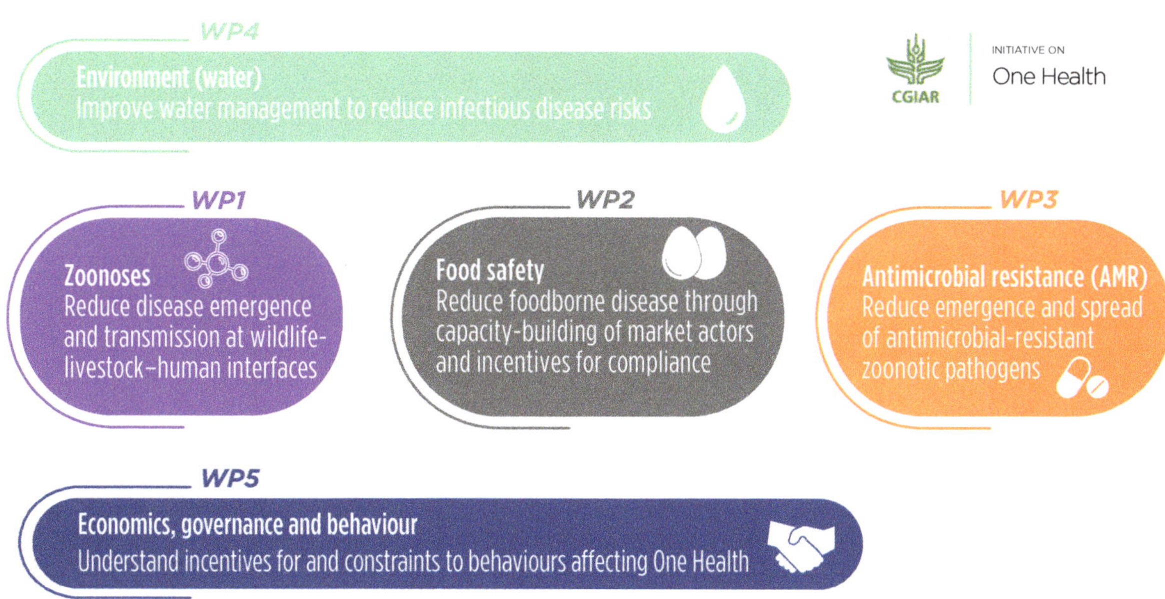

Figure 1. CGIAR One Health initiative project framework broken down by work package (WP), including core components and activities.

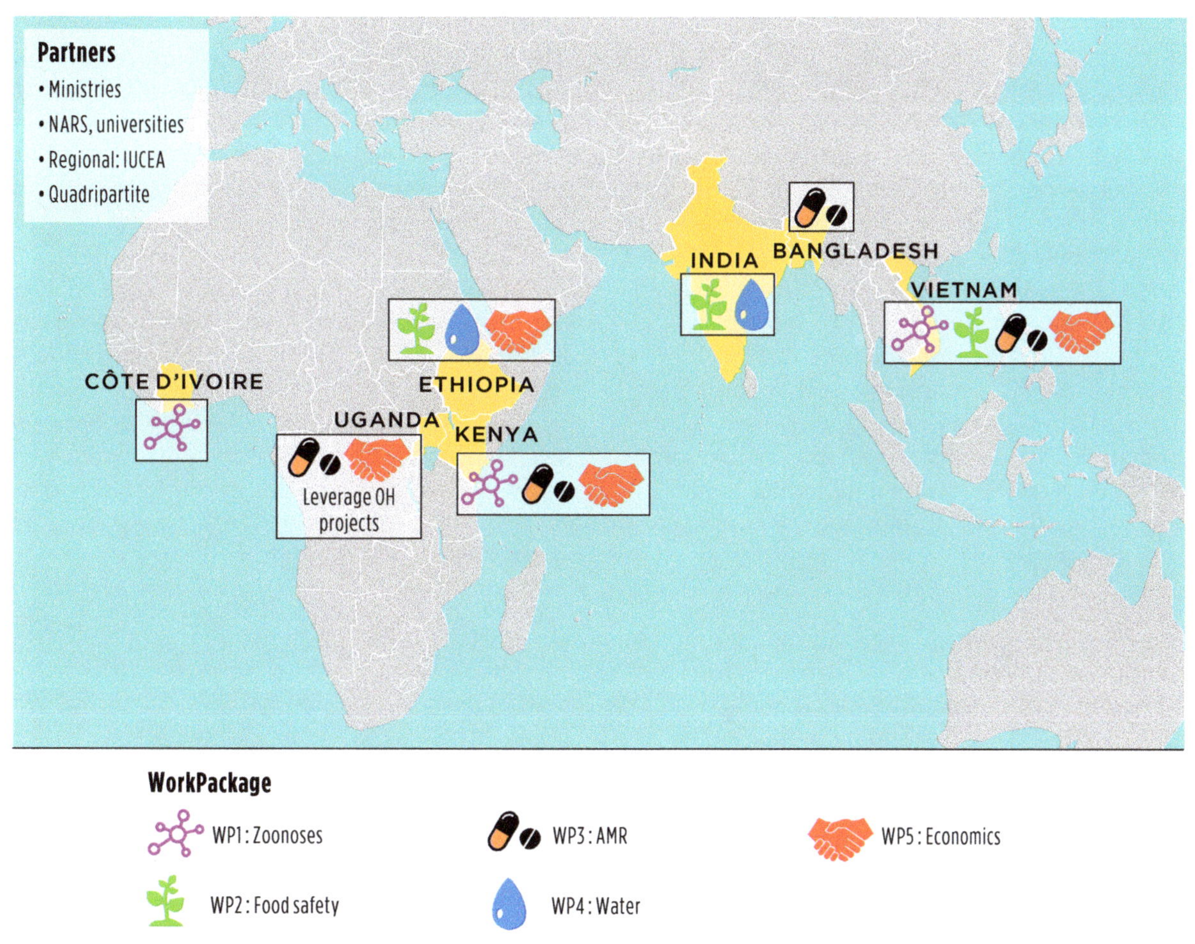

Figure 2. CGIAR One Health initiative study sites and topics.
NARS: National Agricultural Research Systems; IUCEA: Inter-University Council for East Africa; Quadripartite: FAO, WHO, WOAH and UNEP.

Local service providers and communities: key stakeholders for operationalizing One Health

Manuelle Miller,
Esther Schelling,
Eddy Timmermans

The current challenge for One Health is to become operational in the field. This is particularly true at community level in areas where public and private health service delivery is fragile or absent, as in several regions in the Global South, and especially in rural areas. Local civil society and services play a pivotal role at the interface of the three pillars of health: human, animal and environment. Rather than being considered only as the beneficiaries of One Health interventions, these stakeholders should have a more central role in co-identifying public health threats and identifying locally adapted responses.

Among many examples, three projects implemented in Africa by member organizations of the NGO network Vétérinaires Sans Frontières International,[1] in association with human health development stakeholders, underline how local communities and community-based service providers have been involved in development and humanitarian interventions at different stages of the project cycle (Figure 1). For instance, local service providers work closely with decentralized technical services from different ministries (e.g. health posts, animal health services), as are community-based workers specialized in animal health or human health and private sector small-scale businesses.

From the early diagnostic phase through the implementation and evaluation of activities, participatory methodologies are used to generate consensus among local stakeholders on a shared vision of major daily health problems to better manage them with limited resources. In three contexts, from arid pastoral zones (Horn of Africa, program HEAL[2]) to agropastoral areas (Senegal, Thiellal project[3]) and crisis and disaster-affected zones (Niger, action GRSC), local service providers have been jointly trained and supported on different topics identified through participatory diagnostics.

This cross-cutting work helps improve service accessibility and resilience for populations. This is an investment that pays off because services share sparse resources instead of duplicating efforts—e.g. in outreach and information—which also makes them better able to recognize abnormal health events.

Participation of local communities and service providers should be a prerequisite for a sustainable implementation of the One Health approach. Participatory platforms of concerned stakeholders help support communities and can contribute to gender-sensitive strategies for coping with current challenges, including health and environmental threats related to climate change.

1. Vétérinaires Sans Frontières International: www.vsf-international.org
2. https://www.oh4heal.org/
3. https://www.avsf.org/projets/thiellal-rendre-concret-le-concept-one-health-dans-les-territoires-de-haute-casamance/

References

Timmermans E., Rojas Lopez F., Meissner L. 2020. The One Health approach; Improving health risk management in Niger. A new pilot project implemented by Doctors of the World and Vétérinaires Sans Frontières - Belgium. Doctors of the World and Vétérinaires Sans Frontières - Belgium. https://veterinairessansfrontieres.be/wp-content/uploads/2020/11/risques-sanitaires-au-Niger_ENGL_DEF_WEB.pdf

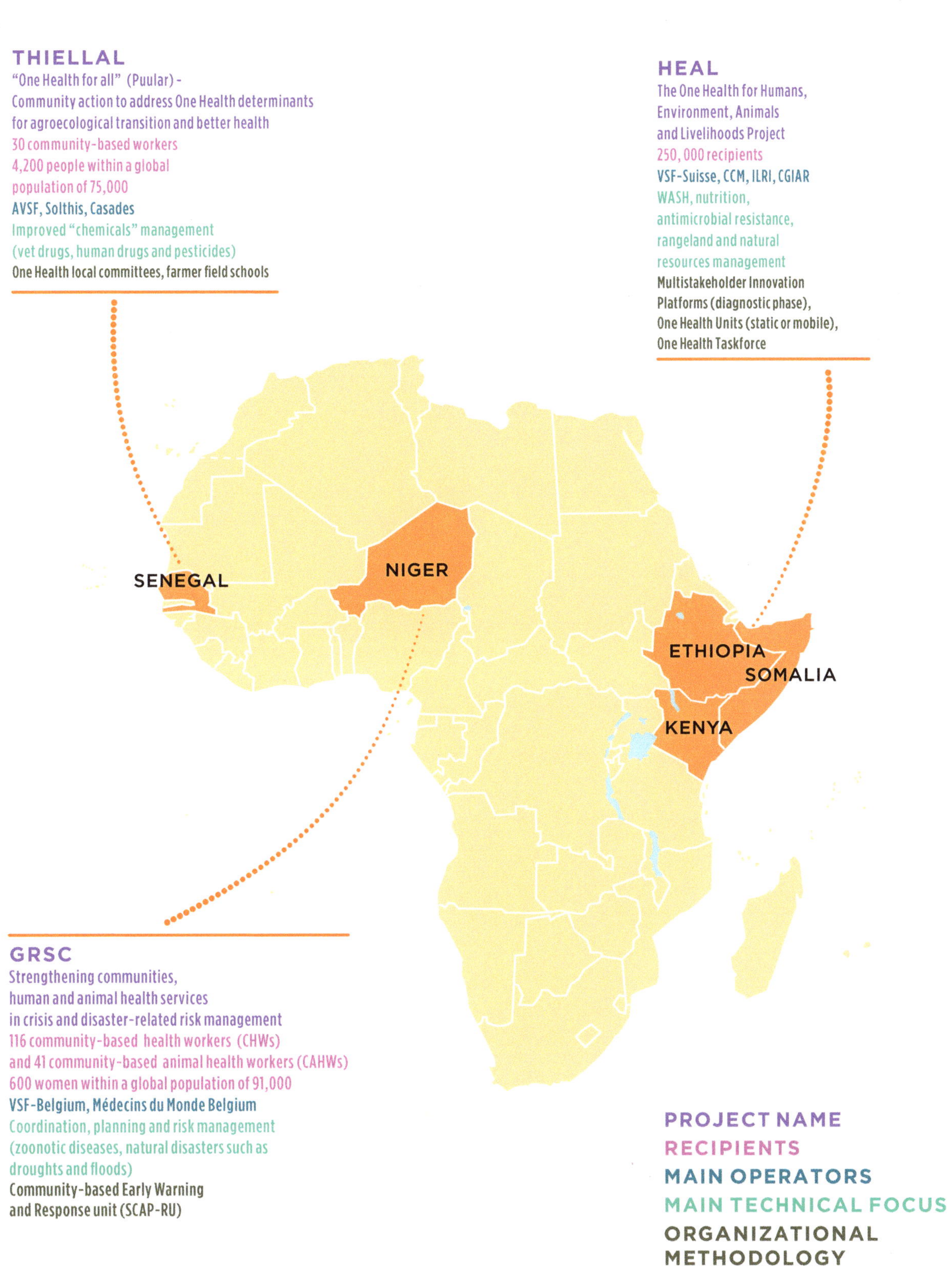

Figure 1. A few examples of projects implemented by three members of the Vétérinaires Sans Frontières – International network, in collaboration with local partners and with active participation from local communities.

Integrated (One) Health service delivery in pastoralist settings in sub-Saharan Africa

Micol Fascendini,
Solveig Danielsen,
Esther Schelling

Although cross-sectoral collaboration is the backbone of One Health, integrated health service delivery remains under-explored and mostly refers to "integrated surveillance". There is growing evidence of wider health benefits from integrating services across sectors and disciplines, particularly in remote low-income settings where public and private services are scarce.

A literature review was carried out on integrated service delivery models for pastoral communities in sub-Saharan Africa, as part of broader research on collaboration and coordination between sectors (public health, veterinary and environmental health) and the establishment of multipurpose delivery points. The search was limited to papers published in English from 1995 to 2022.

Successful published experiences of integrated services in pastoralist communities include examples from Nigeria, Chad, Kenya and Somalia (Figure 1). Integrated services refer to cross-sectoral delivery of services that are jointly planned by at least two health sectors. Cross-sectoral planning leads to the identification of more effective ways to reach the target communities, ensuring services are tailored and relevant to their needs. Initial drivers of these initiatives were low immunization rates or outbreaks; however, once the model was in place, joint services were expanded to also include surveillance, nutrition and health education, for example.

Cross-sectoral collaboration allows for better accessibility and higher service coverage, financial savings, reduced user costs and increased trust in the system. Nevertheless, effective integrated services require a significant financial and non-financial investment mainly in terms of training and supervision of the multidisciplinary team and the creation of multisectoral systems for collaborative planning, working and learning. A solid and durable commitment from the government is essential to guarantee the institutionalization and ownership of any integrated service delivery model (Figure 2).

References

Bomoi I.M., Waziri N.E., Nguku P., Tsofo A. 2016. Integrated human and animal vaccination delivery to Nomadic Fulani communities in Northern Nigeria 2015. *International Journal of Infectious Diseases*, 45, 22. https://doi.org/10.1016/j.ijid.2016.02.082

Danielsen S., Schelling E., Whittaker M. 2020. Reaping One Health benefits through cross-sectoral services. In: *One Health: The Theory and Practice of Integrated Health Approaches* (Zinsstag J., Schelling E. Crump L., Whittaker M., Tanner M., Stephen C., Eds.), pp. 170–183. CABI. https://doi.org/10.1079/9781789242577.0170

Griffith E.F., Kipkemoi J.R., Robbins A.H., Abuom T.O., Mariner J.C. *et al.* 2020. A One Health framework for integrated service delivery in Turkana County, Kenya. *Pastoralism*, 10(1). https://doi.org/10.1186/s13570-020-00161-6

Kamadjeu R., Mulugeta A., Gupta D., Hirsi A.A., Belayneh A. *et al.* 2015. Immunizing nomadic children and livestock – experience in north east zone of Somalia. *Human Vaccines and Immunotherapeutics*, 11(11), 2637–2639. https://doi.org/10.1080/21645515.2015.1038682

Ndiaye S.M., Ahmed M.A., Denson M., Craig A.S., Kretsinger K. *et al.* 2014. Polio outbreak among nomads in Chad: Outbreak response and lessons learned. *Journal of Infectious Diseases*, 210, S74–S84. https://doi.org/10.1093/infdis/jit564

Schelling E., Bechir M., Ahmed M.A., Wyss K., Randolph T.F., Zinsstag J. 2007. Human and animal vaccination delivery to remote nomadic families, Chad. *Emerging Infectious Diseases*, 13(3), 373–379. https://doi.org/10.3201/eid1303.060391

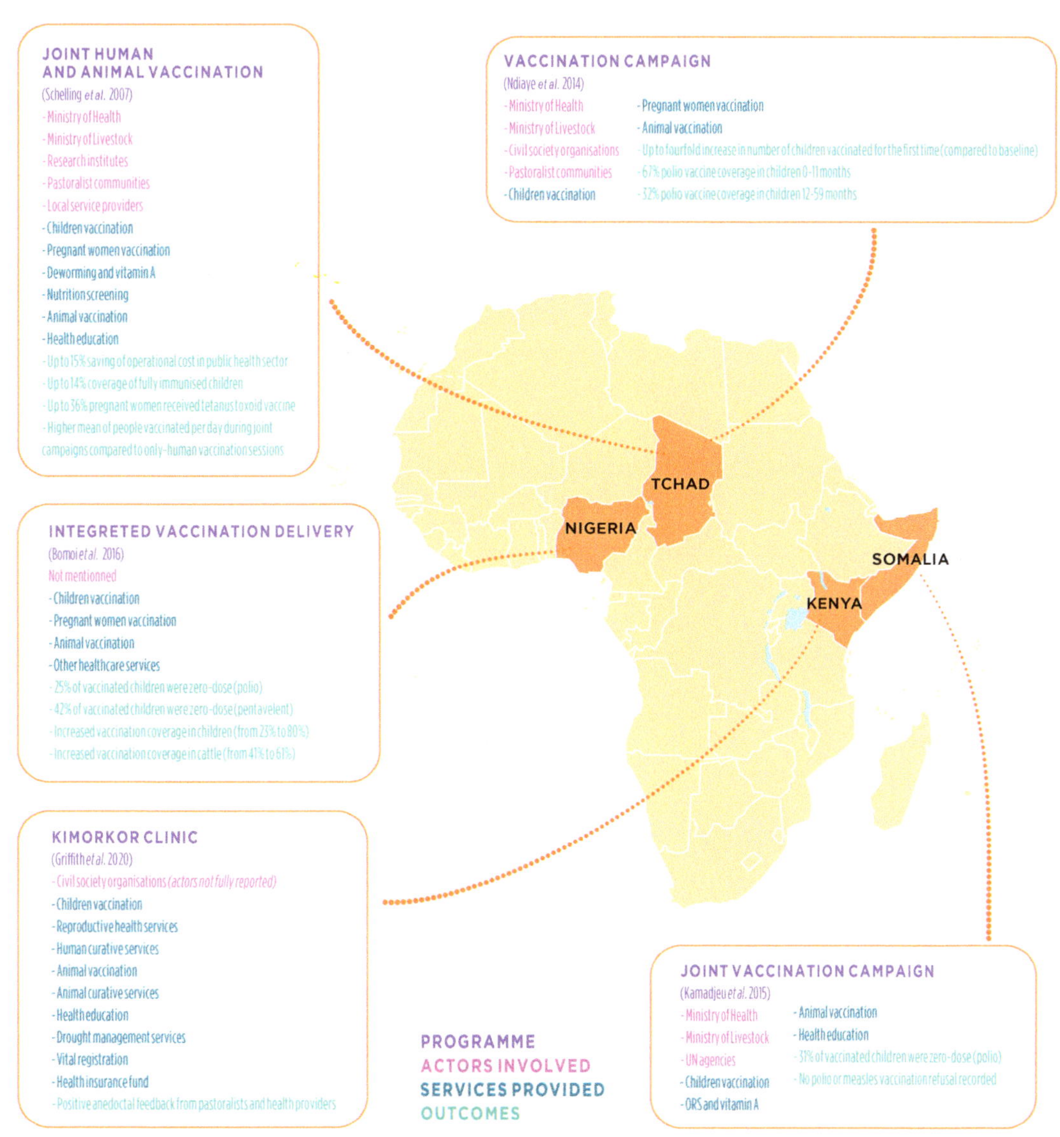

Figure 1. Documented successful experiences of One Health integrated service delivery models among pastoralist communities in sub-Saharan Africa.

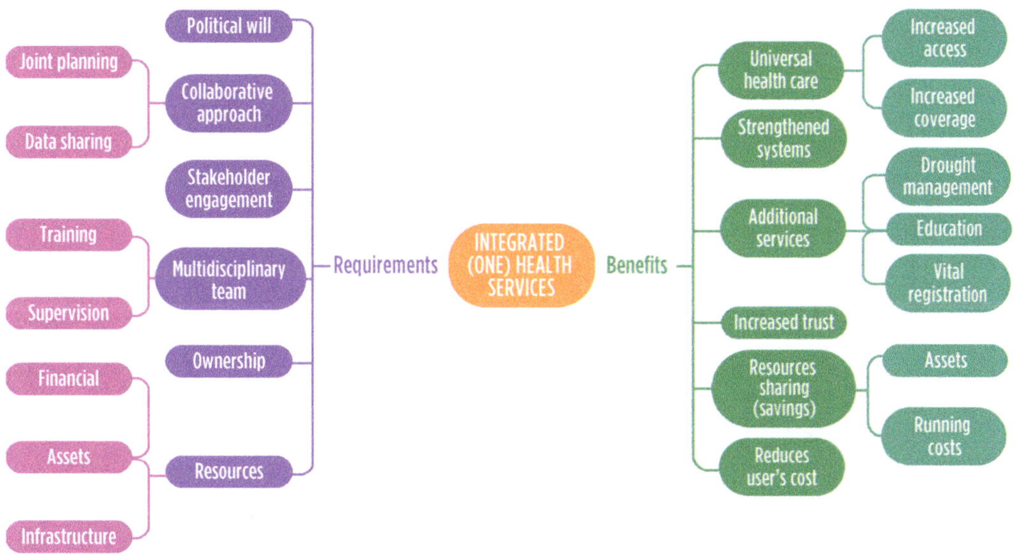

Figure 2. Requirements and benefits of One Health integrated services in pastoralist communities.

Integrated crop-livestock clinics in East Africa

Christine Alokit,
Florence Chege,
Solveig Danielsen

Crop and animal health are intrinsically linked, and they are crucial to agricultural production and livelihoods of smallholder farmers in Sub-Saharan Africa. Poor management of plant and animal health leads to productivity losses and affects the health of humans, animals, plants and the environment in many different ways (Figure 1). For example, mycotoxins in grain make people and animals sick. Misuse of pesticides and veterinary medicines is a major source of contamination, poisoning and antimicrobial resistance. Zoonoses pose a significant threat to public health, and pathogens can be transmitted by water, soil, manure, and plant and animal products. Despite this interconnectedness, farmer advisory services in low-income countries, in addition to having low reach, tend to be organized by sector: crops, animals, human health and the environment are dealt with separately, thus limiting the scope for addressing the One Health issues that connect them.

CABI's work with plant clinics (a plant health advisory service for farmers) over the last two decades has inspired new ideas on how to serve smallholder farmers better. Feedback from farmers and clinic staff laid the groundwork for expanding the clinics to also include animals. With funding from the Biovision Foundation and WTS, integrated crop-livestock clinics ("joint clinics") have been implemented in selected areas in Uganda (six districts since 2021) and Kenya (three counties since 2022), in collaboration with the ministry of agriculture, local government staff from veterinary, agriculture and public health departments (in Kenya) and other stakeholders (Figure 2 and Table 1).

Based on the joint clinic staff's engagement with farmers (documented in clinic records), and other sources of One Health knowledge, project stakeholders have worked together to identify rabies, aflatoxins, worms (Kenya) and misuse of pesticides and veterinary drugs (Uganda) as high-impact areas needing further, cross-sectoral interventions.

Crop-livestock clinics, as a type of cross-sectoral service, are seen as a means to make better use of existing human and financial resources and a promising way to stimulate action and learning across sectors, disciplines and stakeholder groups. Exploration of One Health added values, such as financial, organizational and learning benefits, are ongoing.

References

CABI. Project overview: Joint crop and livestock services for smallholder farmers. https://www.cabi.org/projects/joint-crop-and-livestock-services-for-smallholder-farmers/

CABI. 2019. Plantwise Impact Report. Wallingford, UK: CABI. https://www.plantwise.org/wp-content/uploads/sites/4/2019/04/Plantwise-Impact-Report-2018_WEB.pdf?utm_source=website&utm_campaign=public&utm_content=full-report

Danielsen S., Kajura C., Mulema J., Taylor R., Kansiime M., Alokit C., Tukahirwa B., Schelling E. 2019. Reaching for the low hanging fruits: One health benefits of joint crop-livestock services for small-scale farmers. *One Health*, 100082. https://doi.org/10.1016/j.onehlt.2019.100082

Danielsen S., Alokit C., Aliamo C, Mugambi I. 2022. How crop-livestock clinics are advancing One Health: a pilot case from Uganda. *One Health Cases*. https://doi.org/10.1079/onehealthcases.2022.0002

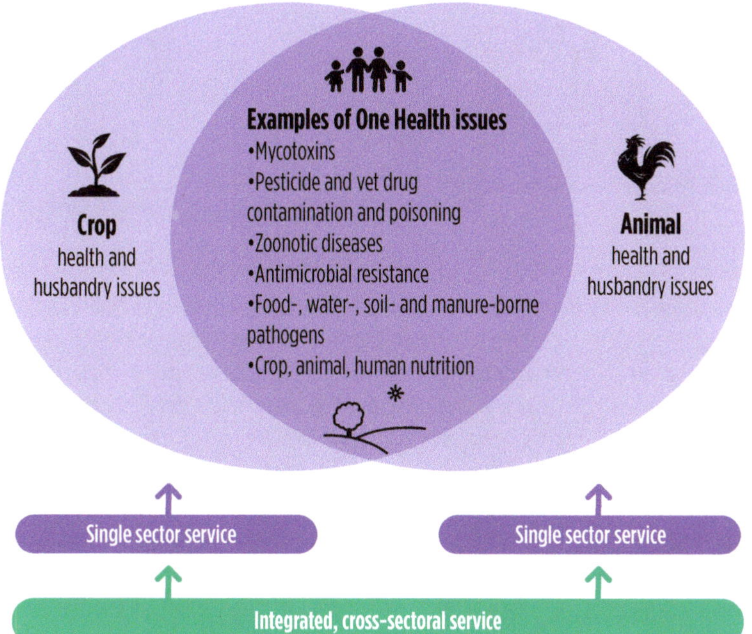

Figure 1. The interconnectedness between crop, animal, human and environmental health and scope of single vs. cross-sectoral services. From Danielsen *et al.* 2022.

Figure 2. Areas where crop–livestock clinics are being implemented: two counties in Kenya and six districts in Uganda.

Table 1. Crop–livestock clinic implementers and collaborators in Kenya and Uganda as of 2022.

Role	Kenya	Uganda
Clinic implementers, oversight	County governments: veterinary, agriculture and public health departments	District local government: agriculture, veterinary departments
Policy guidance, oversight, technical backstopping	Ministry of Agriculture, Livestock, Fisheries and Co-operatives	Ministry of Agriculture, Animal Industry and Fisheries (MAAIF) (various departments)
Training, backstopping, information provision, networking	Infonet Biovision; ILRI; VSF Kenya; CABI; others (evolving)	Makerere University, College of Veterinary Medicine; MAAIF; Infonet Biovision; CABI; others (evolving)

Capacitating One Health in Eastern and Southern Africa (COHESA)

Alexandre Caron, Shauna Richards, Bibiana Iraki, Hélène De Nys, Margaret Karembu, Theo Knight-Jones

The COHESA project[1] aims to generate an inclusive research and innovation ecosystem, facilitating rapid uptake, adaption and adoption of solutions to issues that require a One Health approach in 12 countries in Eastern and Southern Africa (Figure 1). Starting in 2021, the project consortium (ILRI, CIRAD, ISAAA Africentre) has engaged research institutions (aka multipliers; Figure 2) in each country to co-develop and deliver the project's activities. A key theme is creating and/or supporting functional national One Health coordinating bodies, which can foster operationalization of One Health actions. In each country, the activities are flexible enough to adapt to the local context and create synergy with ongoing One Health activities. COHESA thus integrates into the established or developing national One Health movement rather than imposing a pre-established "onesize-fits-all" framework. Furthermore, by connecting the 12 countries, COHESA facilitates inter-country connectivity and experience sharing.

COHESA supports soft-skills development and collective learning, and it recognizes the importance of multi-sectoral and interdisciplinary trainings or workshops to complement the core sectoral competencies (e.g. outbreak investigation, risk-based surveillance) already provided by other initiatives.

Preliminary project outputs indicate that a regional and inter-regional approach provides considerable added value to the project: many Eastern African countries have almost 10 years of One Health engagement and experience that they can share with Southern African countries that are generally only starting to implement One Health. In addition, the flexibility of the project allows the COHESA consortium to respond quickly to requests from countries for activities such as supporting the validation of national One Health Strategic Plans or facilitating alignment with new Quadripartite documents (e.g. Joint Plan of Action). Finally, national One Health stakeholders have provided very positive feedback with regard to soft-skills development and the implementation of participatory tools (e.g. network mapping).

1. www.ilri.org/research/projects/capacitating-one-health-eastern-and-southern-africa-cohesa; https://onehealthobservatory.org/

References

Caron A., de Garine-Wichatitsky M., Figuié M., Meunier J., Mugabe P., Zinsstag J. 2025. Bridging gaps and leveraging opportunities for One Health: Feedback from the 8th World One Health Congress. *CABI One Health*, 4(1), 0003. https://doi.org/10.1079/cabionehealth.2025.0003

Goulet C., de Garine-Wichatitsky M., Chardonnet P., de Klerk L.M., Kock R. *et al.* 2024. An operational framework for wildlife health in the One Health approach. *One Health*, 19, 100922. https://doi.org/10.1016/j.onehlt.2024.100922

FAO, UNEP, WHO, WOAH. 2022. One Health Joint Plan of Action (2022-2026). Working together for the health of humans, animals, plants and the environment. Rome, 86 pp. https://www.who.int/publications/i/item/9789240059139

Richards S., Knight-Jones T., Angombe S., Becker J., Bukachi S.A. *et al.* 2024. Towards institutionalization of One Health in Eastern and Southern Africa. *One Health Cases*, ohcs20240007. https://doi.org/10.1079/onehealthcases.2024.0007

Wako B.Y., Richards S., Grace D., Mutie I., Caron A. *et al.* 2025. Building the future One Health workforce in Eastern and Southern Africa: Gaps and opportunities. *CABI One Health*, 4(1), 0014. https://doi.org/10.1079/cabionehealth.2025.0014

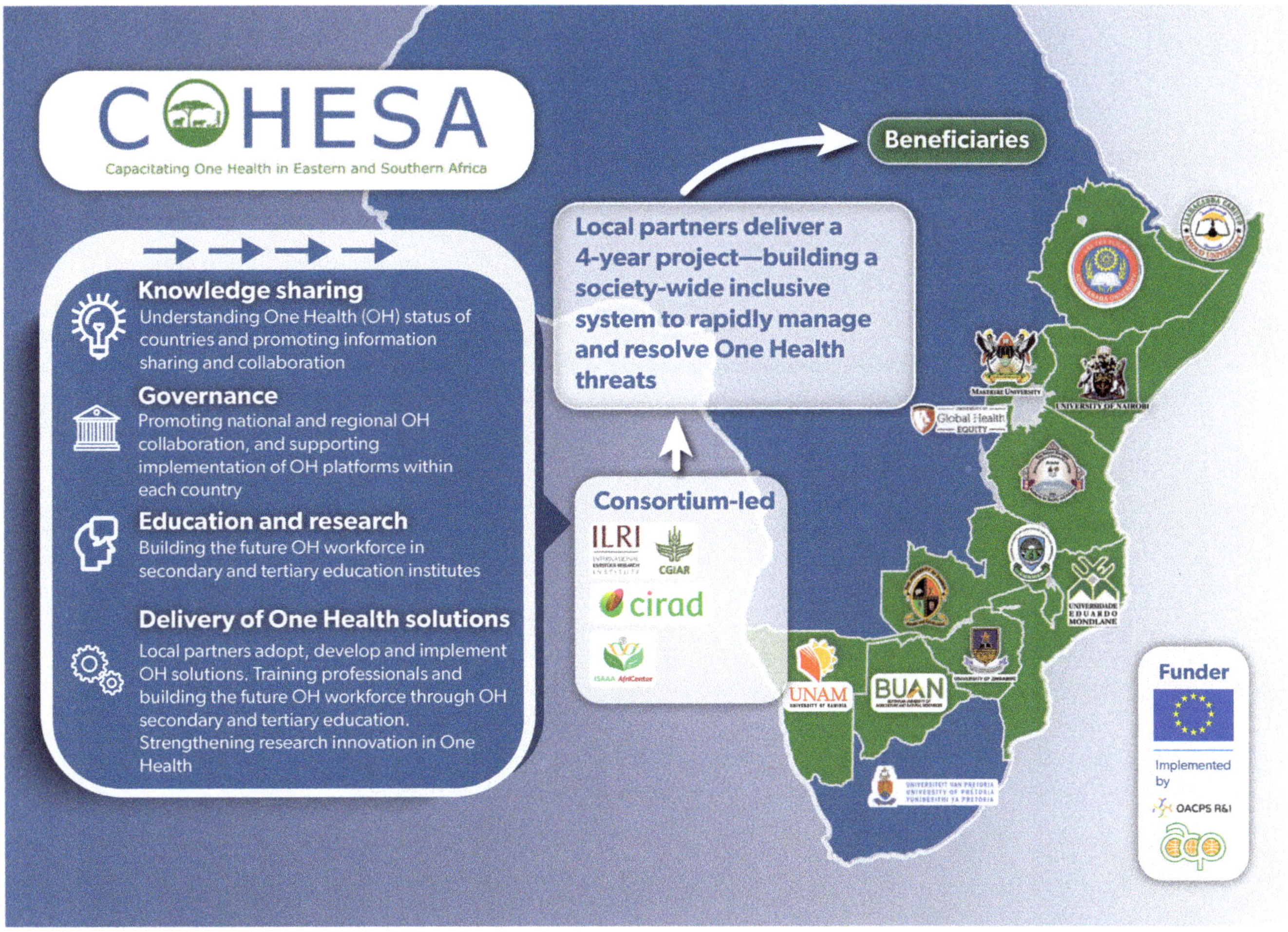

COHESA
Capacitating One Health in Eastern and Southern Africa

Knowledge sharing
Understanding One Health (OH) status of countries and promoting information sharing and collaboration

Governance
Promoting national and regional OH collaboration, and supporting implementation of OH platforms within each country

Education and research
Building the future OH workforce in secondary and tertiary education institutes

Delivery of One Health solutions
Local partners adopt, develop and implement OH solutions. Training professionals and building the future OH workforce through OH secondary and tertiary education. Strengthening research innovation in One Health

Local partners deliver a 4-year project—building a society-wide inclusive system to rapidly manage and resolve One Health threats

Beneficiaries

Consortium-led
ILRI
CGIAR
cirad
ISAAA AfriCenter

Funder
Implemented by
OACPS R&I

UNAM
BUAN
Global Health EQUITY
University of Nairobi
Makerere University
Universidade Eduardo Mondlane

Figure 1. The COHESA project including its Consortium and the 12 associated countries and research institutions (Somalia joined the project in 2022). On the left hand-side, the four main work packages.

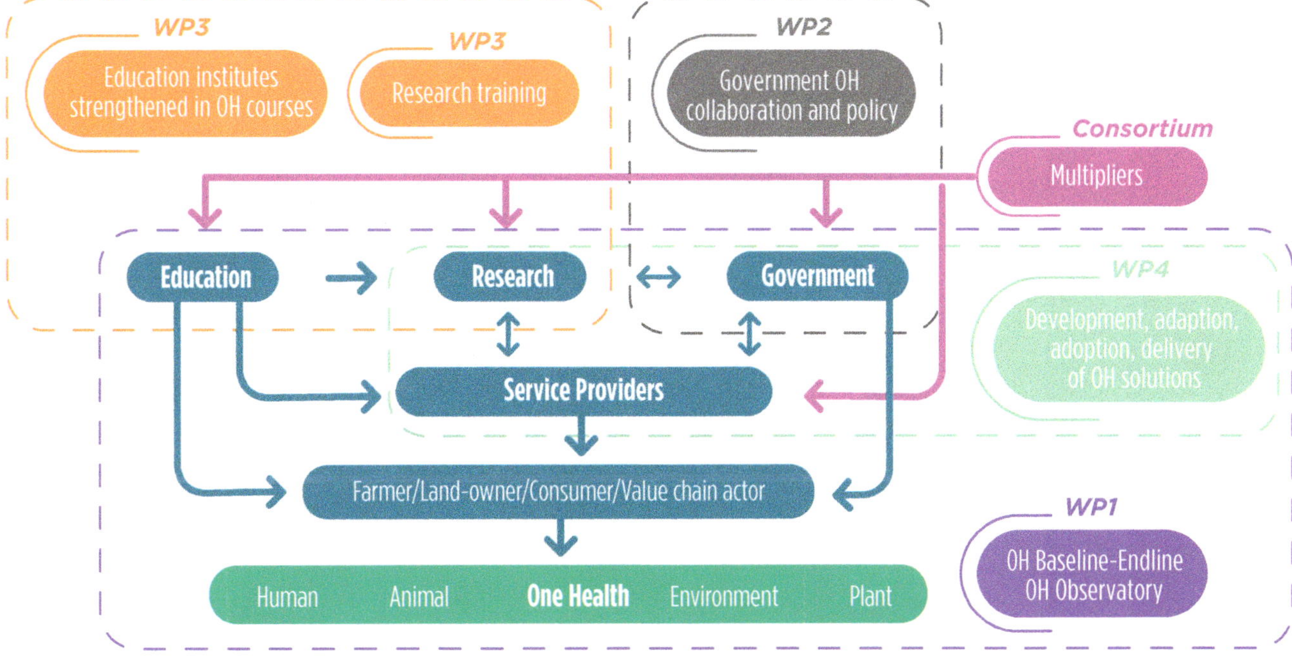

Figure 2. COHESA framework: the consortium supports multipliers in each country to reach out to One Health stakeholders from the government, education and research sectors; these stakeholders and service providers are then capacitated to deliver One Health in action to final beneficiaries (e.g. farmers, consumers).

ILRI One Health research for development portfolio to address food system challenges

Hung Nguyen-Viet,
Steven Lam

For more than 15 years, the International Livestock Research Institute (ILRI) has been striving to understand and address One Health challenges at the intersection of livestock, humans and the environment. We present a synthesis of select ILRI One Health projects implemented with partners across Asia and Africa. Projects were considered relevant if they described the process or outcomes of the initiative; were a relatively large project (defined as having received at least USD 1 million in funding); explicitly used a One Health approach; and were reported on in a journal article or report.

From the 24 projects identified (Figure 1)[1], primarily implemented in Africa and Asia, we analysed processes and outcomes using a realistic evaluation framework (context, mechanisms, outcomes) to present insights within select One Health topic areas such as zoonoses, food safety and antimicrobial resistance (AMR).

ILRI One Health's goal of improving food security and reducing poverty is achieved through three main, interconnected pillars: technical, policy and institutional. ILRI's work has advanced our understanding of challenges and solutions of zoonoses, food safety and AMR.

By challenging conventional wisdom, ILRI revealed that some zoonotic diseases were either more or less common than previously thought, highlighting their impact on human health, wildlife and ecosystems. In food safety, ILRI's decade-long study of informal markets confirmed that food safety is a growing issue affecting smallholder farmers, with significant implications for human health and livestock production. In AMR research, One Health work played a key role by improving knowledge on antimicrobial use, feed quality and governance, while also developing drug quality assessment protocols and influencing AMR policies and interventions at national and regional levels.

Our findings emphasize the need for stronger cross-sectoral collaboration, greater engagement with policymakers to translate research findings into actionable strategies, and the development of adaptable and context-specific interventions.

Ongoing and future research on One Health at ILRI includes One Health projects[2] and being part of the CGIAR Science Programs and Accelerator, in particular Sustainable Animal and Aquatic Foods[3] and Food Frontiers and Security[4].

1. For OHRECA and COHESA projects, see p. 92 and p. 106.
2. https://www.ilri.org/one-health
3. https://www.cgiar.org/cgiar-research-porfolio-2025-2030/sustainable-animal-and-aquatic-foods/
4. https://www.cgiar.org/cgiar-research-porfolio-2025-2030/food-frontiers-and-security/

References

Nguyen-Viet H., Lâm S., Alonso S., Unger F., Moodley A. *et al.* 2025. Insights and future directions: Applying the One Health approach in international agricultural research for development to address food systems challenges. *One Health*, 20, 101007. https://doi.org/10.1016/j.onehlt.2025.101007

ILRI. 2022. *ILRI One Health Strategy: Stopping the global rise of high-impact zoonotic disease, foodborne disease and antimicrobial resistance*, Nairobi, Kenya. https://hdl.handle.net/10568/125264

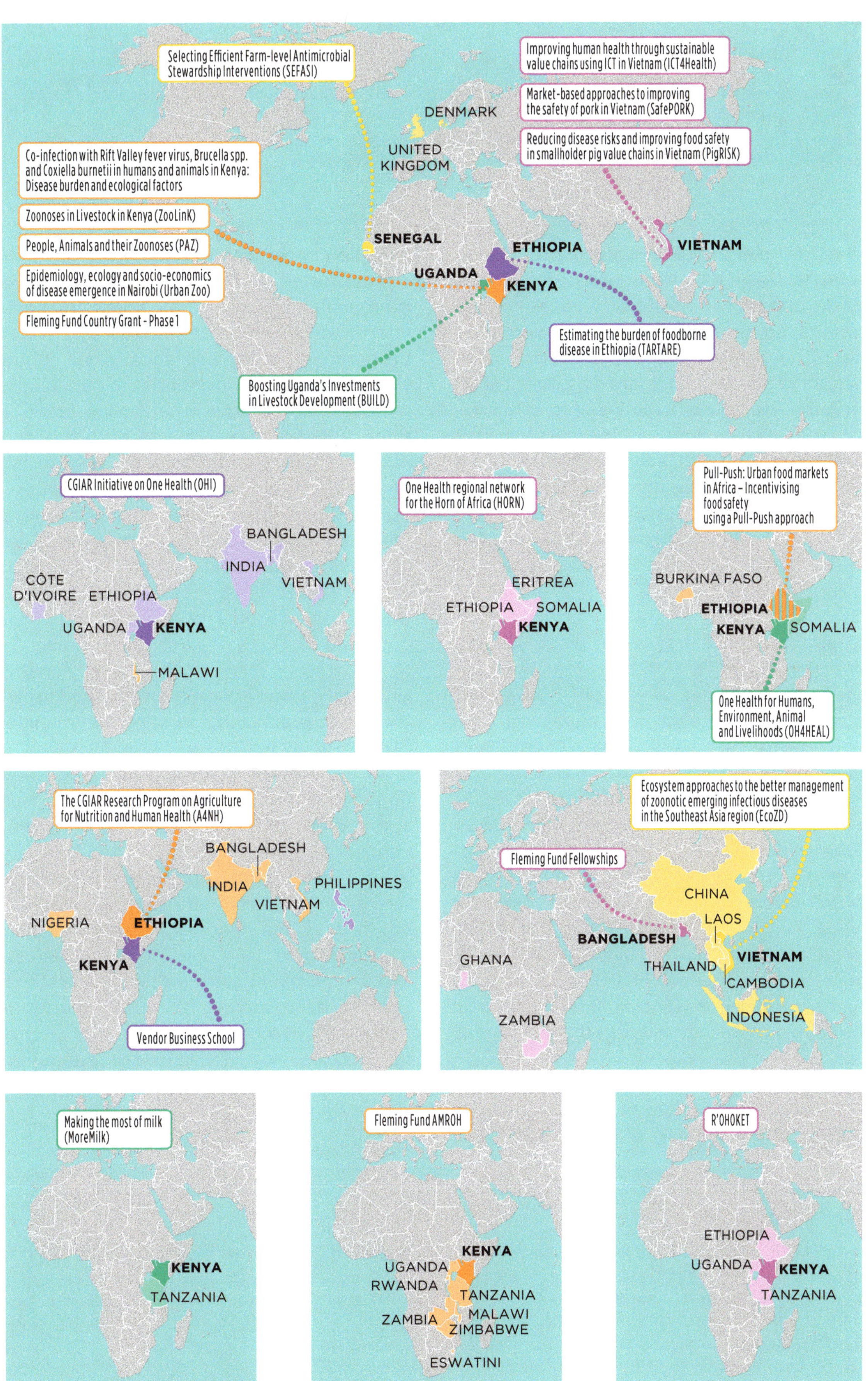

Figure 1. Overview of One Health initiatives at ILRI. Headquarter countries are in bold.

EcoHealth projects in South-East Asia

Steven Lam,
Hung Nguyen-Viet

Like One Health, the EcoHealth approach also emphasizes the interconnectedness of ecosystems and human health but places greater emphasis on ecological processes and their impacts on human health. EcoHealth gained widespread adoption in South-East Asia (SEA) during the late 2000s, largely due to initiatives led by the Canadian International Development Research Centre (IDRC), with a primary focus on mitigating emerging infectious diseases. A scoping review conducted in 2015 identified over 20 EcoHealth programmes, initiatives and projects in SEA following the introduction of this approach.

One notable initiative is the Field Building Leadership Initiative (FBLI), a regional programme spanning five years (2011–2016). FBLI encompassed EcoHealth projects across four countries—Vietnam, Indonesia, China and Thailand (Figure 1). This initiative fostered collaboration among several institutions, including Hanoi University of Public Health, Universitas Indonesia, Kunming Medical University, Mahidol University, World Agroforestry, Vietnam Public Health Association, Vets without Borders Canada, and the International Livestock Research Institute (ILRI). Many of these institutions and key people of FBLI continue to play major roles in advancing One Health and EcoHealth approaches today in the region and globally.

FBLI aimed to conduct EcoHealth research to address the health risks of agricultural intensification, strengthen capacities, and engage communities and other decision-making partners to ensure that research findings inform policy and practice (Figure 2). The FBLI programme resulted in substantial contributions, including 13 international papers, seven national papers, four local policy briefs, and three books[1]. These outputs played a crucial role in informing knowledge dissemination endeavours and community-level interventions. These efforts included using various mediums such as local newspapers, loudspeakers, street theatre, calendars and posters, leading to outcomes such as, in the case of Vietnam, updating the village rules (Hương ước) and the 2020 rural development plan to underscore the importance of sanitation. Ultimately, these interventions helped to shape the health practices of farming communities.

In the region, there is a growing convergence between EcoHealth and One Health approaches, with the latter gaining prominence. This shift is likely due to the heightened focus on zoonotic diseases and pandemic threats. Concurrently, there is rising acknowledgment of the need to incorporate environmental factors into One Health initiatives to tackle emerging health threats and foster sustainable development. Drawing insights from EcoHealth, which has its roots in environmental concerns, could facilitate the integration of sustainability considerations within the One Health framework.

1. For the full list, please visit www.ecohealthasia.net.

References

Nguyen-Viet H., Doria S., Tung D.X., Mallee H., Wilcox B., Grace D. 2015. Ecohealth research in Southeast Asia: Past, present and the way forward. *Infectious Diseases of Poverty*, 4, 5. https://doi.org/10.1186/2049-9957-4-5

Nguyen-Viet H., Pham G., Lam S., Pham-Duc P., Dinh-Xuan T. *et al.* 2021. International, transdisciplinary, and ecohealth action for sustainable agriculture in Asia. *Frontiers in Public Health*, 9, 592311. https://doi.org/10.3389/fpubh.2021.592311

Pham G., Lam S., Dinh-Xuan T., Nguyen-Viet H. 2018. Evaluation of an ecohealth approach to public health intervention in Ha Nam, Vietnam. *Journal of Public Health Management & Practice*, 24, S36–43. https://doi.org/10.1097/PHH.0000000000000732

Figure 1. Geographic distribution of the Field Building Leadership Initiative (FBLI) sites across South-East Asia and Southern China, highlighting the diversity of environmental health challenges addressed through the EcoHealth approach.

These pilot sites reflect the initiative's emphasis on integrating socioecological and health dimensions to tackle complex issues related to agricultural intensification, waste management, and vector-borne diseases. The map also illustrates FBLI's regional scope and the collaborative engagement of academic and policy institutions working across disciplinary and national boundaries to build capacities and inform policy through context-sensitive interventions.

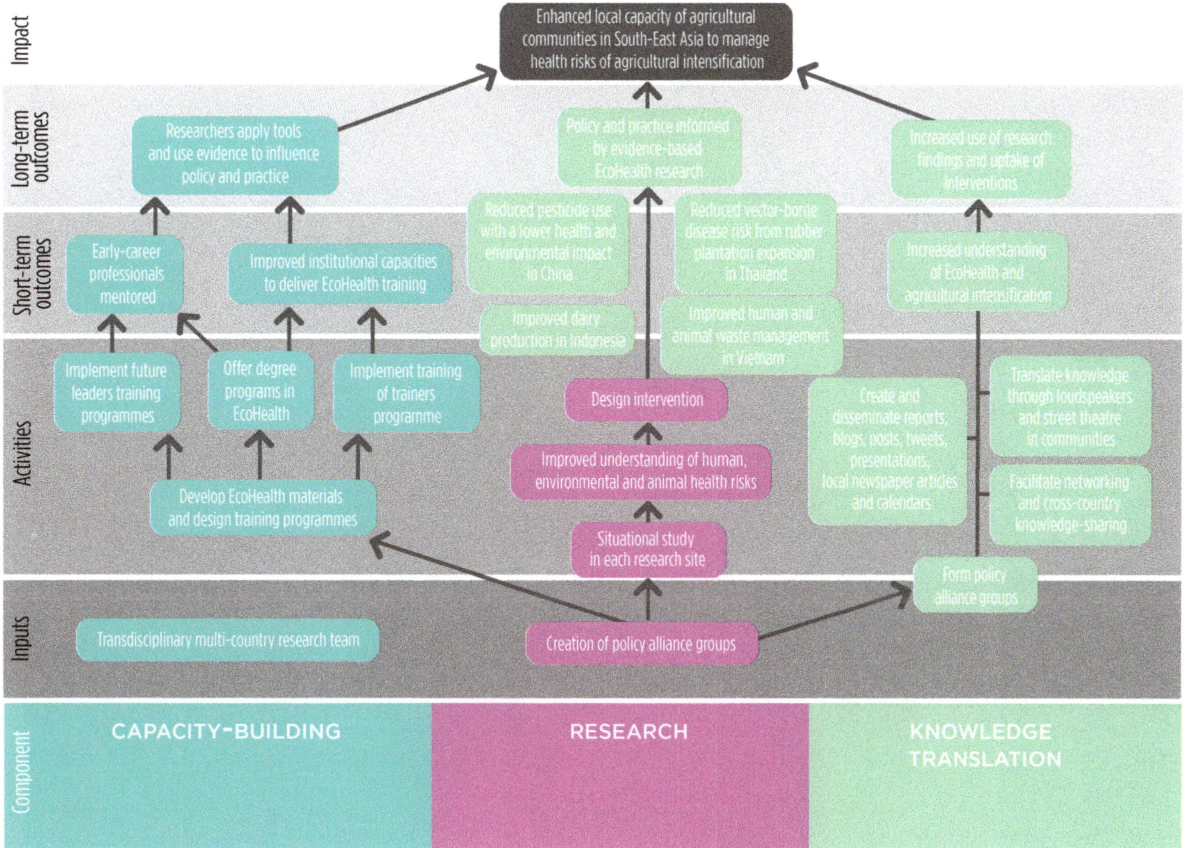

Figure 2. Field Building Leadership Initiative (FBLI) project framework, including core components and activities.

One Health in Vietnam: mapping action

Tu Tu Zaw Win, Phuong Vu Thi
Flavie Goutard

The One Health approach, which integrates human, animal and environmental health to tackle complex global challenges, has become a focal point of Vietnam's efforts to manage zoonotic diseases and pandemics. In response to rising health risks at the human-animal-environment interface, the One Health Partnership (OHP) for Zoonoses (2021–2025) was initiated, involving the Ministry of Agriculture and Rural Development (MARD), the Ministry of Health (MOH) and the Ministry of Natural Resources and Environment (MONRE), along with 33 national and international signatories and other development partners. This study aims to map and evaluate the breadth of ongoing One Health initiatives across Vietnam,[1] analysing their scope, thematic orientation and contributions to national strategic goals.

The research framework utilizes data from the PREZODE dashboard to identify ongoing One Health projects throughout the country. These initiatives primarily focus on zoonotic disease prevention, environmental health and sustainable agricultural practices. A spatial analysis reveals a concentration of activities in ecologically sensitive and highly urbanized regions, such as the Mekong Delta and Red River Delta, areas that are particularly vulnerable to the impacts of urbanization and environmental degradation (Figure 1).

1. https://onehealth.org.vn/en

The evaluation incorporates the ImpresS Change Theory to systematically assess each project across four dimensions: input (activities), output (deliverables), outcome and impact (Figure 2). Thematic analysis shows a strong emphasis on zoonotic disease surveillance, driven by pandemic preparedness. Other prominent themes include the promotion of One Health institutionalization, food safety, animal welfare, efforts to combat antimicrobial resistance (AMR), biosafety and the promotion of sustainable farming practices. Together, these initiatives strengthen Vietnam's One Health capacity by fostering collaboration across sectors and addressing the socioecological determinants of health (Figure 3).

Through this comprehensive study, One Health projects in Vietnam demonstrate their pivotal role in creating a sustainable health ecosystem. By addressing collaboration gaps and ensuring data consistency, the country is poised to significantly advance its One Health agenda. The application of the ImpresS Change Theory facilitates a more rigorous evaluation of these initiatives, ensuring that resources are effectively aligned with tangible outcomes, ultimately fostering impactful and sustainable health solutions.

References

FAO, UNEP, WHO, WOAH. 2023. IHR-PVS National Bridging Workshop – Viet Nam. Hai Phong, Viet Nam. https://extranet.who.int/sph/sites/default/files/document-library/document/NBW%20Vietnam%20-%20Final%20Report.pdf

Ministry of Agriculture and Rural Development (MARD), Ministry of Health (MOH), Ministry of Natural Resources and Environment (MONRE). 2022. Master plan for the One Health partnership framework for zoonoses, 2021–2025 period. https://onehealth.org.vn/upload/2132022%20-%20OHP%20Master%20Plan%20-%20English%20.pdf

Nguyen-Viet H., Lam S., Nguyen-Mai H., Trang D.T., Phuong V.T. *et al.* 2022. Decades of emerging infectious disease, food safety, and antimicrobial resistance response in Vietnam: The role of One Health. *One Health*, 14, 100361. https://doi.org/10.1016/j.onehlt.2021.100361

PREZODE dashboard. 2024. https://loire.shinyapps.io/DashboardPrezodeVN/

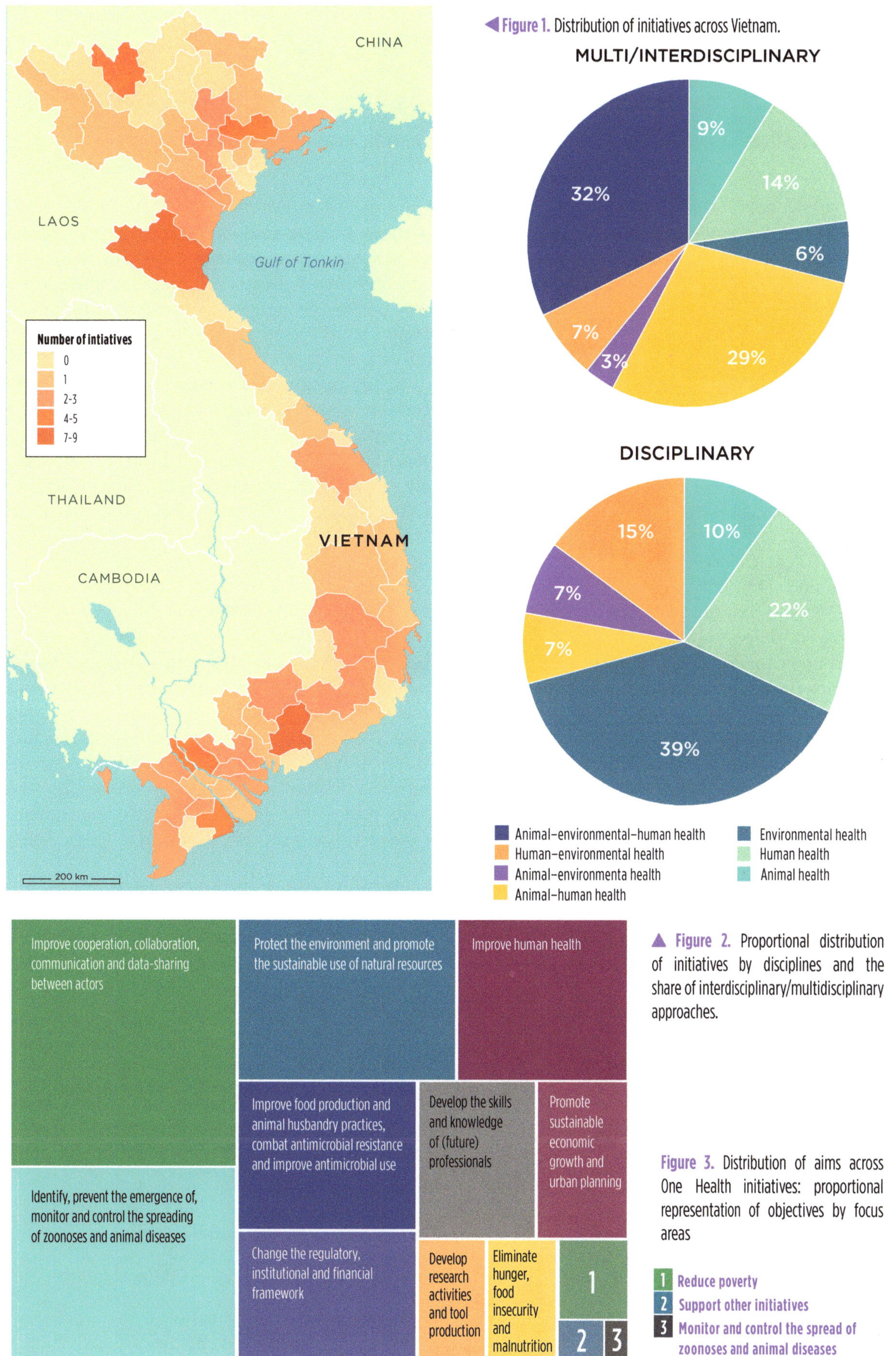

◀ Figure 1. Distribution of initiatives across Vietnam.

▲ Figure 2. Proportional distribution of initiatives by disciplines and the share of interdisciplinary/multidisciplinary approaches.

Figure 3. Distribution of aims across One Health initiatives: proportional representation of objectives by focus areas

Myanmar's One Health surveillance: challenges and next steps

Tu Tu Zaw Win

Myanmar, a country situated in South-East Asia, is actively working to strengthen its One Health surveillance system as a proactive response to the rising threats from zoonotic diseases and antimicrobial resistance (AMR). This initiative recognizes the intricate relationships among human, animal and environmental health, and aims to create a comprehensive framework for tackling public health challenges more effectively.

Myanmar's One Health strategy is structured around a framework designed to enhance disease surveillance, zoonotic disease control and AMR containment through cross-sectoral collaboration. Specifically, the Ministry of Health and Sports (MOHS) is responsible for human health monitoring (Figure 1), while the Livestock Breeding and Veterinary Department (LBVD) of the Ministry of Agriculture, Livestock and Irrigation (MOALI) oversees animal health surveillance (Figure 2), which is critical for detecting and managing zoonotic diseases. Moreover, the Ministry of Natural Resources and Environmental Conservation (MoNREC) manages environmental health aspects like deforestation and pollution. Despite clearly defined roles, certain challenges continue to hinder progress. Enhancing communication between ministries remains a priority, as delays in information-sharing may affect timely responses to emerging health risks. Additionally, better integration of environmental data with health and agricultural sectors is needed to foster a more holistic understanding of the impact of environmental changes on public health. Strengthening coordination across these areas will be key to achieving the full potential of Myanmar's One Health strategy.

A key priority is the development of a real-time data-sharing system to enhance outbreak response. While current passive surveillance methods provide valuable insights, they fall short in supporting timely interventions. Establishing an infrastructure for real-time, cross-sectoral data collection is vital; however, this effort will require both substantial technical expertise and financial investment. Equally important is strengthening Myanmar's One Health framework through targeted capacity-building initiatives. Training professionals in healthcare, veterinary services and environmental monitoring is essential to address operational and technical challenges, particularly in rural areas where access to services remains limited. Additionally, enhancing the legal framework can facilitate stronger collaboration among sectors involved in One Health.

To advance its One Health agenda, Myanmar will benefit from sustained investment in data systems, improved intersectoral coordination, and expanded research efforts. By addressing these critical areas, the country can enhance its ability to protect the health of its people, animals and environment to foster a more resilient and healthy future for all.

References

Global Health Security Index-Myanmar. 2021. https://ghsindex.org/country/myanmar/

MoHS, MoALI, MoNREC. 2019. National One Health Strategic Framework (2019–2023).

Myanmar National Action Plan for Containment of Antimicrobial Resistance. 2017–2022. https://faolex.fao.org/docs/pdf/mya202553.pdf

Win T.T.Z., Campbell A., Soares Magalhaes R.J., Oo K.N., Henning J. 2023. Perceptions of livestock value chain actors (VCAs) on the risk of acquiring zoonotic diseases from their livestock in the central dry zone of Myanmar. *BMC Public Health*, 23(1), 196. https://doi.org/10.1186/s12889-022-14968-y

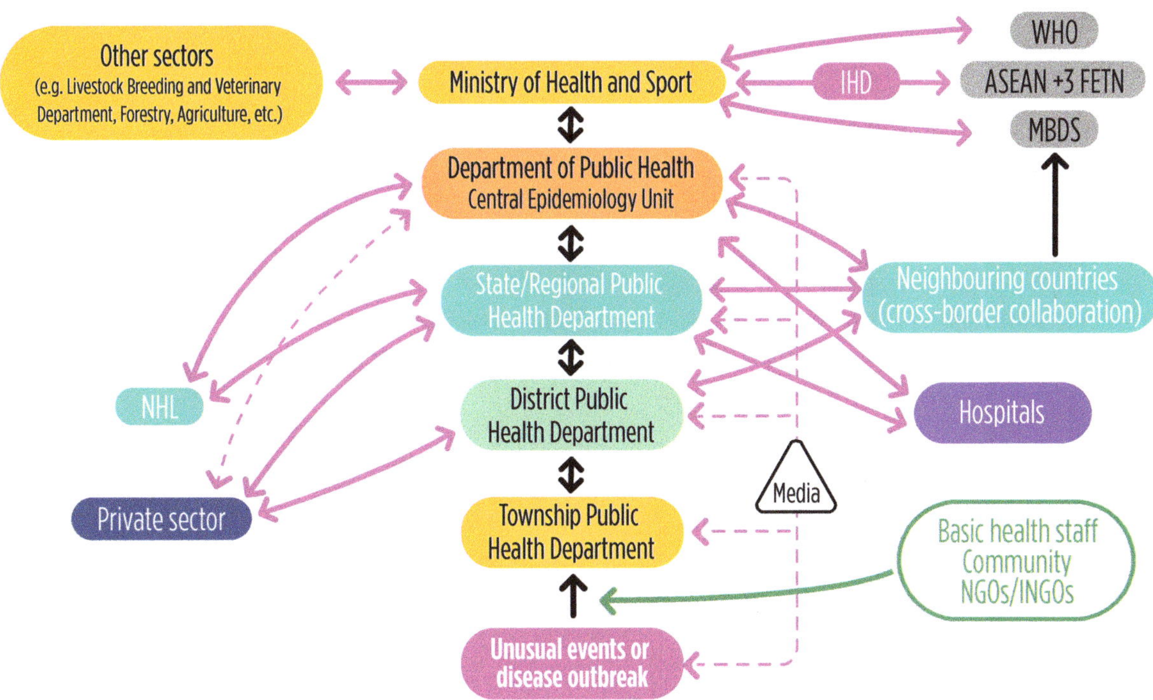

Figure 1. Surveillance system of the Ministry of Health and Sport.

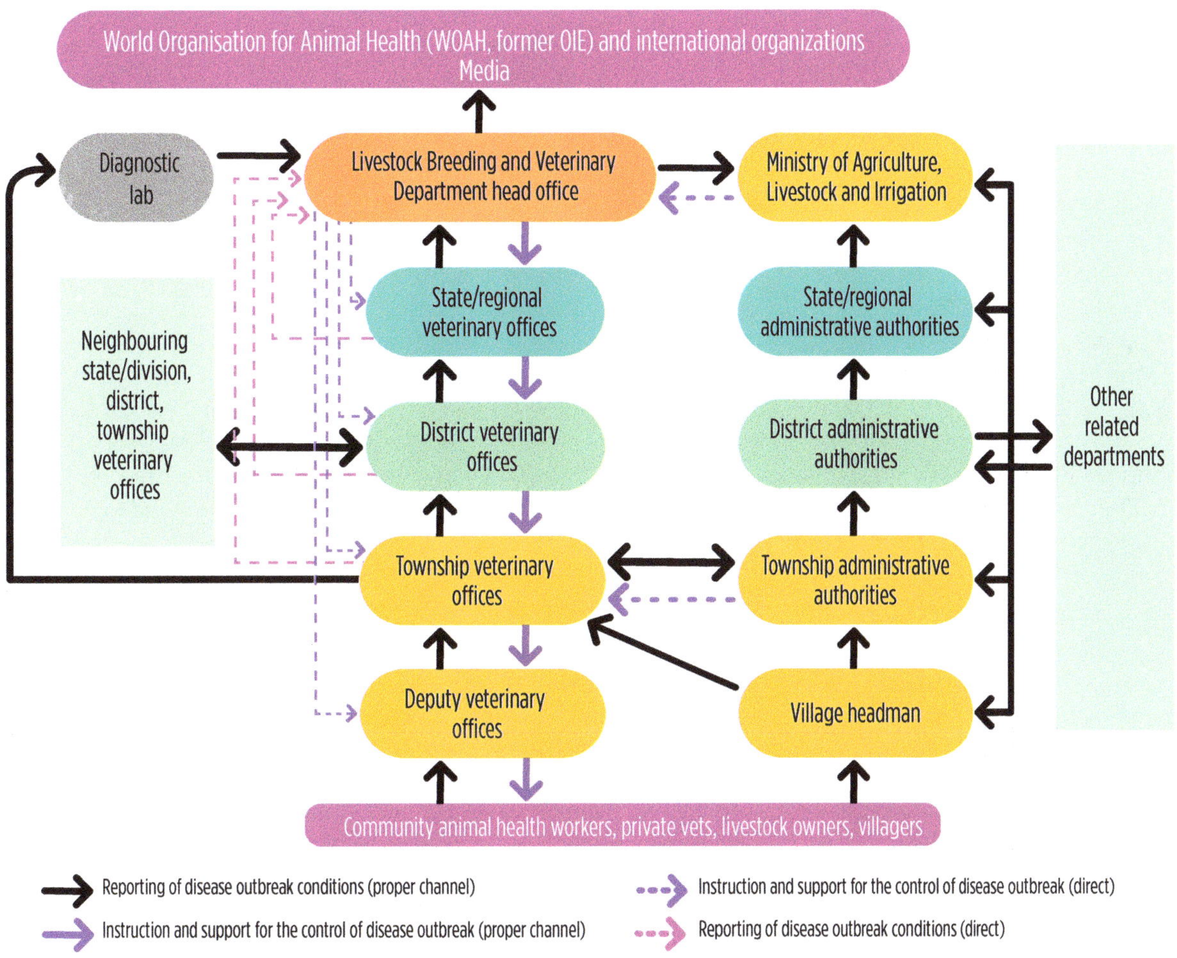

Figure 2. Disease information flow chart and chain of command in the Ministry of Agriculture, Livestock and Irrigation.

Strategizing a collaborative One Health future for the Caribbean

Marie-Jeanne Guenin, Catherine Abadie, Éric Etter

The Caribbean region is threatened by major sanitary risks affecting the health of socio-ecosystems. These risks need to be addressed in a coordinated and collaborative way from the community to the international level, and by adopting an integrated approach recognizing the interdependence between human, animal and environmental health embedded in the One Health definition. Several One Health initiatives promoting intersectoral collaboration supported by various regional actors, networks and organizations from different territories and disciplines have emerged in the Caribbean. Unfortunately, they rarely result in true interdisciplinary activities, and implementing One Health in practice remains challenging due to different obstacles.

To address the challenge of an One Health future in the Caribbean, 44 experts in animal, plant, environmental and human health and the social sciences worked together in a participatory strategic planning process. The aim was to co-build an innovative One Health community of practice (CoP) and collectively set desirable goals to achieve through this CoP (Figure 1). Using this approach, the experts were able to collectively formulate a plausible theory of change (ToC) regarding contextual barriers and levers. An impact pathway illustrates this ToC and its underlying causal relationships between long-term impacts of the CoP and the desirable outcomes. These outcomes may be changes in practices, behaviours and interactions or in knowledge, skills and motivations targeted by the CoP and which result from the appropriation by various actors of the CoP's strategies' outputs (Figure 2). The experts identified seven strategies to contribute to necessary changes at the community and national to transnational levels. The outputs produced within the strategies and the ToC will be consolidated.

The results of this participatory process are avenues for pursuing this change-oriented approach, while the impact pathway will be translated into useful tools for dialogue and monitoring within the CoP. The final objective is for this One Health CoP to be able to identify and adjust these objectives and needs and to create and pinpoint its own opportunities for future One Health activities over time by including other actors who could influence the success of strategies through other One Health projects in the region. This open innovation approach was used to identify synergies between various upcoming projects in the region (AUSCAR, CARIBGREEN, PREACTS 3) and to integrate them within a coherent programme logic that better addresses the challenges raised by the CoP.

References

Pettan-Brewer C., Figueroa D.P., Cediel-Becerra N., Kahn L.H., Francisco Martins A., Welker Biondo A. 2022. Editorial: Challenges and successes of One Health in the context of planetary health in Latin America and the Caribbean. *Frontiers in Public Health*, 10, 1081067. https://doi.org/10.3389/fpubh.2022.1081067

Guenin M.-J., Caribbean One Health community of practice, Thys S., Etter E. 2025. Strengthening one health in the future of the Caribbean through a regional community of practice. *One Health*, 21, 101152. https://doi.org/10.1016/j.onehlt.2025.101152

Guenin M.-J., Etter E., Abadie C., Hasnaoui Amri N., De Romémont A. *et al.* 2024. Proceedings of the ImpresS ex ante and AUSCAR project workshop. Cobuilding a One Health community of practice in the Caribbean. Dominican Republic, 20–23 November, 2023. https://agritrop.cirad.fr/607904/

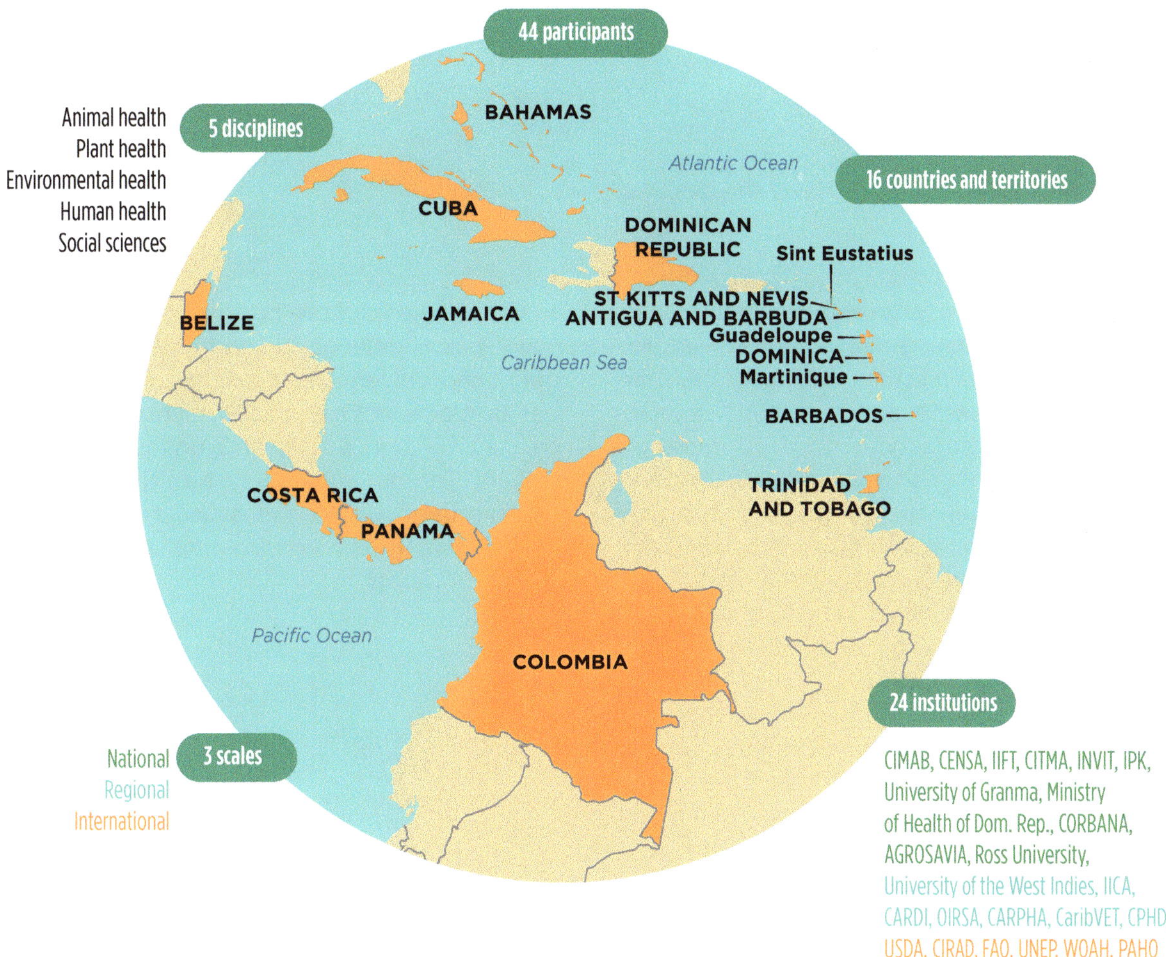

Figure 1. Geographic scope of One Health engagement in the Caribbean.

Figure 2. Pathway from strategies to outcomes and impacts of a Caribbean One Health approach, highlighting actions to strengthen collaboration, education and awareness, and their expected benefits at community and institutional levels.

National mapping of One Health stakeholders in France

Jean-Luc Angot,
Thierry Lefrançois

Implementing an integrated One Health approach requires establishing an appropriate governance system. Because the One Health concept is based on multidisciplinarity and intersectorality, strong cooperation and coordination among the different organizations and bodies operating within the One Health scope are vital. One Health does not necessarily require creating new institutions; rather, it is essentially a question of better organizing dialogue between the different public and private partners. The objective is to break down barriers and combine scientific and administrative cultures. Science and politics must come together, with strong involvement from citizens. On a national scale, many initiatives on both a research and institutional level are required to take up the challenges associated with health crises (Figure 1).

In France, the General Secretariat for Ecological Planning (SGPE), under the authority of the Prime Minister, is responsible for implementing France's Green Nation Action Plan, which includes a health component (Figure 2). The creation of the Committee for Monitoring and Anticipating Health Risks (COVARS), the establishment of inter-ministerial coordination on subjects relating to One Health, the creation of a One Health group for monitoring the 4th National Environmental Health Plan, and the development of the Preventing Zoonotic Disease Emergence (PREZODE) initiative, which has become an international project, are all new steps taken to ensure a transversal One Health approach.

One Health degree courses (Master's, etc.) have been developed and common cores between universities are gradually being put in place. A One Health Institute is being created to train public and private decision-makers. The governance system that is established must consider the issues of prevention, surveillance, research, training and international cooperation.

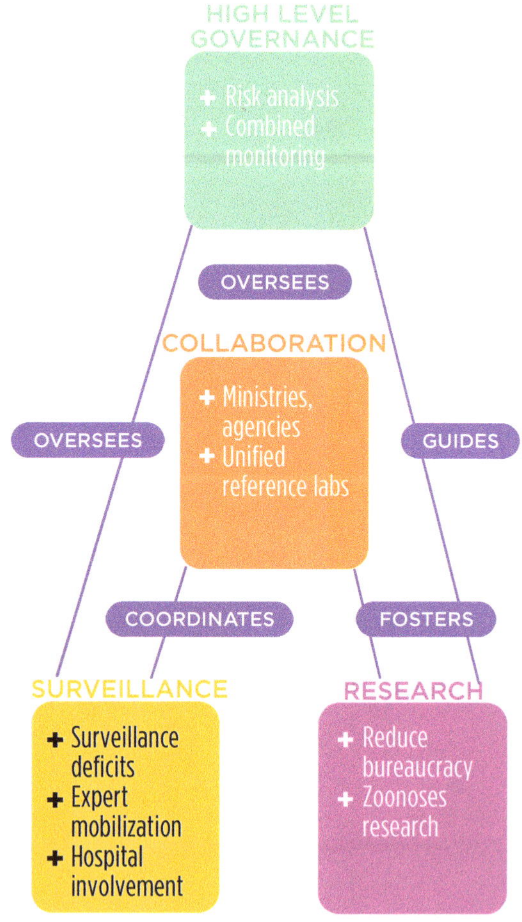

Figure 1. Addressing One Health challenges on a national scale: key strategies and initiatives.

References

Olive M.-M., Angot J.-L., Binot A., Desclaux A., Dombreval L., Lefrançois T., Lury A., Paul M., Peyre M., Simard F., Weinbach J., Roger F. 2022. Les approches *One Health* pour faire face aux émergences : un nécessaire dialogue État-sciences-sociétés. *Natures, Sciences, Sociétés,* 30(1), 72–81. https://doi.org/10.1051/nss/2022023
www.masa

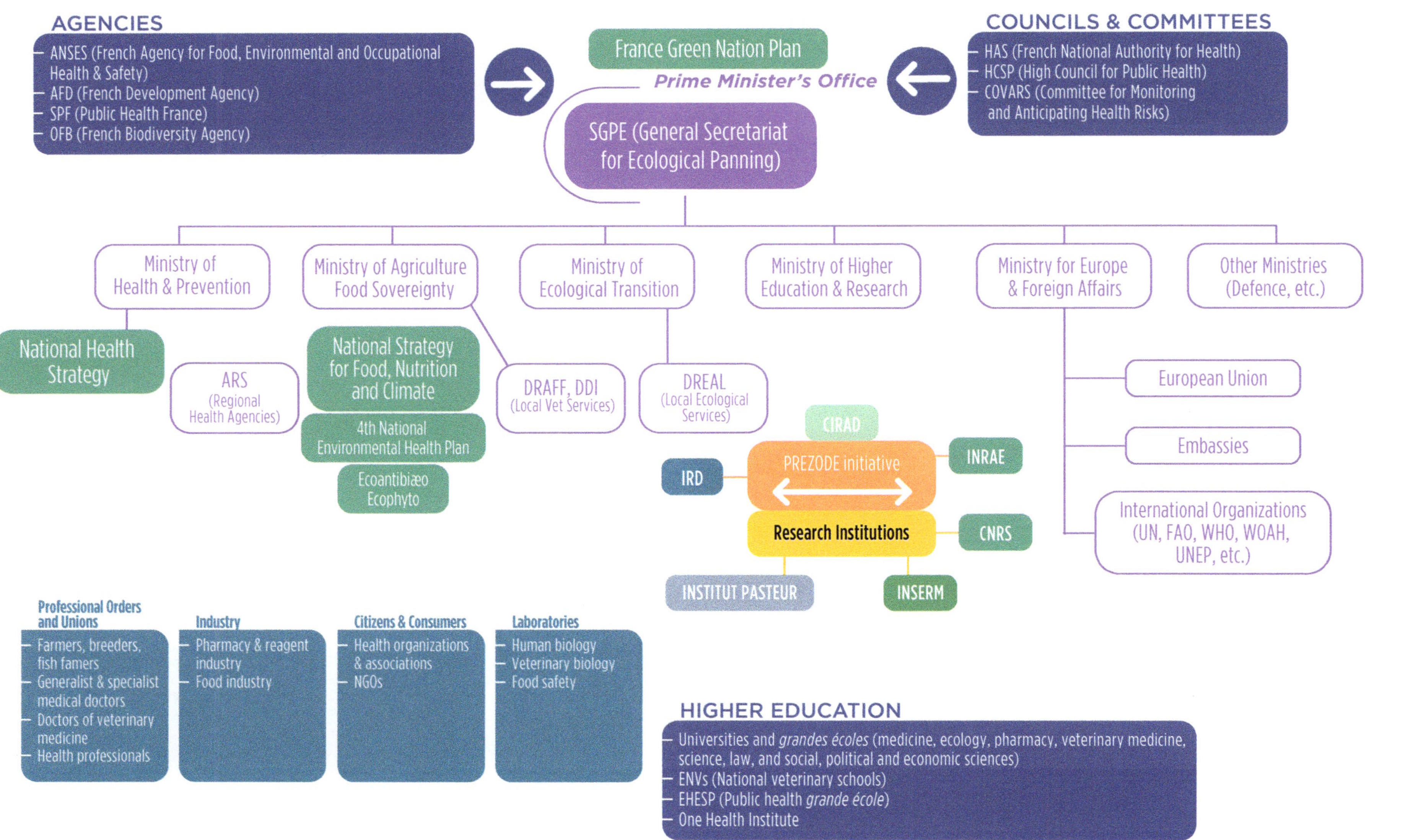

Figure 2. Overview of the organizational structure and key stakeholders involved in France's One Health strategy.

One Health and resilient landscapes: empowering communities through agroecological innovation

Raphaël Duboz,
Océane Wacrenier

The Health & Territories project (known as Santé & Territoires in French[1]) aligns with the global momentum towards agroecological transition by adopting an integrated health approach. The project aims to improve the quality of life for local populations by establishing sustainable agro(food)systems capable of addressing multiple health and environmental challenges. This approach is implemented in the context of health crises and global changes, such as the expansion of cultivated areas, biodiversity loss and climate change. The goal is to develop agricultural practices that minimize health risks while enhancing the productivity and sustainability of production systems.

The project is organized around six living labs, experimental spaces set up in the four partner countries: Benin, Cambodia, Laos and Senegal (Figure 1). These living labs are places where local stakeholders—farmers, researchers and policymakers—collaborate to co-design and test innovative solutions to health and agroecological issues.

The project's One Health approach is manifested through concrete on-the-ground actions such as surveys and diagnostics to assess the impact of agricultural practices on the health of ecosystems, animals and human communities. These assessments are then used to co-develop solutions tailored to local contexts (Figure 2). A central aspect of Health & Territories is the capacity-building of local actors. The project implements training, participatory workshops and knowledge management systems to ensure that agroecological and health innovations are well understood, accepted and adopted by local communities. This capacity-building aims to make territories more resilient to ecological, climatic and social disruptions.

The Health & Territories project is an example of how the One Health approach can be implemented on a territorial scale to address the global challenges of our time, by building bridges between human, animal, plant and environmental health, and placing local communities at the heart of innovation and change. Health & Territories demonstrates that integrating health dimensions into territorial management is not only possible but necessary to achieve sustainable development goals. Ultimately, this ambitious project contributes to an agroecological transition that promotes more resilient, equitable and sustainable production systems while improving the quality of life for local populations.

1. https://www.santes-territoires.org/

References

Lakatos E.S., Pacurariu R.L., Bîrgovan A.L., Cioca L.I., Szilagy A. *et al.* 2024. A systematic review of living labs in the context of sustainable development with a focus on bioeconomy. *Earth*, 5(4), 812-843. https://doi.org/10.3390/earth5040042

Lévesque A., McPhee C., Chrétien F., Gracia-Garza J., Morissette R. *et al.* 2024. Report on the First Forum on Agroecosystem Living Labs (IF-ALL). Adaptation Futures 2023. https://ll-lv.agr.gc.ca/ncloud/index.php/s/EPPJWAxAteLs4Be

LIVING LABS are open innovation ecosystems focused on users, operating in real-world settings to integrate research and innovation within communities. They actively involve users as contributors through participatory methods. The concept originated in the 1990s as a response to the growing demand for more collaborative and user-centered innovation approaches. Over time, living labs have been applied across various sectors, such as environmental sustainability, urban development and health.

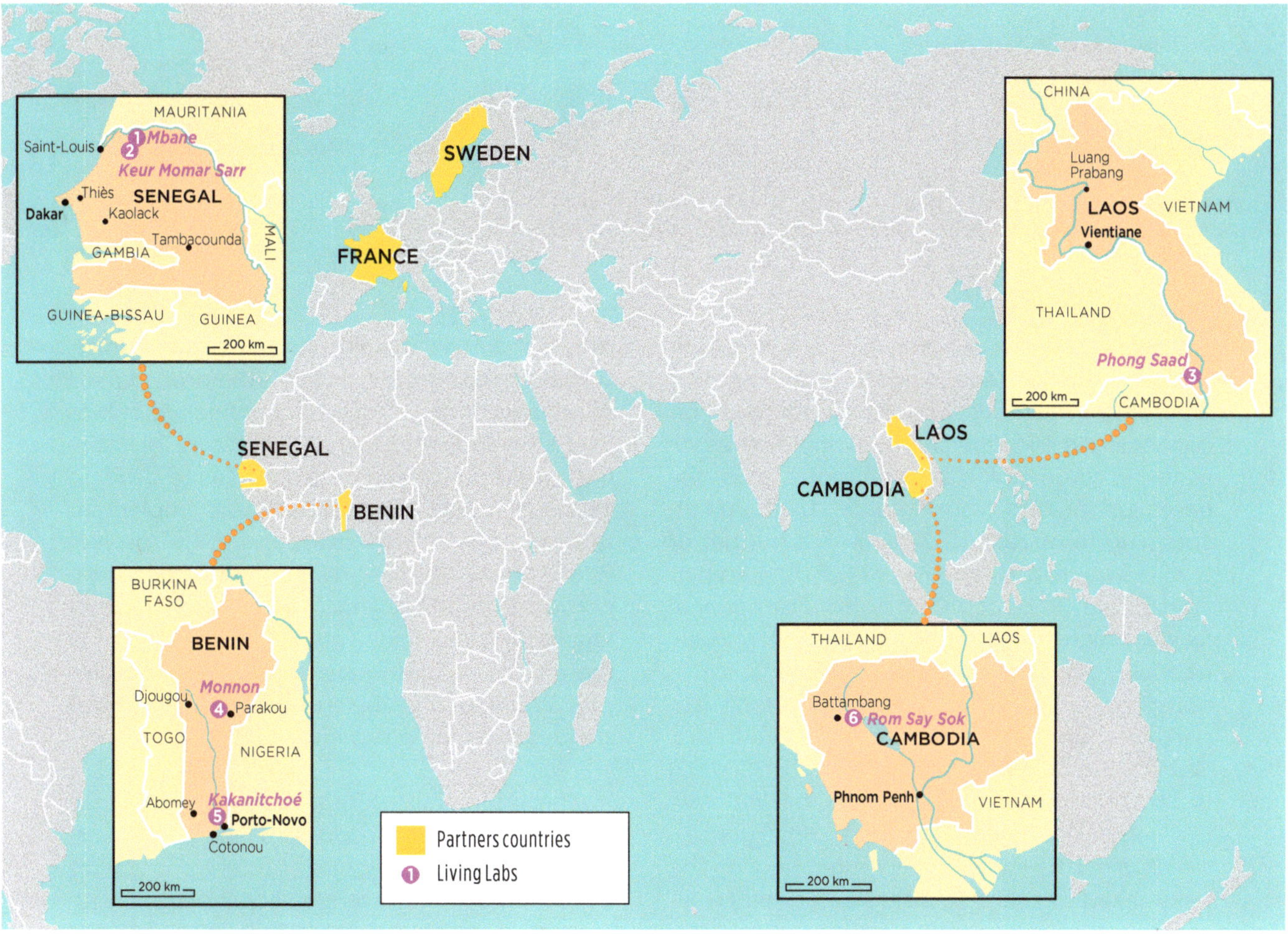

Figure 1. Map of six living labs launched in four countries.

1 SENEGAL Mbane: 18–19 March 2023
2 SENEGAL Keur Momar Sarr: 21–22 March 2023
3 LAOS Phong Saad: November 2022
4 BENIN Monnon: 17 April 2023
5 BENIN Kakanitchoé: 27 March 2023
6 CAMBODIA Rom Say Sok: November 2022

Figure 2. The four pillars of One Health. Source: AI-produced illustration (NightCafé, DALL-E3).

1 ENVIRONMENTAL HEALTH
- Prevention and management of environmental risks (pollution, etc.)
- Balance of biogeochemical cycles
- Natural biodiversity

2 PLANT HEALTH
- Prevention and sustainable management of crop pests
- Adequate plant nutrition
- Agro-biodiversity and balance of populations

3 ANIMAL HEALTH
- Prevention and management of animal diseases
- Adequate animal nutrition
- Genetic diversity of herds

4 HUMAN HEALTH
- Prevention and management of human diseases (hygiene, education)
- Food and nutritional security
- Social diversity

BCOMING: safeguarding biodiversity to prevent the next pandemic

Julien Cappelle,
François Roger

The BCOMING project (Biodiversity Conservation to Mitigate the risks of emerging infectious diseases)[1] focuses on enhancing knowledge and reducing risks associated with emerging zoonotic infectious diseases originating from wildlife. This project, spanning from 2022 to 2026 with a budget of 6 million euros, is led by CIRAD (the French Agricultural Research Centre for International Development) and supported by an international consortium of research institutes, non-governmental organizations and private partners. The target regions include Cambodia, Guinea and Guadeloupe (Figure 1). BCOMING is part of the international PREZODE initiative (Preventing Zoonotic Disease Emergence), which aims to improve understanding of the risks linked to zoonotic disease emergence.

BCOMING's primary goal is to work closely with local communities in biodiversity-rich areas to develop innovative approaches to prevent future epidemics and pandemics. Among its specific objectives, the project aims to advance knowledge on the influence of biodiversity and environmental factors on zoonotic emergence risks. It also seeks to develop efficient surveillance systems and implement tailored prevention strategies. The project intends to establish early warning systems and participatory surveillance to promptly detect emerging outbreaks. This work is grounded in the collection and analysis of biological samples, coupled with the study of ecological and socioeconomic determinants (Figure 2).

Through an integrated interdisciplinary approach, BCOMING adopts a One Health perspective, aspiring to prevent future pandemics by combining biodiversity conservation with the active involvement of local and international communities in establishing effective surveillance systems.

1. www.bcoming.eu

BIODIVERSITY plays a critical role in maintaining ecosystem health and, by extension, human health. Diverse ecosystems provide essential services such as clean air, water and fertile soils, all of which are vital for human well-being. More importantly, biodiversity helps regulate diseases by maintaining balanced ecosystems. For instance, in areas with high biodiversity, predators and competitors help keep populations of disease-carrying species like rodents or mosquitoes in check to reduce the risk of zoonotic spillovers. Conversely, when biodiversity is lost due to deforestation, land-use change or climate change, the natural barriers that prevent the transmission of pathogens from animals to humans are weakened, increasing the likelihood of pandemics. Preserving biodiversity not only safeguards these natural defences but also contributes to long-term human health by supporting healthy ecosystems that can adapt to changes and continue providing essential health-related services.

References

CABI. 2021-2024. Joint crop and livestock services for smallholder farmers. https://www.cabi.org/projects/joint-crop-and-livestock-services-for-smallholder-farmers/

Keesing F., Belden L.K., Daszak P., Dobson A., Harvell C.D. *et al.* 2010. Impacts of biodiversity on the emergence and transmission of infectious diseases. *Nature*, 468(7324), 647–652. https://doi.org/10.1038/nature09575

Plowright R.K., Ahmed A.N., Coulson T., Crowther T.W., Ejotre I. *et al.* 2024. Ecological countermeasures to prevent pathogen spillover and subsequent pandemics. *Nature Communications*, 15, 2577. https://doi.org/10.1038/s41467-024-46151-9

Vora N.M., Hannah L., Lieberman S., Vale M.M., Plowright R.K., Bernstein A.S. 2022. Want to prevent pandemics? Stop spillovers. *Nature*, 605(7910), 419–422. https://doi.org/10.1038/d41586-022-01312-y

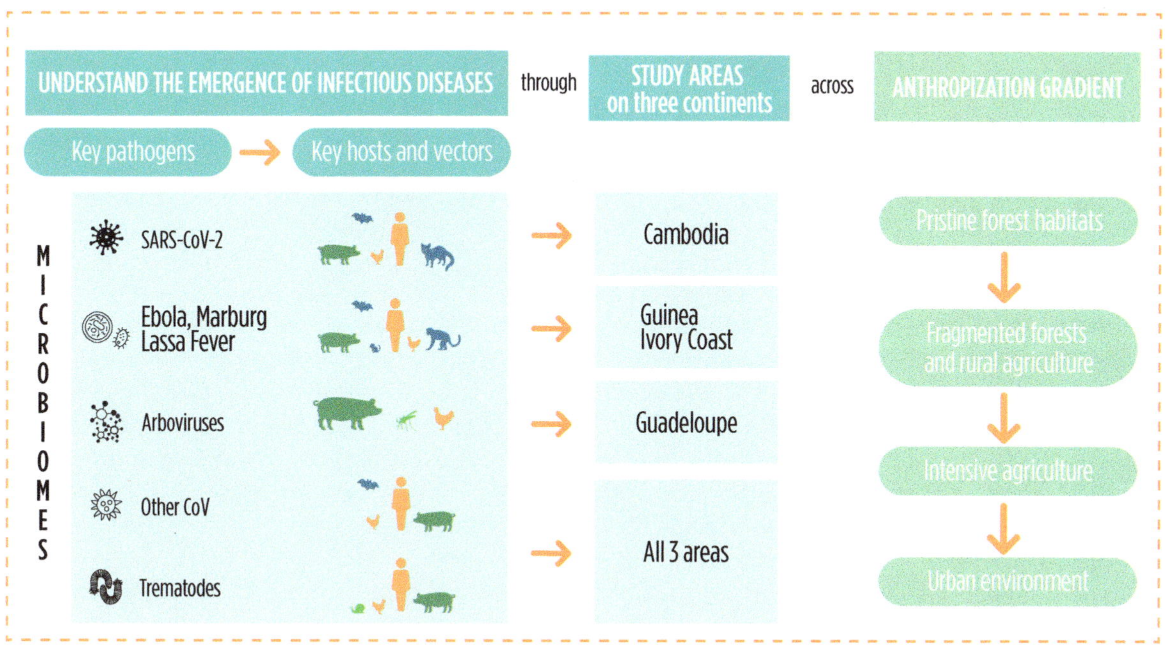

Figure 1. Overview of the main pathogens and hosts targeted at each study site of the BCOMING project across an anthropisation gradient.

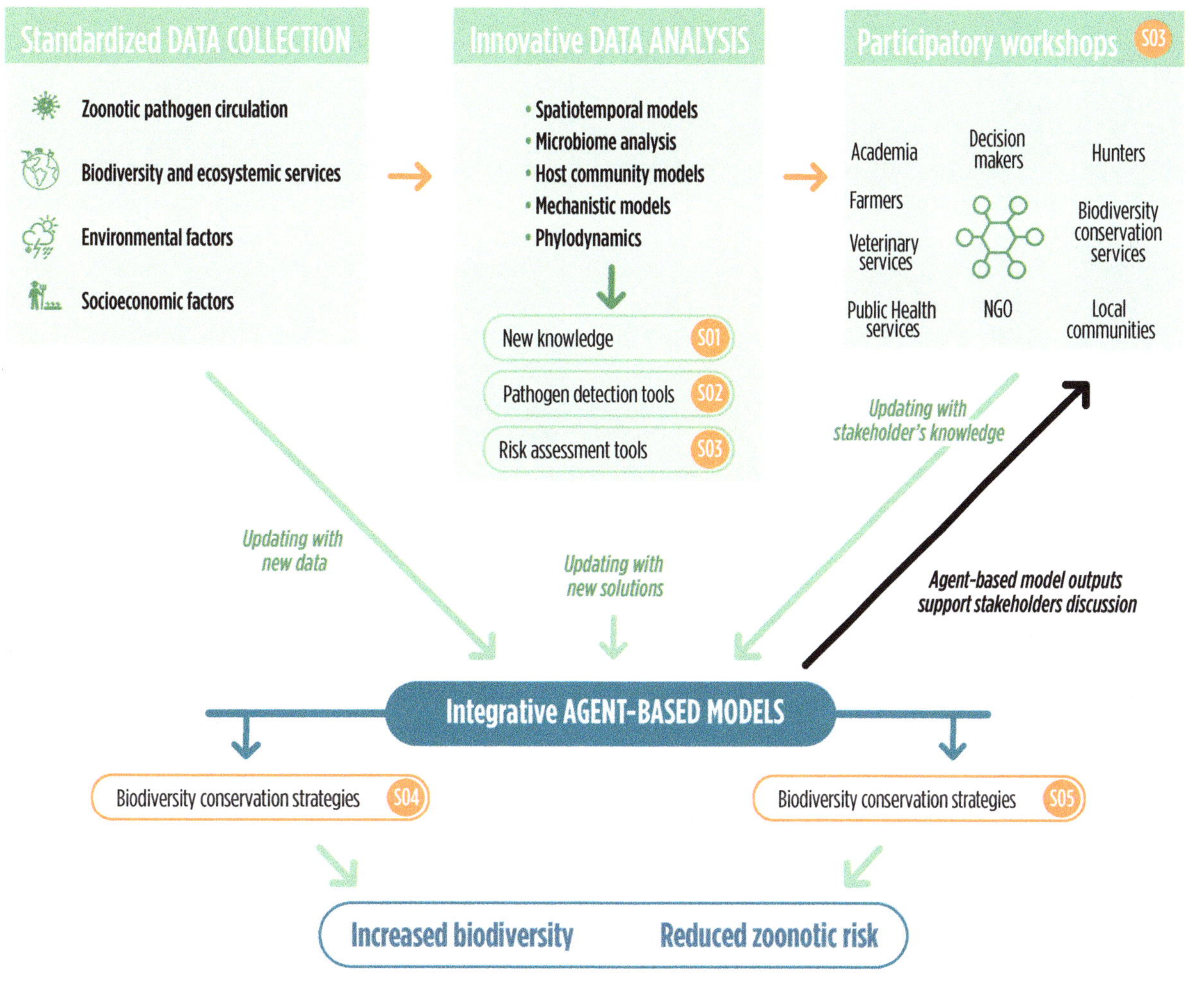

Figure 2. BCOMING project workflow: standardized data collection, advanced analysis and local knowledge integrated into agent-based models to support the co-construction of biodiversity conservation strategies and disease surveillance systems.

ZACAM: One Health long-term socioecological research platform

Nathalie Boutin, Jean-Claude Raynal, Delphine Destoumieux-Garzon, Raphaël Mathevet, Marion Vittecoq, Olivier Boutron, Frédéric Thomas

In France, long-term socioecological research (LTSER) platforms called "zones ateliers", which operate under the aegis of the National Centre for Scientific Research (CNRS) through its Ecology & Environment institute, form an interdisciplinary network dedicated to the study of social-ecological systems and the environment. They aim to answer fundamental ecological questions while considering contemporary societal challenges linked to global change.

The "Zone Atelier Santé-Environnement Camargue" (ZACAM) LTSER platform[1] covers the Rhône delta, a vast coastal area known as the Camargue (Figure 1). ZACAM is the 16th LTSER platform in the national network and the very first dedicated to health-environment issues on the Mediterranean coast.

The Camargue region, where natural areas come in direct contact with anthropogenic pressures, offers an ideal setting for studying long-term disease circulation and emergence, as well as co-constructing solutions for prevention and adaptation to local issues. This vast Mediterranean wetland is a recognized biodiversity hotspot and is home to many species, supporting a range of human activities such as crop and livestock farming and tourism. It is also vulnerable to epidemic risks due to the presence of pathogens, making the prevention of environment-related diseases a crucial concern.

ZACAM is focused on the socioecology of health and takes a "full health" perspective that integrates the One Health approach. It has four priority research themes: 1) zoonotic pathogens, 2) antibiotic resistance, 3) ecotoxicology and 4) environmental mental health (Figure 2). The overall goal is to promote health and environment research in the Camargue through an integrative and interdisciplinary approach to identify concrete solutions for local stakeholders. In this way, one key project called ZOOCAM (PEPR PREZODE programme, Figure 1) will explore the dynamics of zoonotic agents in the Camargue, the human and environmental factors influencing their spread, the modelling of current and emerging zoonotic risks, and the way in which these risks are perceived by local stakeholders.

1. www.zacam.cnrs.fr/en/

References

Destoumieux-Garzón D., Matthies-Wiesler F., Bierne N., Binot A., Boissier J. *et al.* 2022. Getting out of crises: Environmental, social-ecological and evolutionary research is needed to avoid future risks of pandemics. *Environment International*, 158, 106915. https://doi.org/10.1016/j.envint.2021.106915

Boutin N., Mathevet R., Destoumieux-Garzon D., Vittecoq M., Thomas F. In press. La Zone Atelier santé-environnement CAMargue : recherches socio-écologiques selon des approches « une seule santé » et « pleine santé ». *In* Ragueneau O., Lea V., Charpentier I., Tito de Morais C., Bonnefond M. (Eds), *Zones Ateliers, un réseau de co-recherches sur les socio-écosystèmes : historique et perspectives*, London, ISTE Éditions, Collection Sciences, pp. 243-258.

ENVIRONMENTAL MENTAL HEALTH explores how natural and anthropogenic environmental factors influence psychological well-being. Positive aspects include stress reduction in green spaces, connection with nature and interactions with animals. Negatives aspects involve issues like pollution, ecosystem degradation and anxiety driven by climate change or biodiversity loss. In regions like the Camargue, integrating mental health into One Health approaches addresses the psychological impacts of these environmental challenges on local communities.

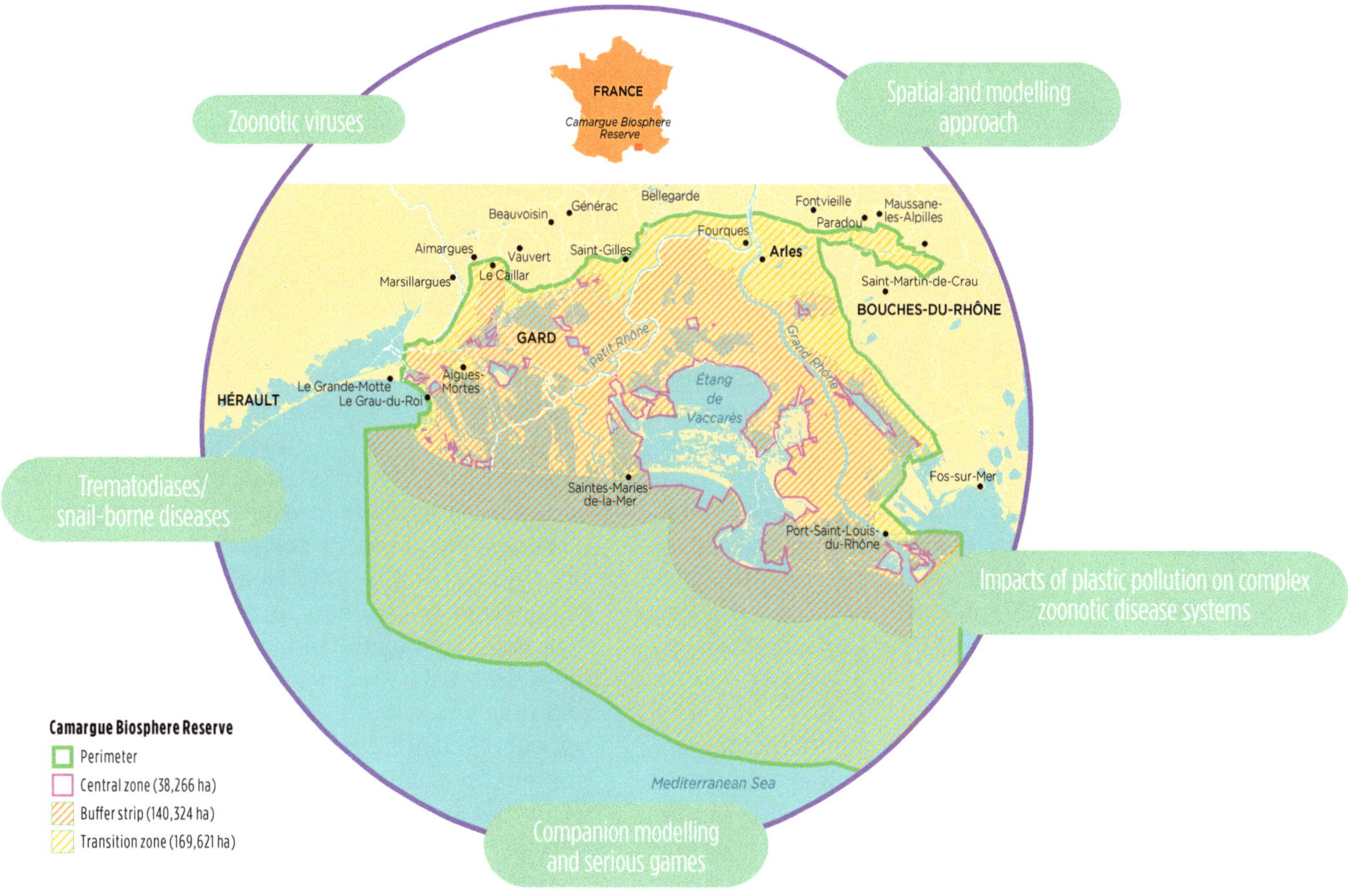

Zoonotic viruses

Spatial and modelling approach

FRANCE

Camargue Biosphere Reserve

Trematodiases/ snail-borne diseases

Impacts of plastic pollution on complex zoonotic disease systems

Companion modelling and serious games

Camargue Biosphere Reserve

Perimeter

Central zone (38,266 ha)

Buffer strip (140,324 ha)

Transition zone (169,621 ha)

Figure 1. The Camargue Biosphere Reserve, where the ZACAM research platform is focusing its One Health research activities.

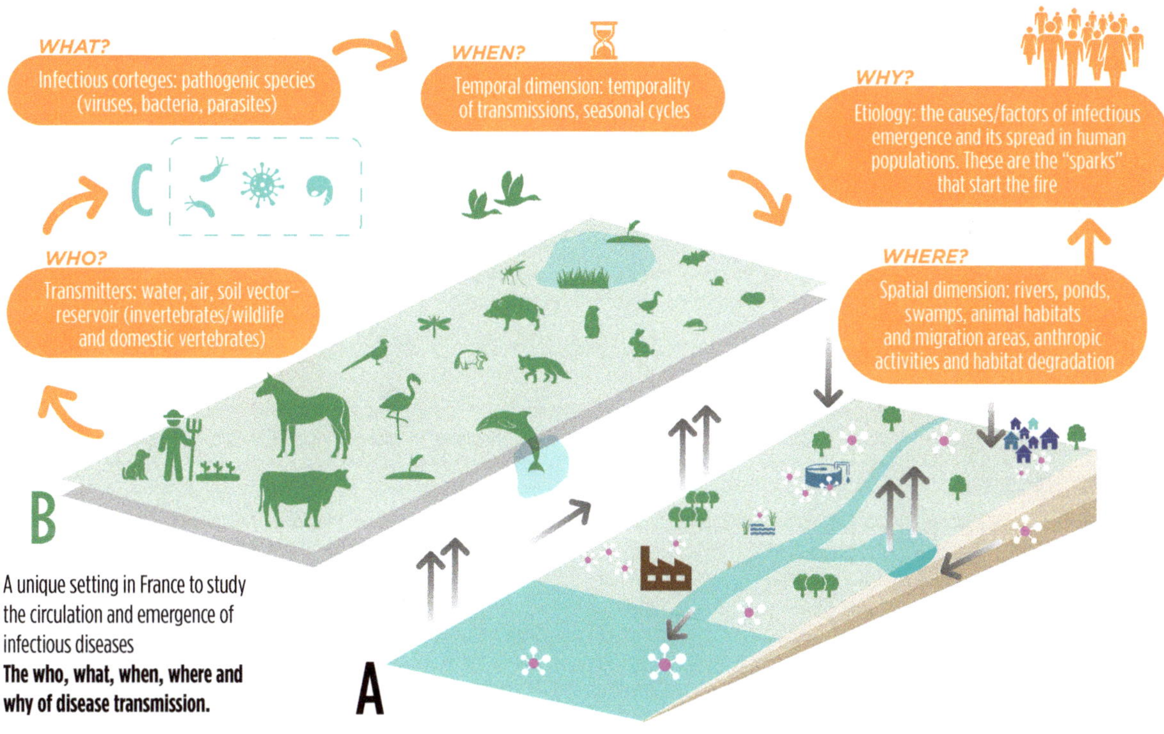

Figure 2. A framework for studying the circulation and emergence of infectious diseases (a priority research area for zoonotic pathogens).
Source: François Renaud.

PREZODE international initiative and its contribution to promoting a One Health approach

Marisa Peyre, Papa Seck, Soawapak Hinjoy, Flavie Goutard, François Roger

One Health is both a scientific concept and a general approach that highlights the links between human, animal and ecosystem health. Despite work on this approach for more than 20 years and a clear scientific consensus, it took a pandemic to achieve meaningful political awareness. Since COVID-19, international initiatives have multiplied to put the One Health approach into practice. PREZODE (Preventing the Emergence of Zoonoses and Strengthening Surveillance Capacities) is one convincing example.

Launched in response to pandemic threats, PREZODE promotes the One Health approach through its scientific strategic agenda and its five pillars of research action (Figure 1). One of PREZODE's main objectives is to proactively manage global health risks to prevent pandemics caused by emerging zoonotic diseases through effective surveillance systems and early detection in high-risk areas. As of 2024, it has nearly 250 institutional members across 80 countries, with 27 government engaged (Figure 2).

By bringing together researchers, decision-makers and local communities, PREZODE plays a key role in transforming the way research is conducted, moving from descriptive research to more participatory action research. This crucial shift encourages the interdisciplinary collaboration that is needed to tackle complex health issues. For example, PREZODE is working to establish a community-based surveillance network in 15 countries to link local actors and actions to national and international networks and strengthen the early detection and prevention of zoonotic diseases. This type of collaboration amplifies the impact of the One Health approach on the ground to promote a rapid, coordinated response to health threats. PREZODE also gives scientists a more active role in advocacy and policy support, helping to bridge the gap between scientific research and policy development. In doing so, the initiative fosters dialogue between communities, researchers and decision makers to develop more effective and relevant public health strategies.

PREZODE offers a structured framework based on key values to facilitate collaboration between sectors and actors in their efforts to take a more integrated approach to health issues. The initiative also raises awareness among decision makers and the general public regarding the importance of the One Health approach for pandemic prevention and ecosystem protection. In this way, PREZODE actively supports the One Health approach through research, surveillance, political engagement and intersectoral coordination.

References

Peyre M., Vourc'h G., Lefrançois T., Martin-Prevel Y., Soussana J.-F., Roche B. 2021. PREZODE: preventing zoonotic disease emergence. *The Lancet*, 397(10276), 792–793. https://doi.org/10.1016/S0140-6736(21)00265-8

PREZODE Strategic Agenda: https://prezode-initiative.org/en/strategic-agenda-and-pillars/

FUNDING AND CATALYTIC ROLE OF PREZODE

The PREZODE initiative is supported by French funding, mainly through the French Development Agency (AFD) and the Priority Research Programme and Equipment (PEPR) co-funded by the French National Research Agency (ANR). Beyond this institutional support, PREZODE also acts as a catalyst for international fundraising by helping mobilize additional resources for research, capacity-building and operational projects aligned with the global agenda of pandemic prevention and One Health integration.

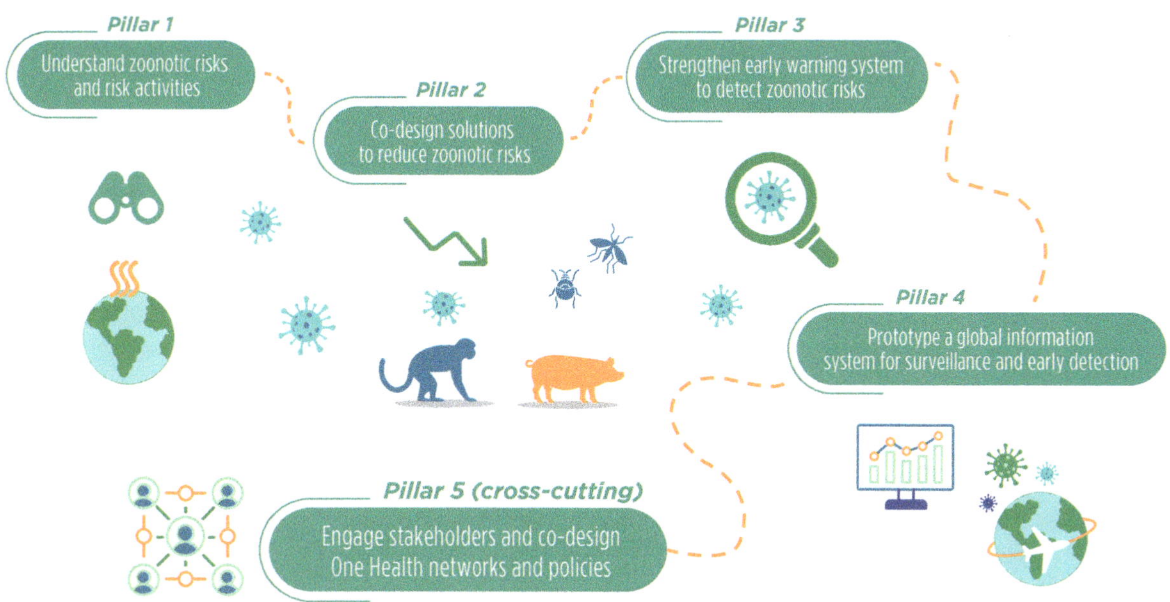

Figure 1. Scientific pillars of the PREZODE initiative: a five-pillar action research framework designed to proactively reduce the risk of zoonotic disease emergence.

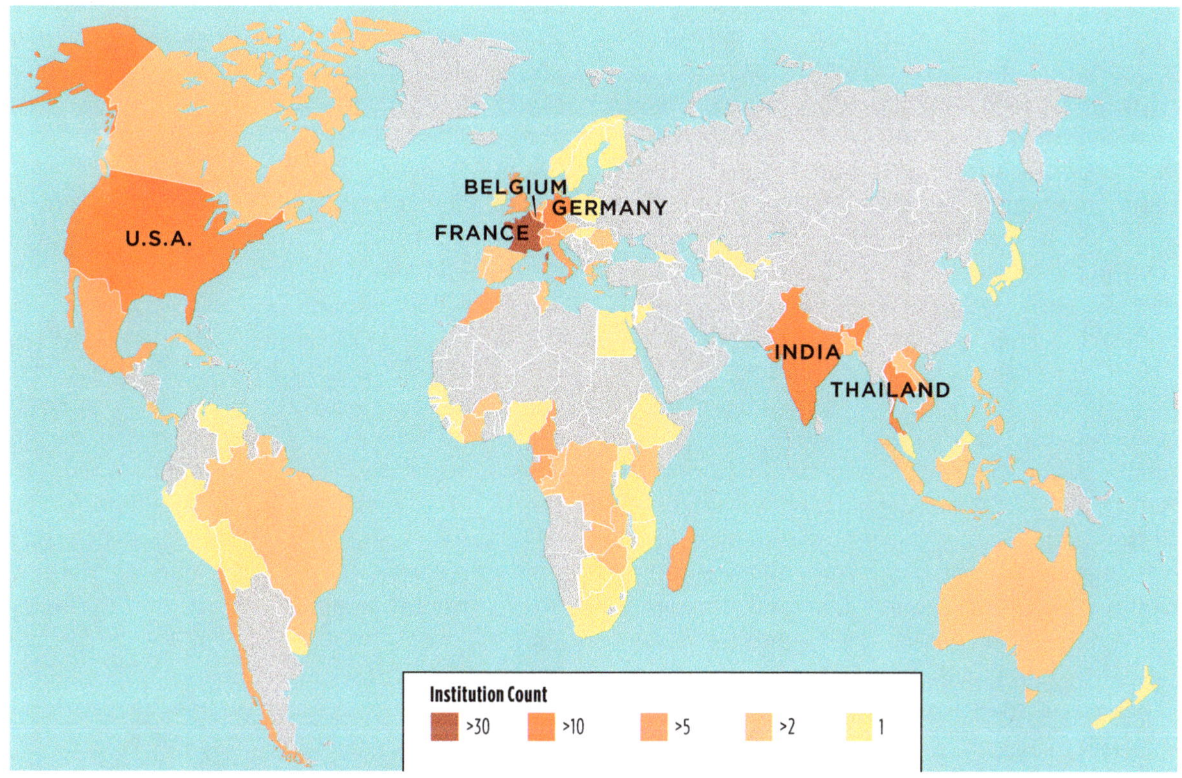

Figure 2. Geographical distribution of PREZODE members and politically engaged countries.
This map illustrates the global reach of the initiative, with nearly 250 institutional members in 80 countries as of 2024, and 27 countries having formalized their commitment to co-developing policies and surveillance systems grounded in One Health principles.

One Health: the ultimate strategy to prevent future pandemics

Marisa Peyre,
Emmanuel Laury, Yoeri Booijink,
Bruno Rosset

The human and economic impacts stemming from the global COVID-19 pandemic were devastating and revealed a critical need for effective pandemic prevention strategies alongside existing preparedness plans. Such efforts should be underpinned by a One Health approach, which experts recognize as key to tackling the complex zoonotic disease outbreaks the world has faced in recent decades, including H5N1 (1997–ongoing), SARS-CoV-1 (2002), H1N1 (2009), H7N9 (2013), Ebola (2014) and MPox (2022).

The COVID-19 pandemic has shown that it would be more cost-effective to invest in prevention than deal with the fallout of a pandemic, but such investments must be relevant, coordinated and sustained over time. In the past 50 years, over USD 1.8 billion has been spent and hundreds of initiatives have been launched internationally, regionally and nationally to officially "prevent, prepare and respond" to zoonotic disease emergence and spread from animals. The One Health concept gained momentum after the H5N1 avian flu pandemic in 2006, leading to more investment in research on zoonotic threats. Unfortunately, efforts tended to focus on crisis response rather than on long-term prevention. Before COVID-19 pandemic (2020), only 30% of the funded projects promoted collaboration between the animal, human and environmental sectors (Figure 1). Around 19% involved animal and human health, 13% environment and animal or human health, and 37% focused only on one sector, mostly human health (Figure 2). These programmes often overlooked biodiversity, early warning systems and local community involvement (< 10% of the number of projects and 1% of the total funding dealt with these issues). Many projects have focused more on capacity-building than empowering local actors and adjusting to community needs, sometimes sacrificing long-term impact in the process. When such key frontline actors are not included, the adoption, impact and sustainability of preventive measures suffers.

An integrated, collaborative and bottom-up approach based on co-creation and One Health principles is needed to prevent future pandemics. Initiatives must factor in local community engagement and empowerment, reduce risk practices, integrate environmental health and preserve biodiversity.

On 20 May 2025, during the 78th World Health Assembly in Geneva, the World Health Organization officially adopted the first ever Pandemic Agreement, endorsed by 124 countries. This legally binding accord aims to strengthen global preparedness and response to future pandemics through improved international coordination, equitable access to health tools and stronger health systems. A key feature of the agreement is the formal integration of the One Health approach, thus recognizing the interconnectedness of human, animal and environmental health and promoting multisectoral collaboration to prevent and mitigate emerging health threats.[1]

1. WHO Pandemic Agreement, Resolution WHA78.1: https://apps.who.int/gb/ebwha/pdf_files/WHA78/A78_10-en.pdf

References

Liu J., Clark H., Kazatchkine M. 2022. Leaders can choose to prevent pandemics. *Nature*, 610(7931), S37. https://doi.org/10.1038/d41586-022-03355-7

Sacco P.L., Valle F., De Domenico M. 2023. Proactive vs. reactive country responses to the COVID-19 pandemic shock. *PLOS Global Public Health*, 3(1), e0001345. https://doi.org/10.1371/journal.pgph.0001345

Traore T., Shanks S., Haider N., Ahmed K., Jain V. *et al.* 2023. How prepared is the world? Identifying weaknesses in existing assessment frameworks for global health security through a One Health approach. *The Lancet*, 401, 673–687. https://doi.org/10.1016/S0140-6736(22)01589-6

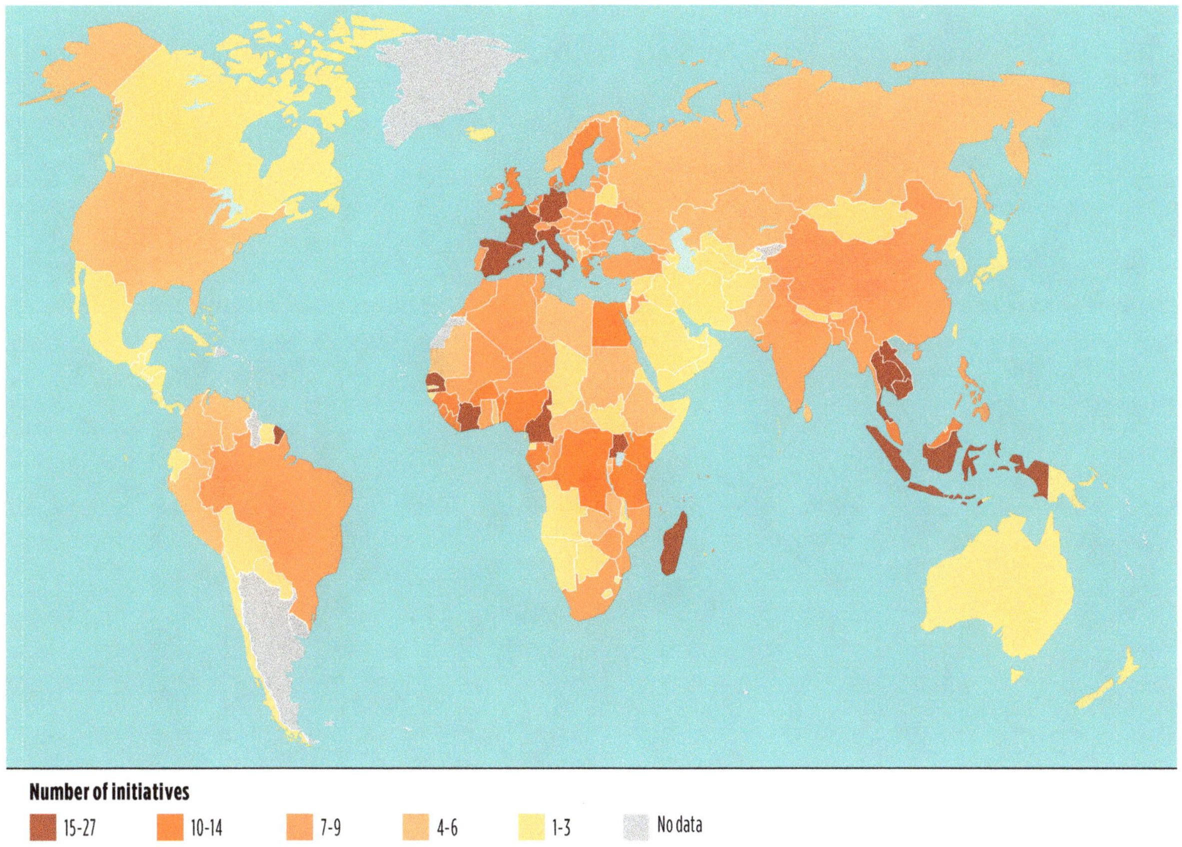

Figure 1. Worldwide map distribution of 115 One Health pandemic prevention, preparedness and response initiatives implemented at regional and international level (i.e. involving at least two countries) since 1975.
Data sources (same for Figure 2): PREZODE initiative mapping platform, a systematic literature review of pandemic prevention initiatives, including the health sector covered by the initiatives (https://prezode-initiative.org/en/); PREZODE co-design workshops (2021–2022); Semi-structured interviews in countries and regions currently ongoing within the PREZODE initiative and the PREACTs program.

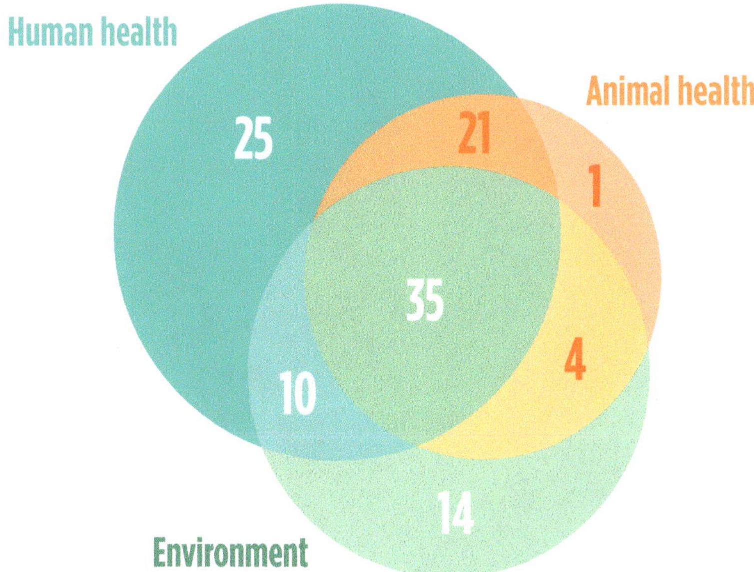

Figure 2. Modified Venn diagram showing the number of initiatives by sectoral involvement. The three colored circles represent the main sectors: human health (blue), animal health (yellow), and environment (green). The overlapping areas indicate intersectoral initiatives, and the numbers within each zone correspond to the number of initiatives involved. The diagram highlights that 37% of initiatives involve a single sector, 32% involve two sectors, and 31% engage all three sectors.

NIKE

4 Future directions and emerging challenges

This section explores the future challenges and innovations within the One Health framework, including emerging trends, unresolved issues and the integration of new tools and approaches. It touches on the impact of climate, the incorporation of the social sciences and equity into One Health strategies. Economic dimensions, the potential of artificial intelligence and the evolution of One Health principles into broader socioecological frameworks like One Welfare are also key focal points. Together, these chapters chart a path for One Health's evolution, addressing both current gaps and future opportunities to build resilient health systems worldwide.

One Health: a widely endorsed but difficult-to-fund approach

Camille Caffier, Franck Berthe, Marisa Peyre

Interest in One Health has risen among health institutions at local, regional and global levels, particularly since the 2019 SARS-CoV-2 pandemic. The One Health approach, which acknowledges the interdependence between human, animal and ecosystem health, is now promoted by the Quadripartite Alliance.[1] It has also been mainstreamed by international financial institutions, such as in the World Bank's One Health Operational Framework and the IDA20 Policy Commitments on One Health. Donors have begun providing funding or financing dedicated to One Health, not only in pandemic prevention, preparedness and response (PPR), but also increasingly in other streams focused on biodiversity, climate change and more.

However, the financing landscape is highly fragmented with many vertical, mostly public health-centric mechanisms (Figure 1). They often react to health crises, rather than reducing risk at the source or breaking the panic-neglect cycle. The COVID-19 crisis triggered a paradigm shift, and specific (although limited) funding dedicated to One Health is now emerging, with two major initiatives worth noting.

In 2020, the French Development Agency (AFD) launched its NGO Sector Innovation Facility (FISONG OH), an innovative financial mechanism to support One Health. Initially intended for general NGO financing, this fund provided EUR 2.5 million for One Health projects in 2020.

In 2022, the World Bank (in collaboration with WHO) established the Pandemic Fund, a Financial Intermediary Fund designed to provide a dedicated stream of long-term financing to strengthen PPR capabilities in low- and middle-income countries. This fund has generated nearly USD 2 billion and supports One Health initiatives in 80% of its projects. The Pandemic Fund managed to raise six US dollars for every dollar in grant funds.

Pragmatic approaches for One Health investments have been proposed through frameworks considering risks, co-benefits and stakeholder financial capacity for a fairer distribution of costs and benefits. The Pandemic Fund's first call for proposals also showed strong demand from beneficiary countries for strengthening public health systems, while raising questions about their capacity to effectively integrate One Health into their health systems. These concerns echo the debates over the inclusion of One Health and financing provisions in the 2025 international Pandemic Agreement, which seeks to strengthen global preparedness for future health crises.

1. The World Health Organization (WHO), the World Organisation for Animal Health (WOAH), the Food and Agriculture Organization of the United Nations (FAO), and the United Nations Environment Programme (UNEP).

References

AFD. "One Health": Responding to Pandemics with a Holistic Approach to Human, Animal and Environmental Health. https://www.afd.fr/en/actualites/one-health-responding-pandemics-holistic-approach-human-animal-and-environmental-health

Berthe F.C.J., Bali S.R., Batmanian G.J. 2022. *Putting Pandemics Behind Us: Investing in One Health to Reduce Risks of Emerging Infectious Diseases.* Washington, D.C.: World Bank Group. http://documents.worldbank.org/curated/en/099530010212241754

WHO, World Bank. 2022. Analysis of Pandemic Preparedness and Response (PPR) architecture, financing needs, gaps and mechanisms. Report prepared for the G-20 Joint Finance & Health Task Force. https://thedocs.worldbank.org/en/doc/5760109c4db174ff90a8dfa7d025644a-0290032022/original/G20-Gaps-in-PPR-Financing-Mechanisms-WHO-and-WB-pdf.pdf

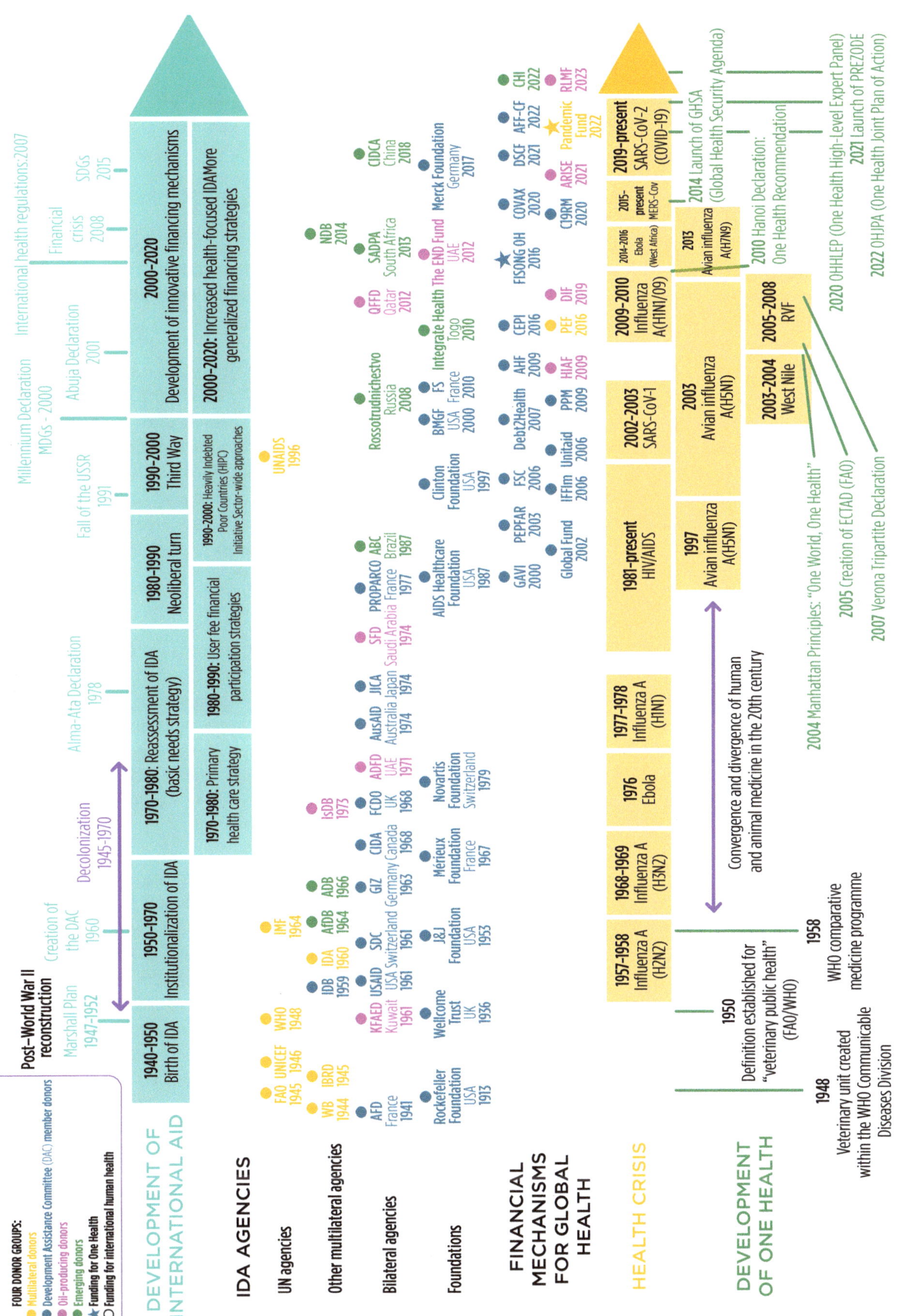

Figure 1. Evolution of international aid and global health financing mechanisms in relation to One Health: from post-World War II development aid to the emergence of dedicated One Health funding streams. The figure highlights the convergence of donor typologies, health crises and institutional initiatives, culminating in recent instruments such as the Pandemic Fund (World Bank).

Investing in One Health: a framework to demonstrate its added value and return on investment

Quadripartite-powered Community of Practice on the Return on Investment for One Health and One Health Investment Planning[1]

The unprecedented set of interconnected global challenges we are facing require fresh solutions that can work in complex systems, where an action can have many consequences, intended or not. One Health recognizes the interdependence of humans, animals, plants and ecosystems and supports systems thinking and collaboration to create sustainable solutions. However, securing sustainable financing for One Health initiatives remains challenging. Demonstrating the return on investment (ROI) of One Health initiatives is essential for justifying investments and guiding decisions on resource allocation. While evidence on the added value of One Health is mounting, several gaps still exist.

To bridge this gap, the FAO and CIRAD, in collaboration with UNEP, WHO, WOAH and the World Bank, supported by experts in the field, have created a framework (Figure 1) to guide economic and financial analyses of One Health initiatives. This framework specifically highlights aspects of particular importance to One Health (Figure 2).

The framework first focuses on the objectives of One Health initiatives and level of intersectoral collaboration by evaluating the strength of the collaboration or integration, and then measuring their added value compared to the outcome without the initiative. The conceptualization phase includes the current situation and context, characterizes the initiative and its goals, describes the scope of the analysis, and defines the perspective and objective. All the costs and benefits are compared to a counterfactual, either business-as-usual (including inaction) or an alternative approach, such as one that is sectoral, unidisciplinary, one-dimensional, insufficiently integrated, or not at scale.

Applying this framework can help build a stronger case for investment by generating economic evidence earlier in the planning process and thereby minimizing poor investment risks. Good practice dictates that investments should include economic and financial analysis at the design stage of an investment, as well as at the completion stage to demonstrate impact.

Discussions with experts and stakeholders indicate that quantitative data on ROI alone is not enough. Effective financial and economic analyses of One Health must also integrate persuasive narratives, including information about societal or community needs, implementation feasibility and resource capacity, combined with economic evidence for different time frames.

The focus of the Community of Practice on the Return on Investment for One Health and One Health Investment Planning, hosted by the One Health Knowledge Nexus in FAO's Virtual Learning Centers, is to generate more evidence and refine methodologies aimed at informing One Health investment planning and leveraging. This community provides a space for interested people to connect, share insights, critique and advance knowledge, and contribute to the long-term economic value of One Health.

1. Text produced collectively by Barbara Häsler, Katrin Taylor, Eleanor Raj, Danny Sheath, Katinka de Balogh, Aashima Auplish, under the leadership of Thanawat Tiensin, in collaboration with Yoeri Booijink, Marisa Peyre, Salman Hussain, Franck Berthe, Sinaia Netanyahu and Chadia Wannous.

References

Auplish A., Raj E., Booijink Y., de Balogh K., Peyre M. *et al.* 2024. Current evidence of the economic value of One Health initiatives: A systematic literature review. *One Health*, 18, 100755. https://doi.org/10.1016/j.onehlt.2024.100755

Quadripartite-powered Community of Practice on the Return on Investment for One Health and One Health Investment Planning: https://virtual-learning-center.fao.org/admin/tool/custompage/view.php?id=50

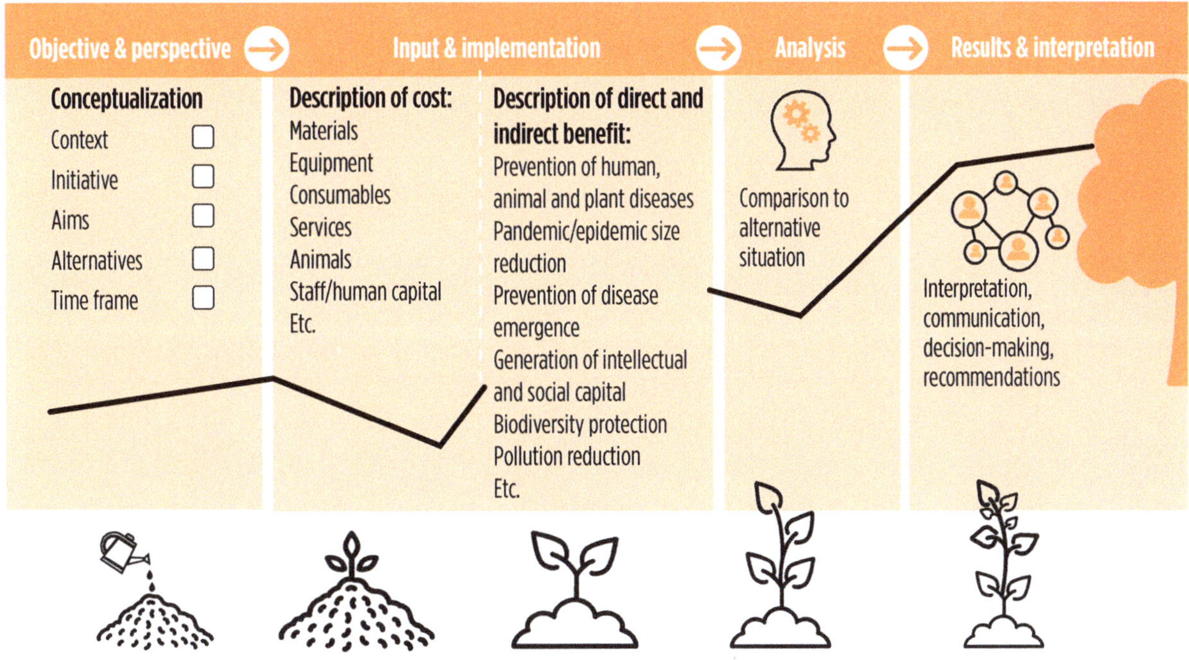

Figure 1. Overview of the methodological framework for the financial and economic analysis of One Health investments.

Figure 2. Ten important aspects to consider when designing and conducting One Health financial and economic analysis (informed by Auplish *et al.* 2024 and the community of practice on One Health return on investment).

Using thematic clustering for data-driven insights on One Health research and knowledge gaps

François Roger,
Marie-Christine Lambert,
Émilie Chirouze

The abstracts from a series of scientific publications related to One Health were analysed and grouped into three clusters based on key terms (Figure 1) to identify major research topics and knowledge gaps.

The first cluster included publications on zoonotic diseases and human-animal interactions, reflecting the importance of the One Health approach, which deals with human, animal and environmental health as a whole. While the research effectively covered virology and epidemiology, perspectives from the social sciences (e.g. efforts to understand cultural practices or socioeconomic factors influencing zoonotic transmission) were limited. Environmental aspects (e.g. habitat alteration, biodiversity loss and ecosystem services) were also underrepresented, despite the central role of the environment in zoonotic disease dynamics.

The second cluster on antimicrobial resistance (AMR) underscored its global relevance, with keywords reflecting resistance mechanisms and bacterial strains like *Escherichia coli*. While this area is well-supported by microbiological and pharmacological research, it lacks substantial input from the behavioural sciences (e.g. how healthcare systems or patient practices influence antibiotic misuse). Crucial topics such as structural inequities in access to antibiotics were not covered, and environmental dimensions, including how AMR is propagated through environmental pathways (e.g. water, soil, agricultural runoff) were insufficiently explored, leaving critical knowledge gaps unaddressed.

The third cluster on global health and climate change included studies on planetary health and climate impacts but lacked robust contributions from the fields of anthropology, political science and ethics, which are vital in addressing global disparities and policy frameworks. Although climate change is an inherently environmental issue, many studies in this cluster did not adequately incorporate environmental data or metrics (e.g. carbon emissions, land-use changes, agriculture or biodiversity indices).

The analysis confirms that research leadership is concentrated in high-income and BRICS countries, while contributions from low- and middle-income countries remain limited (Figure 2).

Another bibliometric analysis supports our findings, highlighting a dominant focus on the human-animal health interface, often at the expense of environmental health. Research remains primarily within natural sciences, with minimal integration of social sciences.

References

Fang Z., Tu S., Huang J. 2024. A bibliometric analysis of One Health approach in research on antimicrobial resistance. *Science in One Health*, 3, 100077. https://doi.org/10.1016/j.soh.2024.100077

Humboldt-Dachroeden S., Rubin O., Sylvester Frid-Nielsen S. 2020. *The state of One Health* research across disciplines and sectors – a bibliometric analysis. *One Health*, 10, 100146.https://doi.org/10.1016/j.onehlt.2020.100146

Miao L., Li H., Ding W., Lu S., Pan S. *et al.* 2022. Research priorities on One Health: a bibliometric analysis. *Frontiers in Public Health*, 10, 889854. https://doi.org/10.3389/fpubh.2022.889854

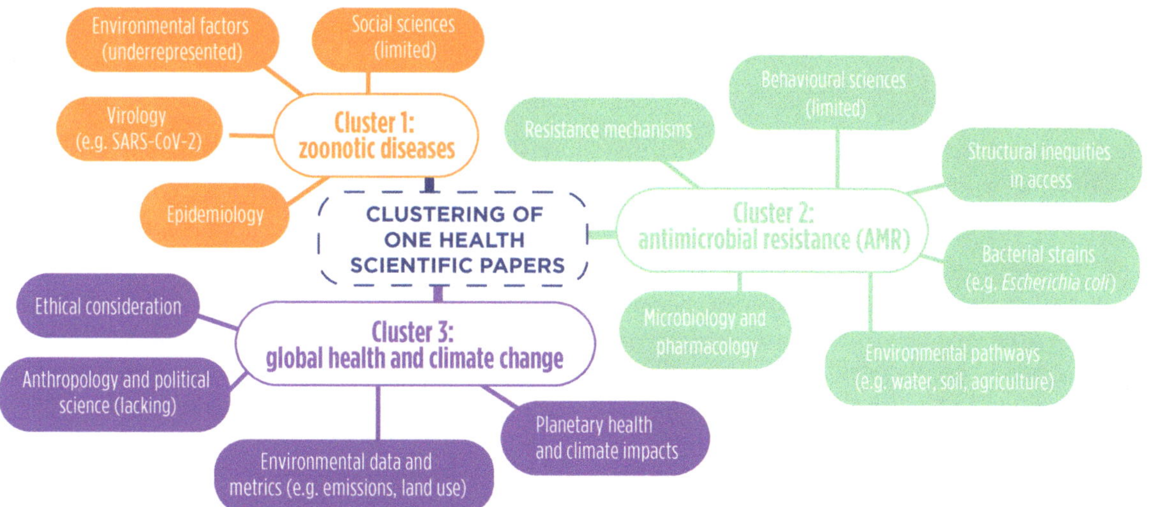

Figure 1. Mindmap of 3 clusters found after analysis of a database of 3,498 articles which were selected from the Web of Science (WoS) and MEDLINE databases (query date: 22 February 2023) using the terms "One Health", "EcoHealth" and "Planetary Health" based on keyword searches by authors, medical subject heading (MeSH) terms from MEDLINE and WoS algorithm-generated keywords.
The clustering was done by using TF–IDF (term frequency–inverse document frequency) and k-means: the abstracts were preprocessed to remove stop words and convert them into numerical vectors using the TF–IDF method. Dimensionality reduction was performed with PCA (principal component analysis) to project the data into two dimensions for visualization. The *k*-means clustering algorithm (which partitions data into k-groups, or clusters, by minimizing intra-cluster variance) was then applied to separate the abstracts into clusters.

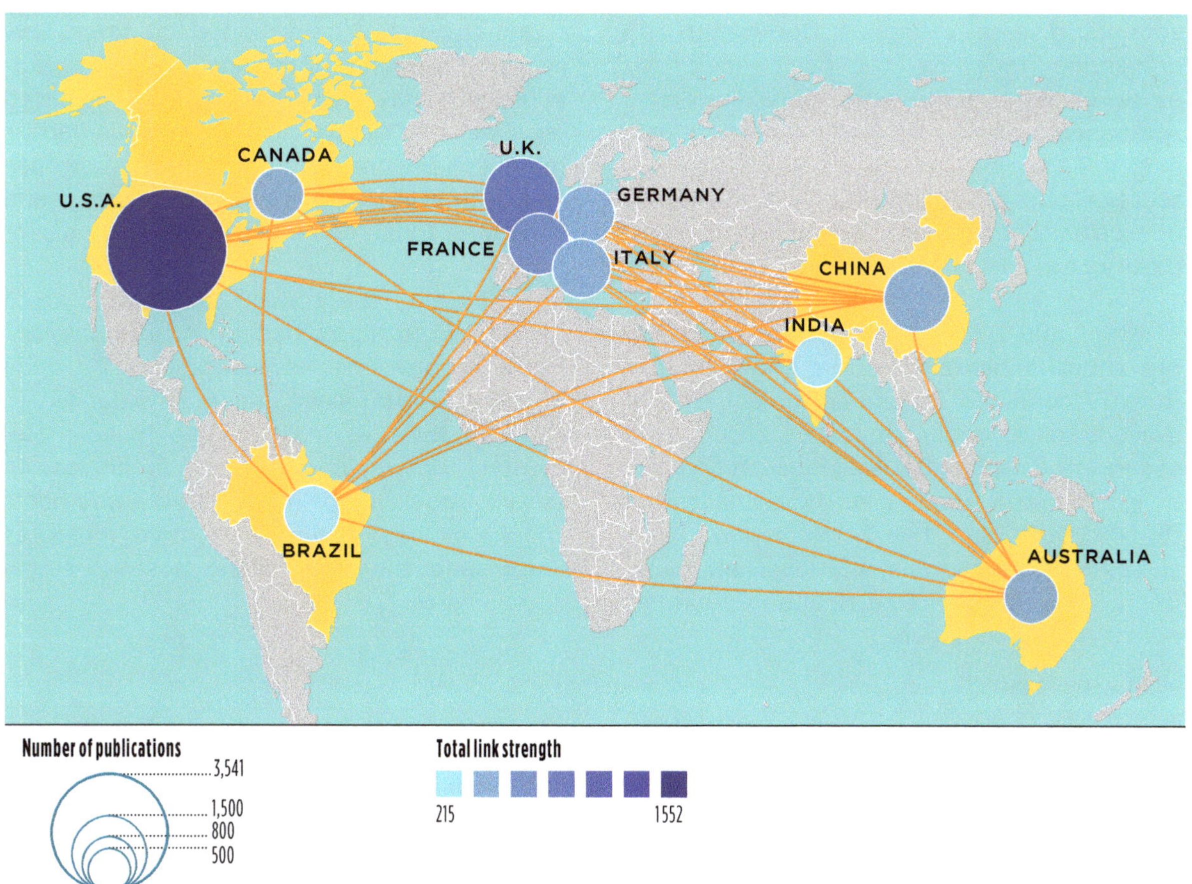

Figure 2. Top 10 national collaborative networks on One Health, 2003–2021. From Miao *et al.* 2022.
The United States dominates global scientific production and is highly connected to other countries. The intensity of a country's collaborations is represented by the color of its circle. The connections indicate where collaborations exist but do not provide a precise measure of the volume of co-authored publications between two countries.

One Health governance: implementation bottlenecks in the Global South

Julie Meunier, Musso Munyeme, Soawapak Hinjoy, Servane Baufumé, Papa Seck, Marisa Peyre

Given the rising threat of zoonotic diseases to global health systems, the One Health strategy has become increasingly crucial as an innovative, cost-effective and holistic approach to dealing with public health emergencies. This approach recognizes the interconnectedness of humans, animals and the environment to fully integrate all three of these aspects.

Implementing One Health can be particularly challenging in the Global South due to limited resources and inadequate healthcare infrastructure in some places. The COVID-19 epidemic highlighted the urgent need for comprehensive health plans that incorporate zoonotic disease prevention. One Health and prevention principles have been recognized as key elements in the international Pandemic Agreement adopted by WHO member states in May 2025. However, some countries remain sceptical about the feasibility and affordability of such an accord given the diversity and complexity of the areas it concerns.

As of mid-2024, the PREZODE pandemic prevention initiative[1] has over 240 members from 77 countries. Its primary objective is to prevent animal-origin pandemics, ensure food security and protect the livelihoods of communities. To better understand these challenges, the PREZODE initiative performed a global survey to evaluate the institutionalization and operationalization of One Health in countries across the world (Figure 1).

1. https://prezode.org/

Surveys, interviews with experts and literature reviews revealed that while some nations have formally structured One Health governance and integrated One health strategy into their national health plans, these efforts often remain separate and operate parallel to mainstream health governance, which can limit resource allocation and long-term collaboration (Figure 2). International agendas that focus on pandemic threat preparedness and response have influenced the operationalization of One Health over time, but do not always address national priorities, which in turn leads to fragmented efforts and outcomes.

Initiatives must be designed to address new zoonotic dangers while meeting national needs in order to truly prevent pandemic risks based on One Health principles. These initiatives must also provide a comprehensive approach to health that benefits all by supporting and improving fundamental healthcare systems.

Addressing the specific challenges faced by Global South countries is crucial in international negotiations. These challenges include financial, institutional and structural barriers to adopting the One Health approach and implementing upstream preventive initiatives. To ensure global well-being, effective agreements should include robust preventive measures and respect the needs of the most vulnerable communities.

References

Meunier J., Munyeme M., Hinjoy S., Baufumé S., Seck P. *et al.* 2024. One Health governance and implementation at country level: An overview of the situation worldwide including the Global South [Abstract No. 2620]. *In Abstract Book of the 8th World One Health Congress*, Cape Town, South Africa, 20–23 September 2024. Global One Health Community.

Peyre M., Vourc'h G., Lefrançois T., Martin-Prével Y., Soussana J., Roche B. 2021. PREZODE: preventing zoonotic disease emergence. *The Lancet*, 397(10276), 792–793. https://doi.org/10.1016/s0140-6736(21)00265-8

Schwalbe N., Hannon E., Gilby L., Lehtimaki S. 2024. Governance provisions in the WHO Pandemic Agreement draft. *The Lancet*, 403(10434), 1333–1334. https://doi.org/10.1016/s0140-6736(24)00585-3

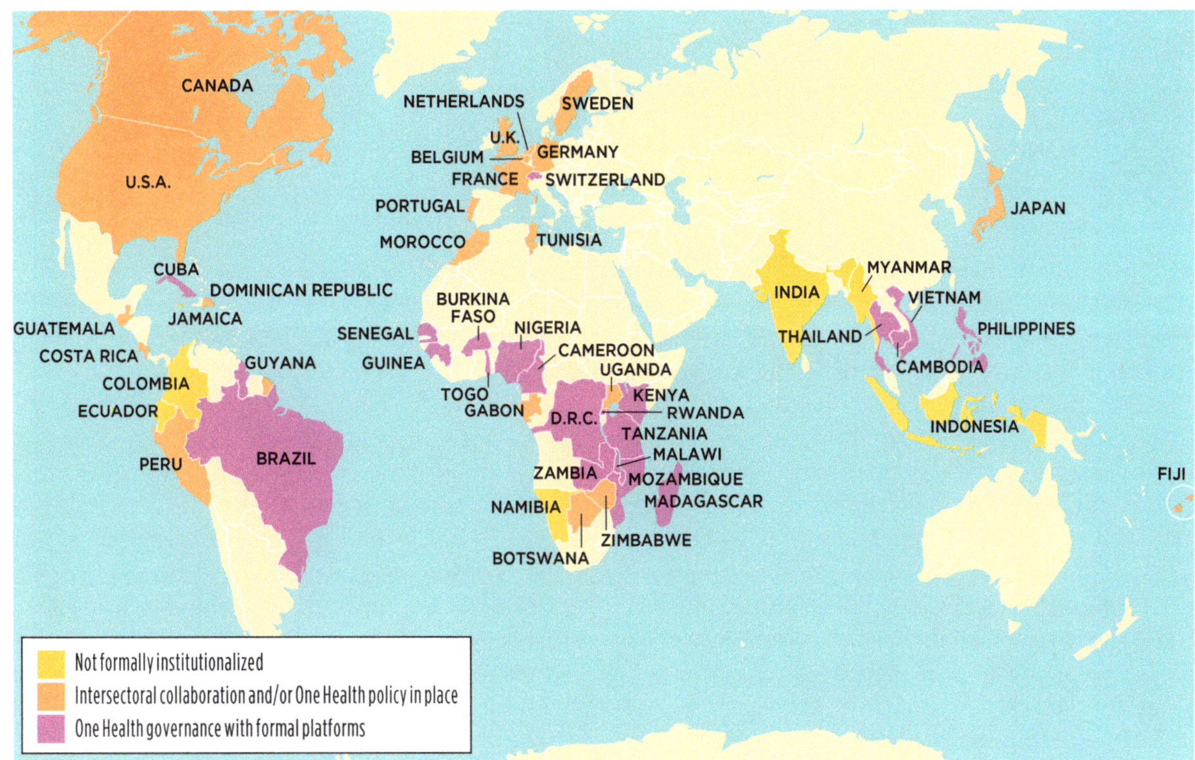

Figure 1. World mapping of One Health (OH) institutionalization status at country level: OH governance with formal platforms ; intersectoral collaboration and/or OH policies in place ; not formally institutionalized.

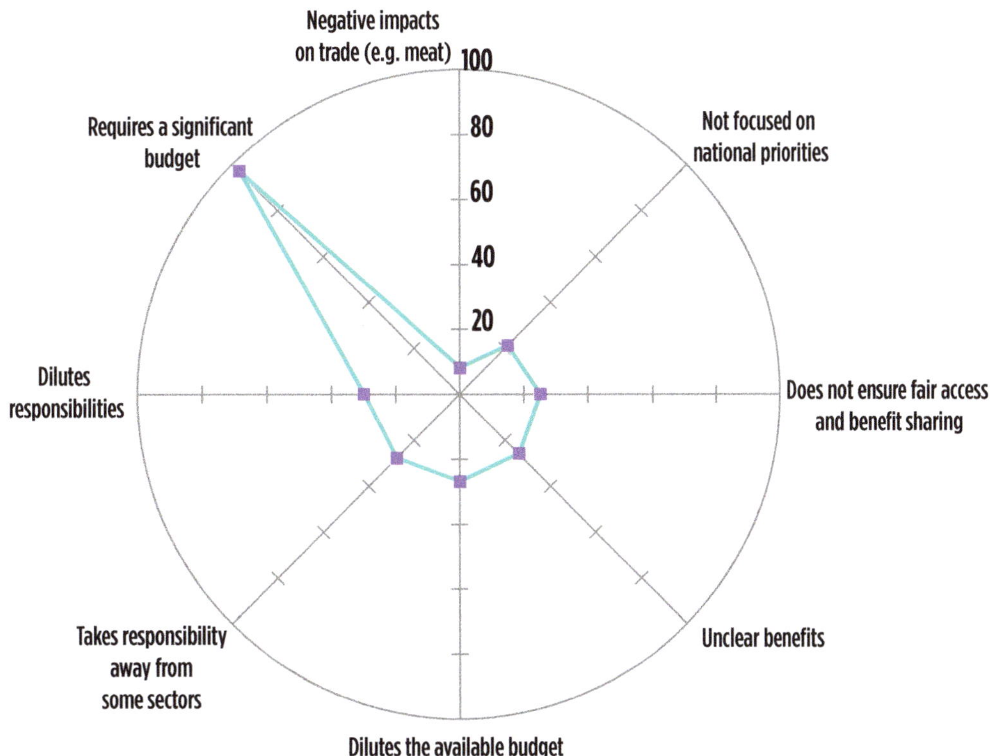

Figure 2. Relative importance of the main limits of One Health operationalization according to respondents' perceptions.

Networks shaping the future of One Health: focusing on the Afro-Eurasian landmass

François Roger

Global health challenges are increasingly complex, and mainly driven by environmental, climatic, economic, political and social factors. Globalization also accelerates the international spread of emerging diseases and antimicrobial resistance. Dealing with these issues requires One Health networks, such as the regional platforms developed by CIRAD to enhance research capacities, facilitate intersectoral exchanges, generate scientific knowledge and translate it into practical applications. These platforms also promote stakeholder relations, development and implementation of specialized public policies, and training of a new generation of health researchers (Figure 1 and Table 1).

These platforms catalyse collaboration among local, regional and global actors. Existing One Health networks, however, face significant challenges. Distribution is essentially towards Europe and North America, while networks in the Global South are underrepresented. In particular, there is need for more equitable governance structures within networks to ensure better inclusion of low- and middle-income countries. Many challenges often arise at the local stakeholder level: limited human and financial resources, translation of health standards from international bodies down to local political authorities, and intersectoral partnerships between all kinds of actors involved in complex issues.

Improved communication on the co-creation of effective local health policies involving researchers, policymakers and field operators is key. This calls for engaging communities in One Health initiatives by actively involving local communities, especially those who are most affected, in managing health threats to ensure research is relevant and cultural appropriateness is respected. Most networks are primarily concerned with new infectious diseases and pandemic threats at the expense of handling other pressing issues, such as environmental concerns. Real progress in health will be achieved when more inclusive, equitable and sustainable networks are established that link local, regional and global actors while being driven by the needs of vulnerable communities. Doing so would allow regional health networks to share methodologies and tools and create a global platform to stimulate comparative research, strengthen scientific advocacy with policy decision makers and improve global research coordination. These regional networks include links with international organizations, which already exist but need to be strengthened if international research and related recommendations are to be appropriately adapted to a national context.

References

Mwatondo A., Rahman-Shepherd A., Hollmann L., Chiossi S., Maina J. *et al.* 2023. A global analysis of One Health Networks and the proliferation of One Health collaborations. *The Lancet*, 401 (10376), 605-616. https://doi.org/10.1016/S0140-6736(22)01596-3

Roger F., de Romémont A., Binot A., Loire É., Girard P. 2019. Strengthening the global health dialogue: linking research networks in the Global South. *Perspective*, 53, 1-4. https://doi.org/10.19182/perspective/31833

ONE HEALTH COMMISSION (OHC)

The OHC is a US-based, non-profit organization established in 2009 to advocate globally for the One Health approach. Rather than being a field-based network, OHC serves as an international platform for coordination, advocacy and education, connecting stakeholders across human, animal, environmental and plant health domains. OHC plays a key role in facilitating awareness, dialogue and resource sharing across sectors and geographies. It convenes thematic communities and working groups—such as the One Health Education Task Force, the One Health Social Sciences Initiative and student-led alliances like ISOHA—to promote systems thinking and collaborative action. Through webinars, campaigns (e.g. One Health Awareness Month) and global partnerships, OHC helps integrate One Health principles into policy, research and education worldwide.
www.onehealthcommission.org

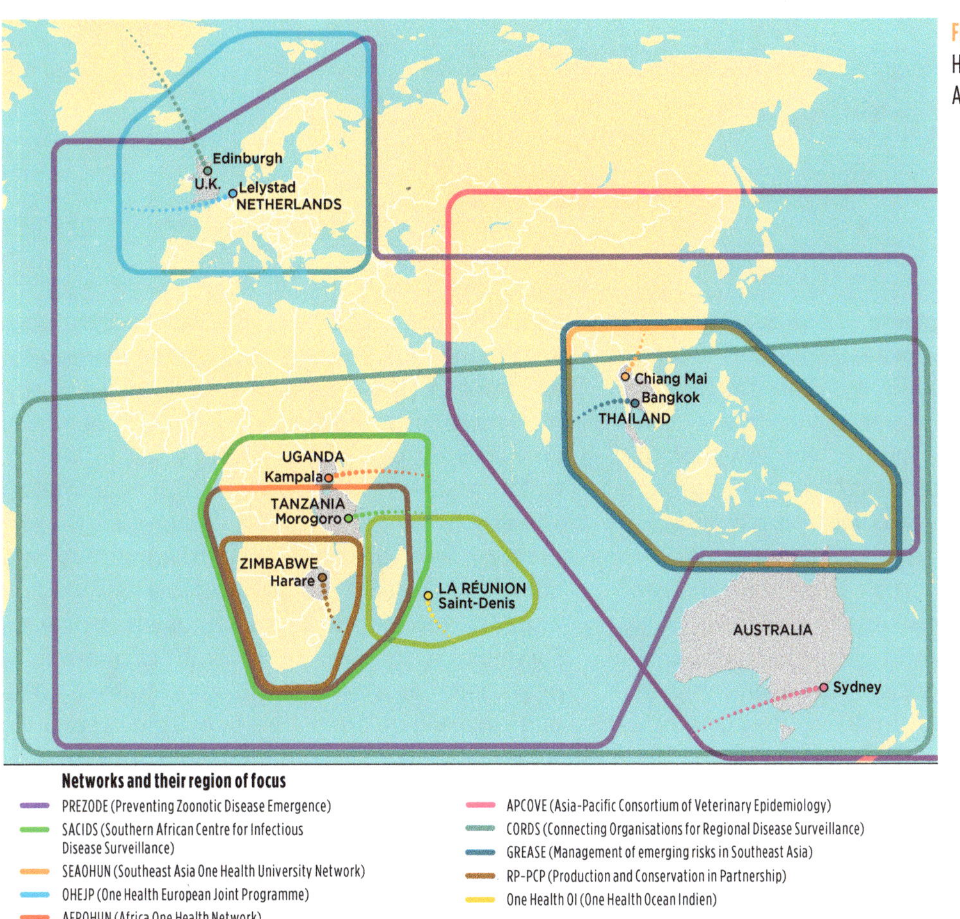

Figure 1. Mapping of key One Health networks across the Afro-Eurasian landmass.

Table 1. Comparative overview of One Health network strategies, activities and objectives.

STRATEGIES	PRIMARY ACTIVITIES	KEY OBJECTIVES
PREZODE (Preventing Zoonotic Disease Emergence)	Research, capacity-building and collaboration	To reduce the risk of zoonotic disease emergence by strengthening global surveillance and prevention networks
SACIDS (Southern African Centre for Infectious Disease Surveillance)	Research, capacity-building and surveillance	To enhance Africa's capacity to detect and respond to infectious diseases through a One Health approach
SEAOHUN (Southeast Asia One Health University Network)	Education, capacity-building and research	To strengthen the capacity of universities in Southeast Asia to address health challenges through a One Health approach
One Health European Joint Programme (OHEJP)	Research, training and policy development	To promote a collaborative One Health approach to foodborne zoonoses, antimicrobial resistance and emerging threats
Africa One Health Network (AFROHUN)	Education, capacity-building and research	To promote One Health education and practices across African universities and institutions
Asia-Pacific Consortium of Veterinary Epidemiology (APCOVE)	Research, capacity-building and collaboration	To improve veterinary epidemiology and One Health practices in the Asia-Pacific region
CORDS (Connecting Organisations for Regional Disease Surveillance)	Research, capacity-building and collaboration	To promote interdisciplinary research and the implementation of One Health approaches particularly in low- and middle-income countries
GREASE (Management of emerging risks in Southeast Asia)	Managing emerging health risks at the human–animal–environment interface, research, capacity-building and policy advising	To improve the surveillance and control of zoonotic diseases and manage emerging health risks in Southeast Asia through a One Health approach
RP-PCP (Production and Conservation in Partnership)	Health, sustainable rural development and biodiversity conservation	To address health as part of the sustainable development of rural populations and biodiversity conservation through collaborative partnerships
One Health OI (One Health Océan Indien)	Controlling animal and human infectious diseases, capacity-building and policy advising	To address animal and human health challenges in the Indian Ocean islands through a collaborative One Health approach

Using PADI-web to monitor animal and plant diseases in digital media sources

Mathieu Roche, Carlène Trevennec, Isabelle Pieretti, Elena Arsevska

New and re-emerging disease outbreaks have become increasingly common in recent decades, driven by climate change, anthropization of natural environments and contact with wildlife due to human mobility and animal trade activities. A global approach to monitoring public, animal and plant health is needed to address this issue. Monitoring the news across digital media can provide relevant information on disease outbreaks and make disease surveillance and pandemic preparedness more comprehensive. However, manually extracting relevant information from unofficial digital news sources is time-consuming.

The Platform for Automated Extraction of Disease Information from the web (PADI-web1[1]) was designed to make this task easier. PADI-web has been used since 2016 as part of the epidemic intelligence activities of the French National Animal Surveillance Health Plateform. This tool crawls Google news related to public, animal and plant health, including zoonotic diseases. It offers a multilingual approach, automated information classification and extraction modules, a notification tool configurable according to end-user needs, and maps for animal and plant health monitoring. The PADI-web pipeline involves five steps (Figure 1), and the algorithms associated with steps 3 and 4 use text mining and artificial intelligence approaches that are fine-tuned for animal and plant diseases. PADI-web for animal disease surveillance focuses specifically on zoonotic diseases such as avian influenza, which enables cases detected in mammalian species to be investigated (Figure 2). Human vector-borne diseases (e.g. chikungunya and dengue) have also been added to a new version of PADI-web, and the version for plant disease surveillance focuses on several EU quarantine pests, such as *Xylella fastidiosa* (Figure 3). The platform will be enhanced to enable syndromic surveillance for the early detection of emerging epidemics and new host plants.

PADI-web has been integrated into international projects, including the Monitoring Outbreaks for Disease Surveillance in a data science context (MOOD) project (H2020 2020–2024). The MOOD project seeks to develop innovative tools and services for the early detection, assessment, and monitoring of current and future infectious disease threats in the context of continuous global, environmental and climatic change. As part of the MOOD project, different case studies have focused on widespread diseases (e.g. avian influenza and West Nile) as well as neglected diseases (e.g. leptospirosis).

1. https://www.padi-web-one-health.org

References

Arsevska E., Valentin S., Rabatel J., de Goër de Hervé J., Falala S. *et al.* 2018. Web monitoring of emerging animal infectious diseases integrated in the French animal health epidemic intelligence system. *PLoS ONE*, 13(8), e0199960. https://doi.org/10.1371/journal.pone.0199960

Dupuy C., Locquet C., Brard C., Dommergues L., Faure E. *et al.* 2024. The French National Animal Health Surveillance Platform: an innovative, cross-sector collaboration to improve surveillance system efficiency in France and a tangible example of the One Health approach. *Frontiers in Veterinary Science*, 11, 1249925. https://doi.org/10.3389/fvets.2024.1249925

Roche M., Rabatel J., Trevennec C., Pieretti I. 2024. PADI-web for plant health surveillance. In: *Intelligent Information Systems: CAiSE* (Islam S., Sturm A., eds.). Lecture Notes in Business Information Processing, 520, pp. 148–156, Springer, Cham. https://doi.org/10.1007/978-3-031-61000-4_17

Valentin S., Arsevska E., Rabatel J., Falala S., Mercier A. *et al.* 2021. PADI-web 3.0: A new framework for extracting and disseminating fine-grained information from the news for animal disease surveillance. *One Health*, 13, 100357. https://doi.org/10.1016/j.onehlt.2021.100357

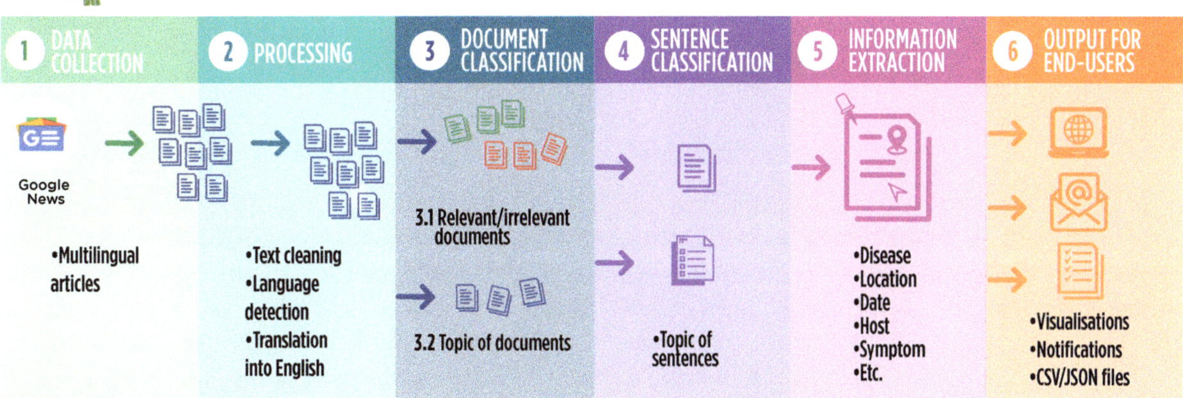

Figure 1. PADI-web pipeline.

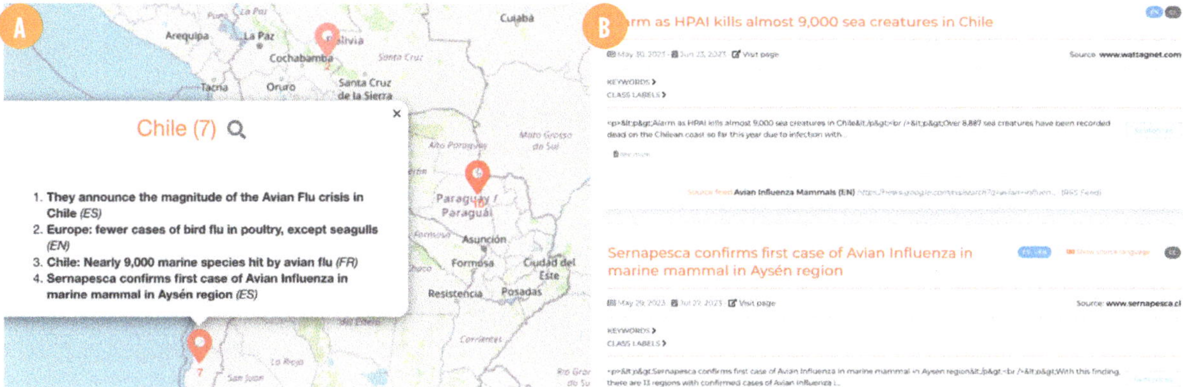

Figure 2. Avian influenza (animal health) (01/05/2023–01/07/2023). (A) List of texts automatically classified as outbreak declarations focusing on Chile. (B) Examples of texts collected in English, French and Spanish that are formatted with PADI-web. Note that the first events of detection of avian influenza on mammals were identified in Chile with PADI-web on 25 May 2023 vs. 7 June 2023 with the World Animal Health Information System (WAHIS).

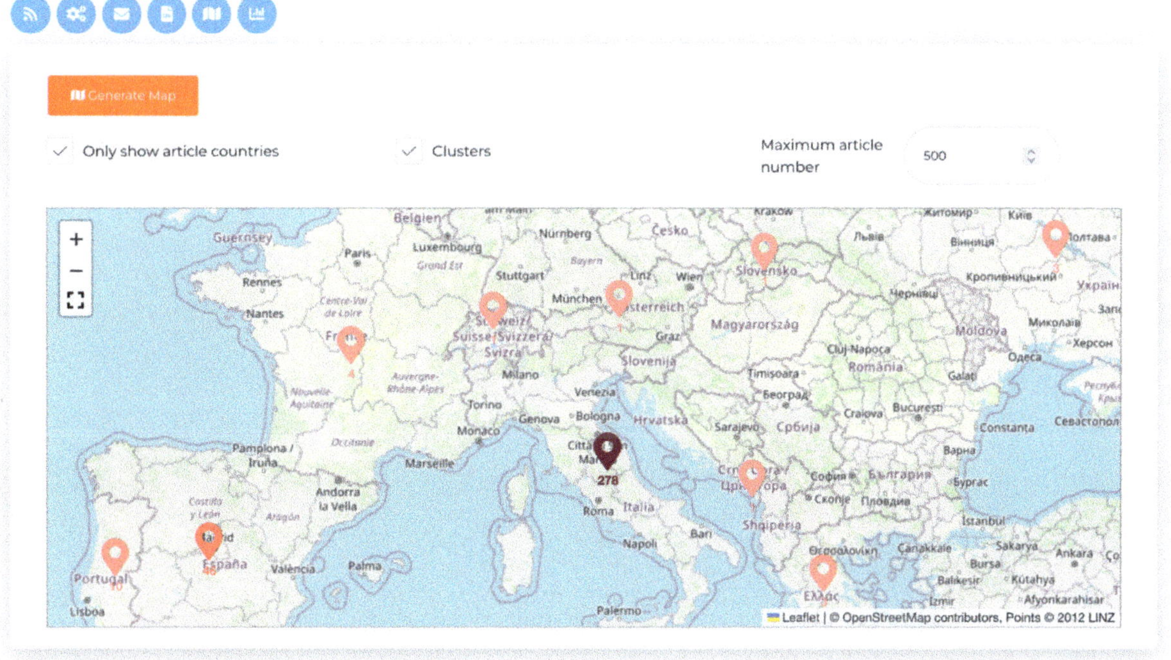

Figure 3. *Xylella fastidiosa* (plant health) (01/03/2024–01/09/2024). Geolocation of news collected and normalized with PADI-web and automatically classified as relevant using artificial intelligence approaches – 363 texts.

The scope of public-private partnerships in One Health

Daan Vink, Bruno Rosset, Rahul Srivastava, Isabelle Dieuzy-Labaye, Marisa Peyre

Complex One Health challenges, such as zoonotic disease control, food safety and antimicrobial resistance (AMR), require multi-faceted approaches. Public-private partnerships (PPPs)—i.e. collaboration between public institutions, private enterprises and civil society to achieve common goals—are inherently suited to the intersectoral activities required for One Health implementation. PPPs provide frameworks which accommodate government agencies (responsible for regulation and health policy), private companies (often leaders in innovation, research and development) and civil society organizations (which can provide broad-based community support). Such frameworks provide the scope to integrate efforts across human, animal and environmental health sectors. By aligning interests and coordinating across sectors, pooling expertise, and combining resources and operational capacities, well-implemented PPPs can amplify the reach and cost effectiveness of interventions and allow for greater and lasting impact. They can also offer flexibility and responsiveness to emerging One Health risks and threats such as pandemics, AMR and food safety crises.

From an operational perspective, PPPs are a practical solution for creating resilient One Health systems, particularly in low-resource settings where public agencies may lack funding or technical capacity. They can ensure that the One Health approach is not only theoretical but actionable and scalable. For example, GALVmed collaborates with private veterinary companies and public sector institutions to develop and distribute affordable livestock vaccines and diagnostics in sub-Saharan Africa and South Asia.

A structured questionnaire sent to 181 World Organisation for Animal Health (WOAH) Member Countries and 47 private contacts in 2017 collected information on 99 different PPPs operating in 76 countries. These were built into a publicly-accessible database. The PPPs had a global reach, with up to four described per country (Figure 1). Four were self-classified as including One Health-related activities, and another five described activities with One Health relevance (Figure 2). However, of these nine PPPs, only three were implemented collaboratively by animal health and human health agencies and so could be considered "true" One Health PPPs. The key operational areas included AMR, One Health surveillance, food safety and communication. Eight of the nine PPPs were set up in higher income countries. While these data may not be comprehensive or up to date, they show that the number of PPPs initiated as a One Health activity remains limited, particularly in the Global South.

References

Galière M., Peyre M., Muñoz F., Poupaud M., Dehove A. *et al.* 2019. Typological analysis of public-private partnerships in the veterinary domain. *PLoS ONE*, 14(10), e0224079. https://doi.org/10.1371/journal.pone.0224079

GALVmed. Global Alliance for Livestock Veterinary Medicines. https://www.galvmed.org

The WOAH Database of Public-Private Partnerships. https://bulletin.woah.org/?panorama=04-3-1-2023-1_ppp-database

The WOAH PPP Handbook: Guidelines for Public-Private Partnerships in the veterinary domain. https://www.woah.org/fileadmin/publicprivatepartnerships/ppp/EN/Handbook_EN.html

World Organisation for Animal Health. Public-Private Partnerships in the veterinary domain. https://www.woah.org/en/what-we-offer/improving-veterinary-services/pvs-pathway/public-private-partnerships-in-the-veterinary-domain

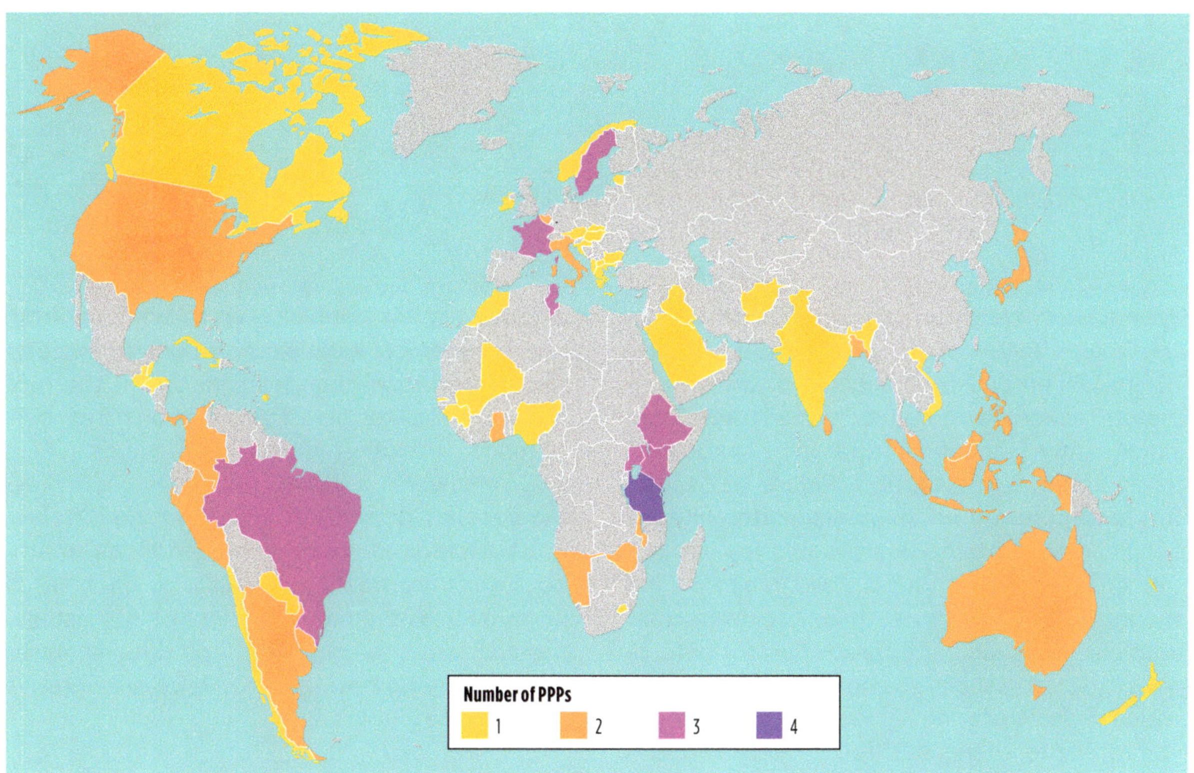

Figure 1. World map showing country-level number of PPPs recorded in the World Organisation for Animal Health (WOAH) PPP database (n = 99).

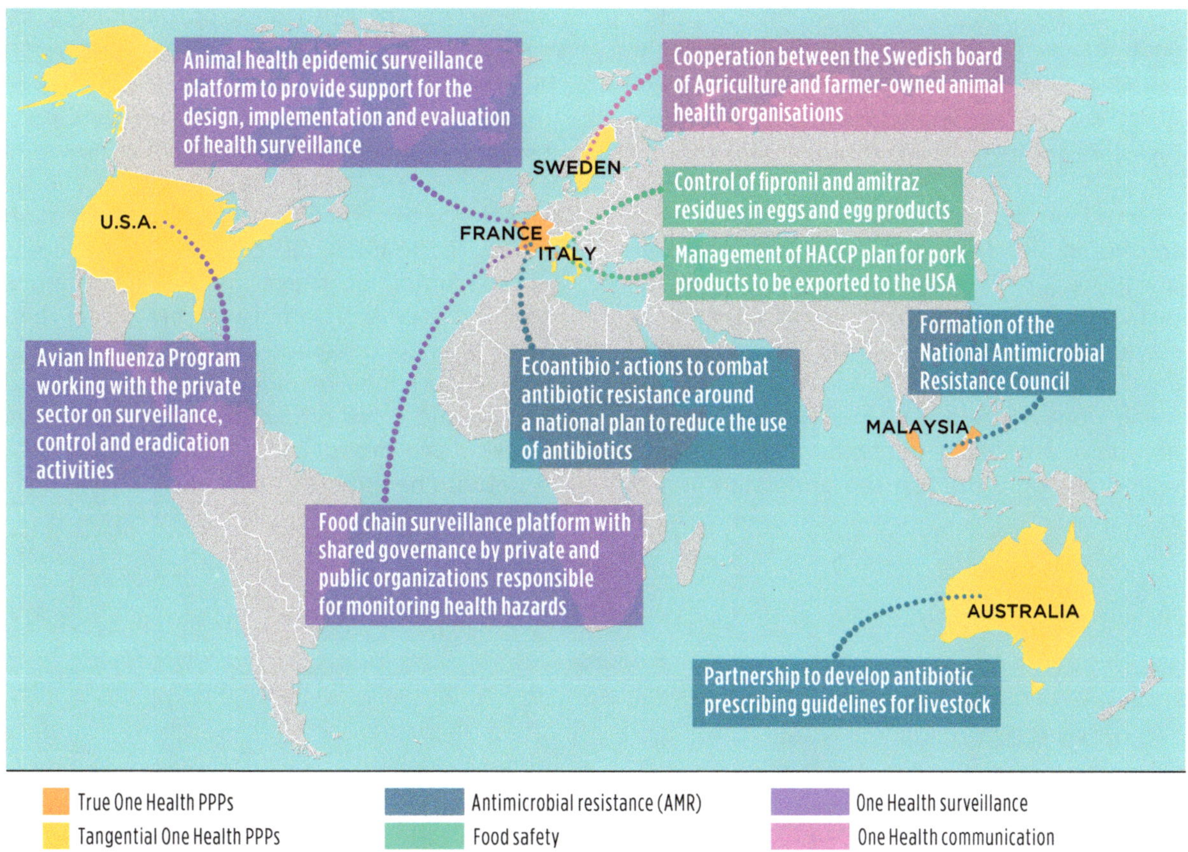

Figure 2. Subset of the WOAH PPP database showing countries implementing PPPs with One Health relevance (n = 9), categorized by type. True One Health PPPs are ones where both animal and human health agencies are involved in implementation; tangential PPPs are ones where One Health activities are implemented, but without evidence of intersectoral collaboration. HACCP: Hazard Analysis and Critical Control Point.

No One Health without a true integration of social sciences

Anne Conan, Séverine Thys, Sergio Guerrero Sanchez, Guillaume Fournié

The latest definition of One Health by the One Health High-Level Expert Panel (OHHLEP) emphasizes "society" as a key component, acknowledging that human behaviours significantly influence pathogen exposure, transmission and evolution (Figure 1). Despite this recognition, epidemiologists often employ social science methods as tools they are not trained for and without a solid grasp of the theoretical foundations. This gap is further amplified by a lack of appreciation for heterogeneity among social science fields and for the wide range of academic disciplines that are as varied as the interests of humankind activities and organizations (anthropology, economics, education, history, law, political science, psychology, sociology). Finally, when social science researchers are invited to One Health projects, their participation is included as an afterthought in the projects, relegated to the role of a "bolt-on" to the research process.

True interdisciplinarity at the social-biological scientific interface demands that research questions are framed by both epidemiologists and social scientists, going beyond the mere identification of human behaviours as risk factors. It requires an exploration of how these behaviours are shaped by social, economic and political contexts. This approach would not only deepen the understanding of disease ecology but also enable a shift from purely technical local interventions to society-based structural changes that target the root causes of pathogen transmission and persistence and other One Health issues. There are, however, multiple challenges in achieving such an integration, including the differing epistemological frameworks and the methodological approaches specific to each discipline, such as the nature of data of interest and the fundamental objectives of data analyses while following both qualitative or quantitative approaches (Figure 2).

Overcoming this divide requires mutual respect and an acknowledgement of the complementarity of both sets of disciplines by being engaged in reciprocal data-sharing and joint analysis. For instance, social scientists can investigate the drivers of risk behaviours identified by epidemiologists, providing nuanced perspectives on how these behaviours can be altered as well as how control interventions can be adapted to the local context. Subsequently, the outcomes of these behaviour changes must be studied from both biological and social science perspectives, thus fostering a more comprehensive understanding of disease ecology and more effective control strategies targeting the broader societal structures that perpetuate health risks.

References

Barnett T., Pfeiffer D.U., Hoque M.A., Giasuddin M., Flora M.S. *et al.* 2020. Practising co-production and interdisciplinarity: Challenges and implications for One Health research. *Preventive Veterinary Medicine*, 177, 104949. https://doi.org/10.1016/j.prevetmed.2020.104949

Brown H., Nading A.M. 2019. Introduction: Human animal health in medical anthropology. *Medical Anthropology Quarterly*, 33(1), 5–23. https://doi.org/10.1111/maq.12488

OHHLEP, Adisasmito W.B., Almuhairi S., Behravesh C.B., Bilivogui P. *et al.* 2022. One Health: A new definition for a sustainable and healthy future. *PLOS Pathogens*, 18(6), e1010537. https://doi.org/10.1371/journal.ppat.1010537

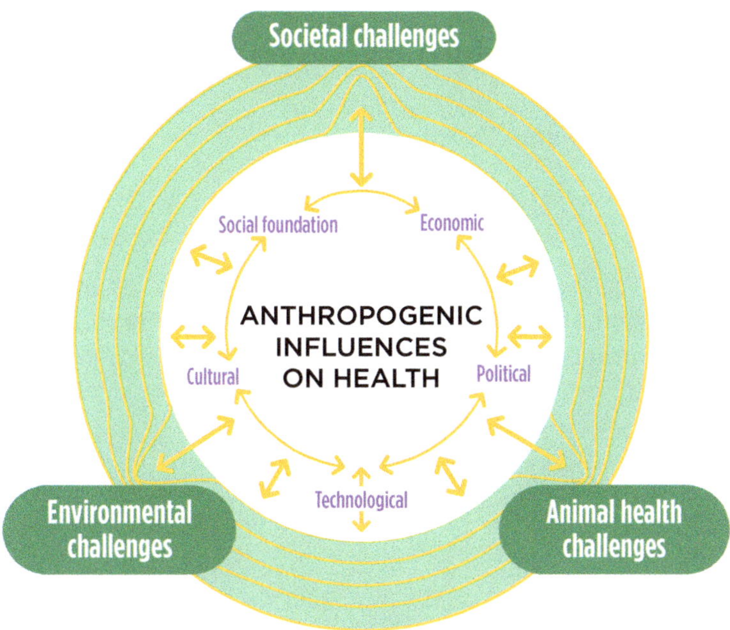

Figure 1. Problem statement identified by the One Health High-Level Expert Panel (OHHLEP) in their One Health Theory of Change. One Health Theory of Change: "Working toward a world better able to prevent, predict, detect, and respond to health threats and improve the health of humans, animals, plants, and the environment while contributing to sustainable development." Source: https://cdn.who.int/media/docs/default-source/one-health/ohhlep/ohhlep--one-health-theory-of-chance.pdf

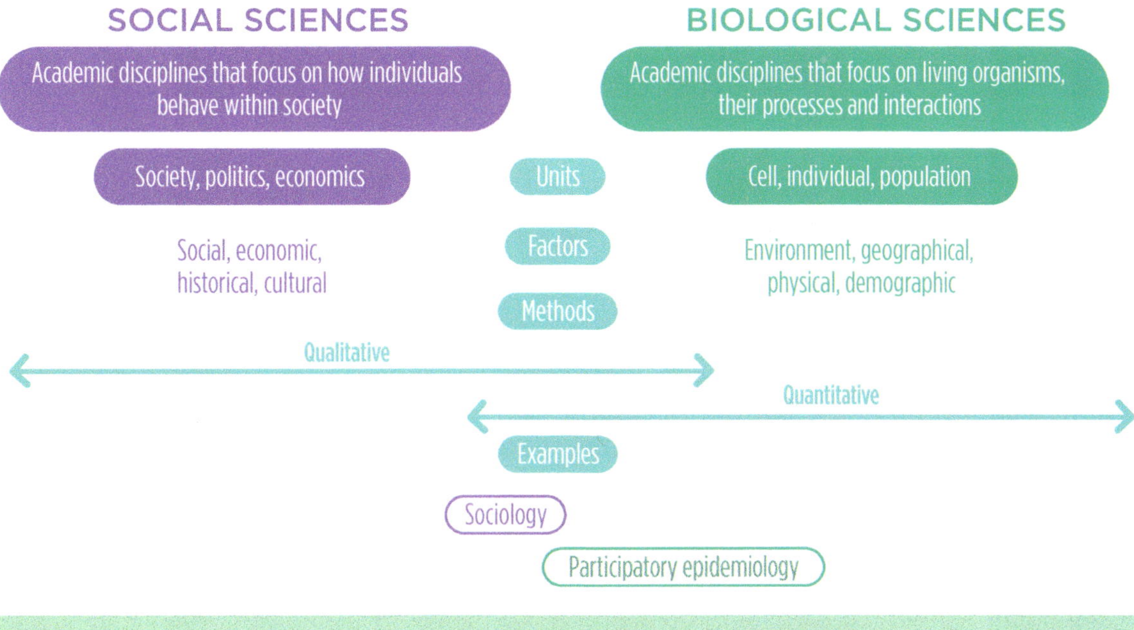

QUALITATIVE	QUANTITATIVE
Interpretation based Describe the reality as seen by individuals (inductive approach) No predefined sample strategies (e.g. purposive, convenience, snowball sampling) Interviews/focus group Discussion/participant observations Individuals as unique and complex cases Flexible, dynamic process	Significance based Test a hypothesis (deductive approach) Probabilistic sampling strategies (e.g. randomized, stratified clustered sampling) Closed questionnaires (including knowledge attitudes and practices)/biological samples Individuals as part of a population with cases that can be generalized Predefined study framework

Figure 2. Differences and similarities between the social sciences and the biological sciences.

How economics connects food, health and the environment

Alexis Delabouglise

Economics studies how humans make use of their resources to produce goods and services. Economic processes and production and consumption choices made by food-system actors determine the impact of animal infectious diseases on human health, welfare and the environment.

Livestock diseases undermine the production efficiency of goods and services, which increases the environmental impacts of livestock. The high mortality due to the most lethal diseases limits the availability of animals for sale and consumption, while other diseases undermine growth, milk and egg production as well as draft animal power (Ethiopia). Livestock diseases affect poor households' nutrition and finances, with negative impacts on health and access to education. Improved management of livestock diseases on family farms is expected to indirectly improve health and human capital in low-income countries (Kenya).

Domestic animal populations and mobility are largely driven by the demand for livestock products and the environmental conditions that determine feed availability. Consumer demand varies over time, and livestock movements tend to be concentrated at specific times of the year, exacerbating the spread of pathogens affecting livestock and humans (Senegal and Vietnam). Spontaneous responses of livestock producers to disease outbreaks are essentially aimed at reducing disease costs and include increased trade activities or antimicrobial usage, with implications for disease control and human health (South Vietnam, figure bottom right).

Economics contributes to understanding the ways human, animal and environmental health are interconnected. Examples of important future research areas include:

- Future adaptations of producers to an increasingly risky environment (climate change, sanitary risks). Such adaptations will affect livestock mobility and stocking density as well as the risk of disease emergence and antimicrobial resistance.
- Indirect beneficial effects of improved livestock health on human health and well-being (better nutrition, improved access to education) and the environment (resource use). These indirect effects depend on household responses to increased livestock productivity and must be further studied.
- Consumption of wildlife products. In many countries of sub-Saharan Africa and South-East Asia, households consume wildlife products as an alternative to livestock. Increasing wildlife consumption may threaten ecosystem preservation and raise the risk of disease emergence. We need to better understand to what extent households substitute wild and domestic animal products in their diets in response to economic or sanitary shocks.

References

Ciss M., Giacomini A., Diouf M.N., Delabouglise A., Mesdour A. *et al.* 2023. Description of the cattle and small ruminants trade network in Senegal and implication for the surveillance of animal diseases. *Transboundary and Emerging Diseases*, 2023, 1-13. https://doi.org/10.1155/2023/1880493

Delabouglise A., Choisy M., Phan T.D., Antoine-Moussiaux N., Peyre M. *et al.* 2017. Economic factors influencing zoonotic disease dynamics: demand for poultry meat and seasonal transmission of avian influenza in Vietnam. *Scientific Reports*, 7, 5905. https://doi.org/10.1038/s41598-017-06244-6

Delabouglise A., Thanh N.T.L., Xuyen H.T.A., Nguyen-Van-Yen B., Tuyet P.N. *et al.* 2020. Poultry farmer response to disease outbreaks in smallholder farming systems in southern Vietnam. *Elife*, 9.

Otiang E., Yoder J., Manian S., Campbell Z.A., Thumbi S.M. *et al.* 2022. Vaccination of household chickens results in a shift in young children's diet and improves child growth in rural Kenya. *Proceedings of the National Academy of Sciences*, 119(24), e2122389119. https://doi.org/10.1073/pnas.2122389119

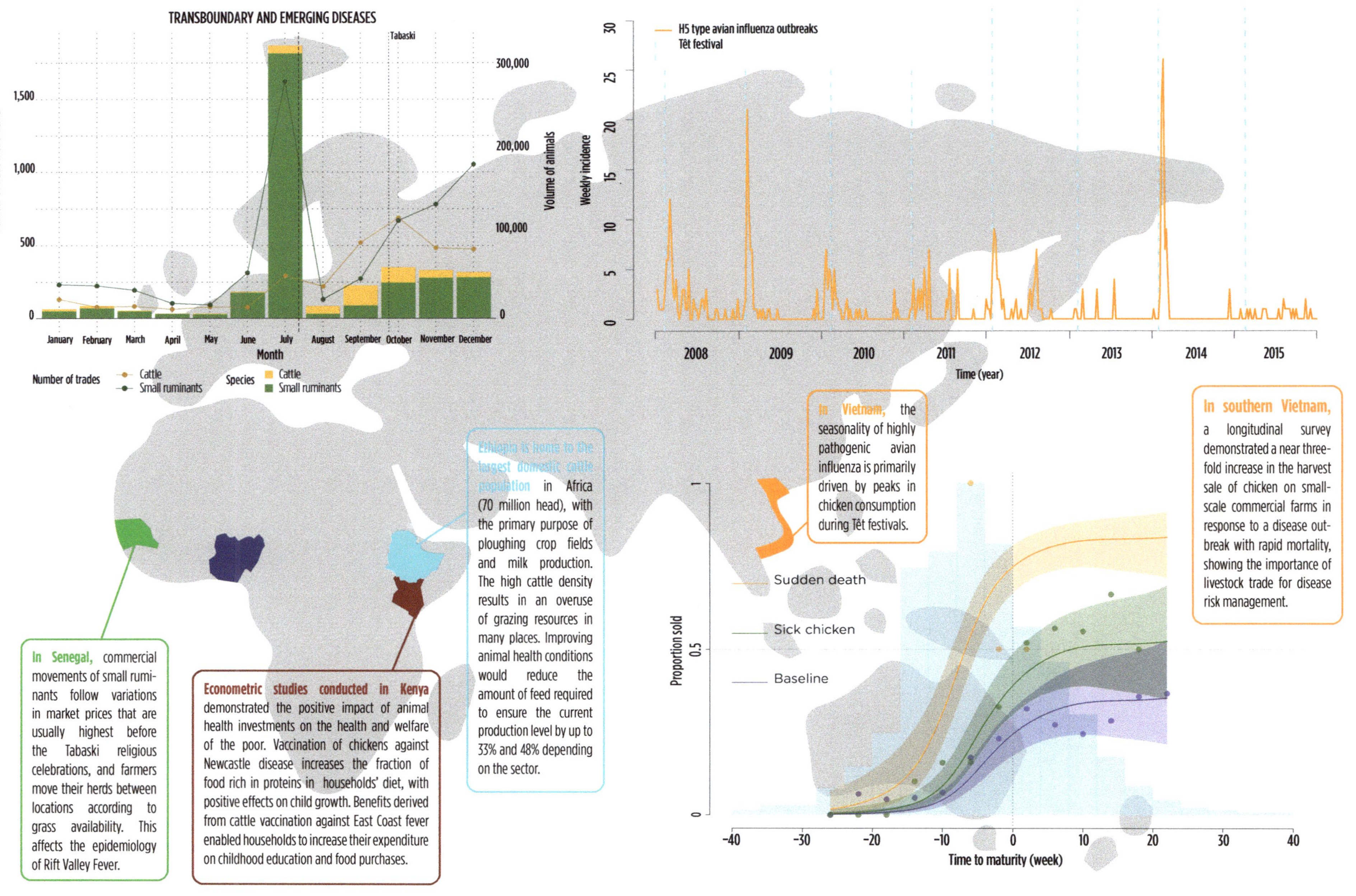

TRANSBOUNDARY AND EMERGING DISEASES
Number of trade movements
1,500
1,000
500
0
Tabaski
300,000
200,000
100,000
0
Volume of animals
January February March April May June July August September October November December
Month
Number of trades
Cattle
Small ruminants
Species
Cattle
Small ruminants
Weekly incidence
30 25 20 15 10 5 0
H5 type avian influenza outbreaks
Têt festival
2008 2009 2010 2011 2012 2013 2014 2015
Time (year)
In Senegal, commercial movements of small ruminants follow variations in market prices that are usually highest before the Tabaski religious celebrations, and farmers move their herds between locations according to grass availability. This affects the epidemiology of Rift Valley Fever.
Econometric studies conducted in Kenya demonstrated the positive impact of animal health investments on the health and welfare of the poor. Vaccination of chickens against Newcastle disease increases the fraction of food rich in proteins in households' diet, with positive effects on child growth. Benefits derived from cattle vaccination against East Coast fever enabled households to increase their expenditure on childhood education and food purchases.
Ethiopia is home to the largest domestic cattle population in Africa (70 million head), with the primary purpose of ploughing crop fields and milk production. The high cattle density results in an overuse of grazing resources in many places. Improving animal health conditions would reduce the amount of feed required to ensure the current production level by up to 33% and 48% depending on the sector.
In Vietnam, the seasonality of highly pathogenic avian influenza is primarily driven by peaks in chicken consumption during Têt festivals.
In southern Vietnam, a longitudinal survey demonstrated a near three-fold increase in the harvest sale of chicken on small-scale commercial farms in response to a disease outbreak with rapid mortality, showing the importance of livestock trade for disease risk management.
Proportion sold
1
0.5
0
Sudden death
Sick chicken
Baseline
-40 -30 -20 -10 0 10 20 30 40
Time to maturity (week)

Participatory approaches and One Health: the example of community-based surveillance systems

Maxime Tesch, Marion Bordier, Alexis Delabouglise, Marie-Marie Olive

Participatory approaches are now being developed in many fields, including for the prevention and control of infectious diseases. These approaches, mostly applied to research action, promote a bottom-up process that allows stakeholders—especially those who are usually excluded from decision-making—to participate in knowledge production, problem definition and solutions development. Stakeholder participation varies in terms of intensity, as illustrated by the ladder of citizen participation developed by Arnstein (Figure 1).

Participatory epidemiology, the application of these inclusive approaches to epidemiology, is a discipline that was developed in the 2000s, to conduct surveillance of the last cases of rinderpest in the world and to finally eradicate it. Participatory epidemiology has since expanded to include an integrated vision of health, mainly by enabling exchanges between organizational silos, in line with the One Health approach. These methods have also made it possible to involve representatives from all levels, in particular non-institutional stakeholders such as communities, in the design and implementation of surveillance activities. For instance, a number of community-based surveillance systems (CBSs) have been developed thanks to these approaches (Figure 2). Local stakeholders at the frontline of emergence, such as hunters, livestock keepers and community health workers, are involved in the design and operationalization of these systems.

In Gabon, as part of the Sustainable Wildlife Management (SWM) programme, hunter communities contributed to a CBS of zoonotic diseases at the interface between wild and domestic animals and humans.[1] They were involved in three ways: (1) first, an elicitation process was implemented to identify signs of diseases in wildlife observable by hunters and local stakeholders who could act as liaisons between the local and central level, (2) representatives of hunter communities participated in designing the surveillance system, and (3) hunters and other relevant community members were trained in disease surveillance implementation in wildlife, livestock and human, enabling a timely notification of early signs of the presence of zoonotic diseases. This example highlights the central role of participatory approaches for creating CBSs tailored to the context. Local stakeholders create their own tools to detect zoonotic diseases and thus limit the risk of epidemic transmission within their community. Gabon's Ministry of Health has also expressed an interest in strengthening and scaling up this approach and community event-based early warning systems. There is still room for progress in the way researchers mobilize participation in epidemiology and surveillance, to involve at an earlier stage stakeholders in the development of the problem statements.

1. www.cirad.fr/en/cirad-news/news/2023/community-based-surveillance-to-combat-pandemics

References

Arnstein S.R. 1969. A ladder of citizen participation. *Journal of the American Institute of Planners*, 35(4), 216–224. https://doi.org/10.1080/01944366908977225

Catley A., Alders R.G., Wood J.L. 2012. Participatory epidemiology: Approaches, methods, experiences. *The Veterinary Journal*, 191(2), 151–160. https://doi.org/10.1016/j.tvjl.2011.03.010

Mariner J.C., House J.A., Mebus C.A., Sollod A.E., Chibeu D. *et al.* 2012. Rinderpest eradication: Appropriate technology and social innovations. *Science*, 337(6100), 1309–1312. https://doi.org/10.1126/science.1223805

McNeil C., Verlander S., Divi N., Smolinski M. 2022. The landscape of participatory surveillance systems across the One Health spectrum: Systematic review. *JMIR Public Health and Surveillance*, 8(8), e38551. https://doi.org/10.2196/38551

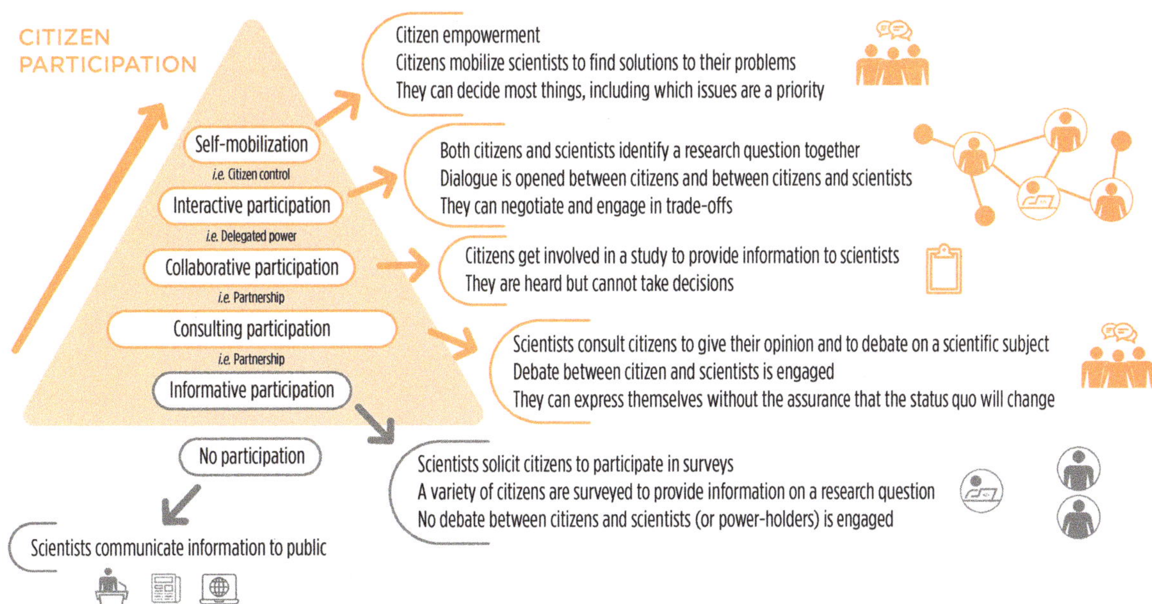

Figure 1. The pyramid of participation adapted from Arnstein's ladder of citizen and policy-maker participation.

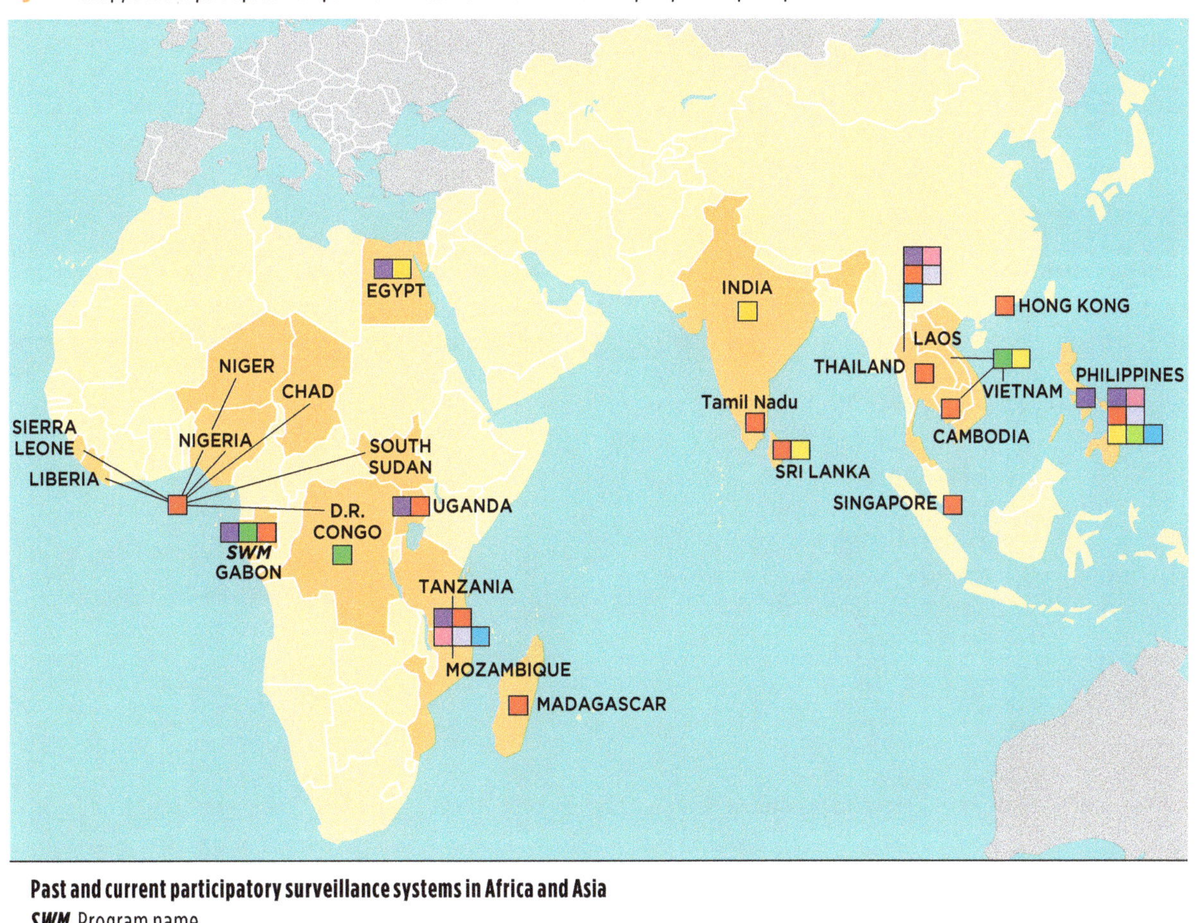

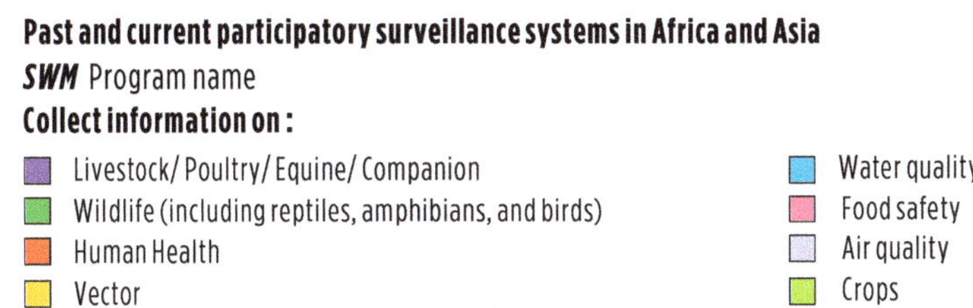

Figure 2. Community-based surveillance system and a level of One Health integration. Adapted from the Ending Pandemics database: https://endingpandemics.org/participatory-surveillance-programs-map/.

Bridging gaps: gender equity as the neglected link in One Health implementation

Dien Nguyen Thi,
Séverine Thys

Gender, referred to as the social construction of norms, roles and responsibilities associated with perceived differences between the biological sexes, shapes the behaviours, practices and related health risks of the women, men and gender-diverse people in a given society. Diseases and disasters threatening human, animal and environmental health are all gender-based risks due to the different tasks, priorities, resources and decisions taken. The One Health approach must take gender into consideration to design effective, sustainable interventions in human, animal and environmental health tailored to the specific socioecological and cultural context and to ensure benefits are equitably shared among stakeholders.

The anthropocentric vision of health that neglects socioecological determinants and the power dynamics that limit the participation of marginalized groups (particularly women) impede the integration of the principles of diversity, equity and inclusion into One Health. The unequal representation of women in leadership, decision-making and education in the One Health arena create gaps in efforts to mainstream gender-related concepts in human, animal and environment health and in the interdisciplinary One Health debate. More generally, gender inequities remain a neglected aspect of the One Health agenda and activities.

Several initiatives seek to address growing demand from policies, programmes and funding agencies for appropriate frameworks and systematic tools and methodologies to support gender mainstreaming in One Health (Figure 1). These initiatives include the Women for One Health Network[1], the Gender Working Group of the Network for Ecohealth and One Health[2] and a framework to identify key gender considerations in One Health research for development (Figure 2), with a focus on low- and middle-income countries.

Bringing a gender-sensitive approach to One Health requires assessing gender needs; recognizing the reality of gender negotiations in different socioeconomical and cultural contexts over time; analysing gender inequalities, roles, norms and power relations at the human–animal–environment interface; and addressing the deep drivers of the interconnected crises.

1. https://wfoh.org/
2. https://www.ecohealthinternational.org/regional-chapters/europe/

References

Cataldo C., Bellenghi M., Masella R., Busani L. 2023. One Health challenges and actions: Integration of gender considerations to reduce risks at the human-animal-environmental interface. *One Health*, 16, 100530. https://doi.org/10.1016/j.onehlt.2023.100530

Friedson-Ridenour S., Dutcher T.V., Calderon C., DiPrete Brown L., Olsen C.W. 2019. Gender analysis for One Health: Theoretical perspectives and recommendations for practice. *EcoHealth*, 16, 306–316. https://doi.org/10.1007/s10393-019-01410-w

Garnier J., Savic S., Boriani E., Bagnol B., Häsler B., Kock R. 2020. Helping to heal nature and ourselves through human-rights-based and genderresponsive One Health. *One Health Outlook*, 2, 22. https://doi.org/10.1186/s42522-020-00029-0

Léchenne M., Cediel-Becerra N., Cailleau A., Greter H., Yawe A. *et al.* 2024. Toward social and ecological equity: A feminist lens on One Health. *CABI One Health*, 3, 1. https://doi.org/10.1079/cabionehealth.2024.0002

Robbiati C., Miarisoa Andriamandroso A., Auerswald H., Cediel Becerra N., Dente M.G. *et al.* 2024. Diversity, equity and inclusion in One Health could crucially support prevention of health threats, but a change in mind-set is needed. *Cambridge Open Engage*. https://doi.org/10.33774/coe-2024-4zf5g-v3

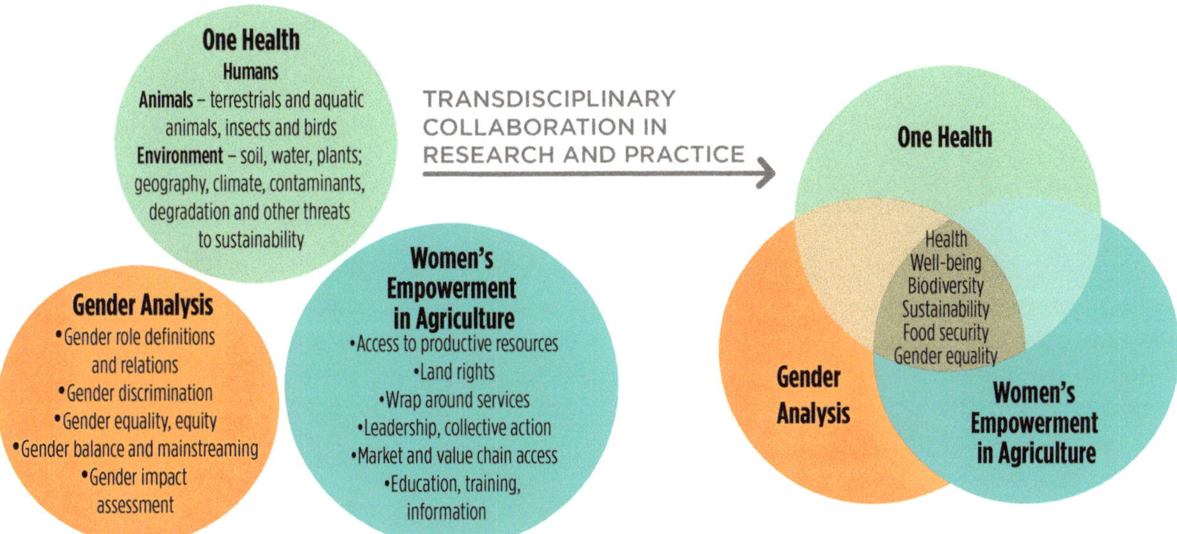

Figure 1. Toward a framework for gender and one Health. From Friedson-Ridenour *et al.* 2019.

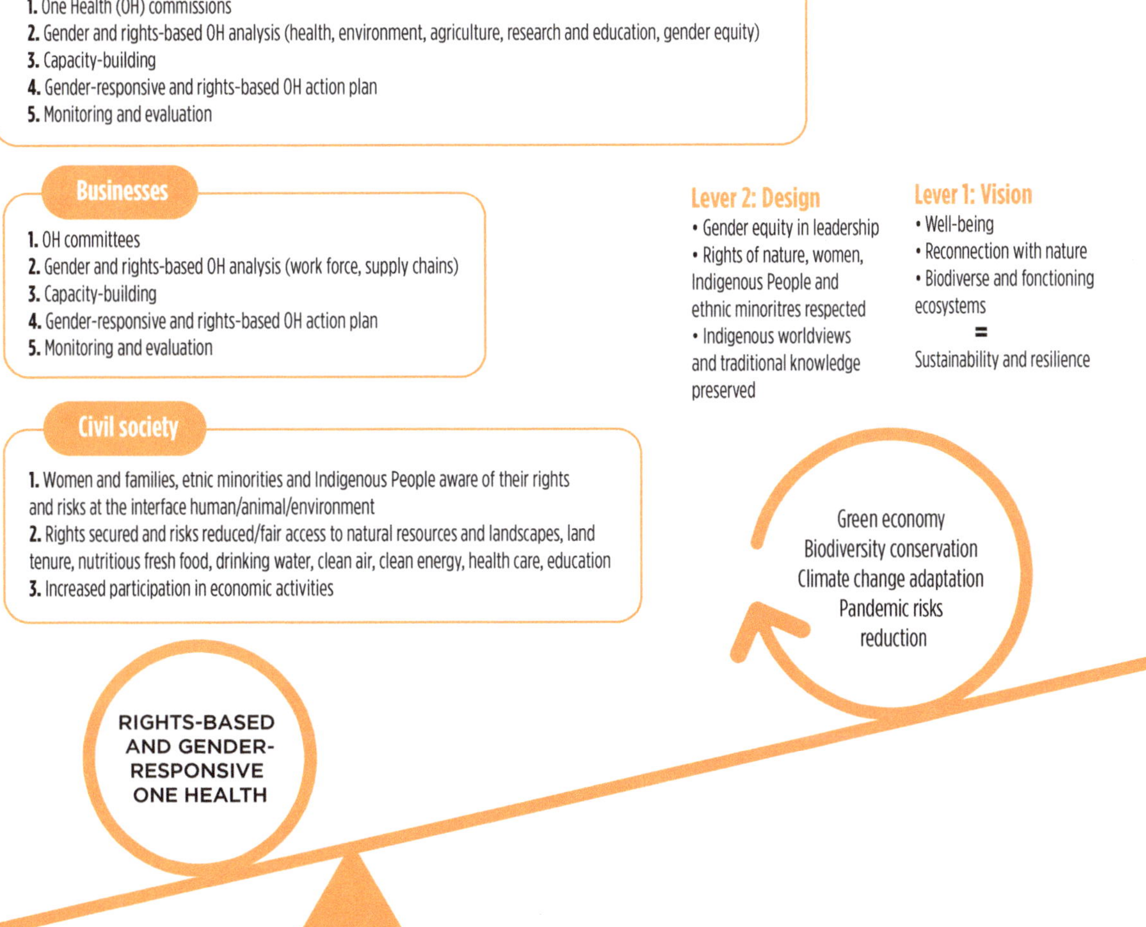

Figure 2. A framework for mainstreaming gender-responsive and rights-based One Health to deliver improved well-being for all and healing of nature. From Garnier *et al.* 2020.

One Welfare: integrating animal welfare into One Health

Jean-Luc Angot

The notion of animal welfare encompasses not only the health and physical well-being of an animal but also its psychological well-being and the possibility of expressing behaviours specific to its species. It is intrinsic to human-animal relationships. Animal welfare is based on "Five Freedoms" (Figure 1), originally drawn up by the UK Farm Animal Welfare Council (FAWC). The World Organisation for Animal Health (WOAH) has been developing international standards grounded in science since 2005 (Figure 2).

Animal welfare is completely in line with the agroecological transition and the United Nations Sustainable Development Goals. It has multiple dimensions—scientific, philosophical, ethical, cultural, sociological, religious, political and economic—and therefore involves several disciplines. Animal welfare should not be considered a constraint but rather a way to make livestock farming better and support sustainable development. It concerns livestock farming conditions and production systems, transport and slaughter as well as international trade.

Animal welfare is inseparable from the notion of human well-being—for livestock farmers, veterinarians and their assistants, animal transporters, slaughterhouse employees and anyone in contact with animals. The well-being of humans in relation to animals is inherently reciprocal; the improvement of one depends on the improvement of the other.

Treating animals well is a factor in human well-being. More generally, respecting the environment and living in harmony with nature is a source of well-being. The well-being of humans and other animal species depends on biodiversity and the environment in which they exist—in other words, the holistic One Welfare concept.

According to the World Health Organization (WHO), well-being is a component of health. Furthermore, the definition of One Health adopted by the WHO, WOAH, Food and Agriculture Organization (FAO) and United Nations Environment Programme (UNEP) refers specifically to well-being ("The One Health approach mobilizes multiple sectors, disciplines and communities at varying levels of society to work together to foster well-being."). One Welfare is therefore an integral part of One Health (Figure 3), especially since both concepts require a multidisciplinary and intersectoral approach and permanent dialogue with stakeholders to support the co-construction of actions (Figure 4).

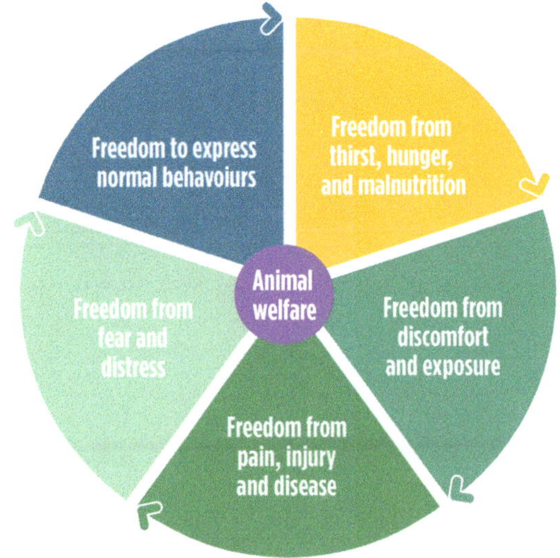

Figure 1. Animal welfare: the five freedoms (FAWC, 1965).

References

WHOA standards (Terrestrial and Aquatic Animal Health Code) since 2004: https://www.woah.org/en/what-we-do/animal-health-and-welfare/animal-welfare/development-of-animal-welfare-standards/

ANSES: https://www.anses.fr/fr/thematique/sant%C3%A9-et-bien-%C3%AAtre-des-animaux and https://www.cnr-bea.fr/2024/02/12/animal-welfare-basis-one-health-un-convention-solution-human/

Physical and mental state of an animal in relation to the conditions in wich it lives and dies.

WOAH Terrestrial and Aquatic Animal Health Code

The positive mental and physical state linked to the satisfaction of [an animal's] physiological and behaviourial needs as well as its expectations. This state varies depending on the animal's perception of the situation.
ANSES, 2018

Figure 2. Definitions of animal welfare.

Figure 3. One Health, One Welfare. From WOAH 2024.

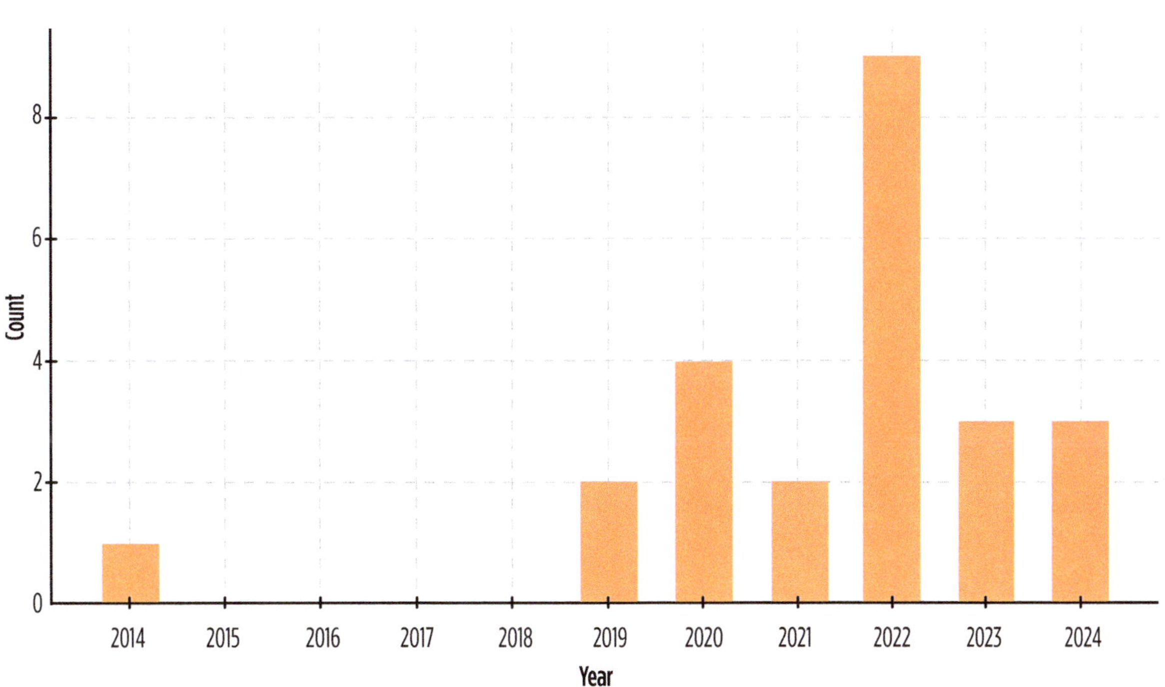

Figure 4. Number of publications recorded in PubMed related to "One Welfare" and "One Health" from 2014 to 2024.

Dogs: key players in the One Health approach

François Roger

Because of their close relationships with humans and ability to adapt to different environments (Figures 1 and 2), dogs hold a critical place within the One Health approach. They can have both positive and negative impacts, either directly or indirectly, on human and environmental health (Figure 3).

Dogs can be carriers of zoonotic diseases, infecting humans with diseases such as rabies through a bite. Systematic vaccination of dogs is a key part of One Health strategies aimed at keeping both human and animal populations safe from these zoonotic diseases. Dogs can also be used as "sensors" to provide early warnings of disease. Their excellent sense of smell can detect many human diseases (including different cancers, COVID-19 and bacterial infections) well before symptoms manifest. This ability makes them important public health collaborators and could support efforts for earlier interventions that could help lower the risk of infectious agents. One major contribution that dogs make to human health goes beyond the field of disease diagnosis: they are also a significant factor in improving people's mental health. Dogs are known to decrease stress, anxiety and depression simply through their companionship. Because of these positive effects, they are widely used in animal-assisted therapy programmes to support patients in overcoming psychological disorders and trauma.

Dogs play an important role in rural ecosystem management by protecting livestock from predators and facilitating the coexistence of wildlife and human activities. However, their predatory behaviour must be carefully managed to avoid disrupting local wildlife populations. Additionally, the global pet food industry has a broader negative impact on the environment and the ecosystems through significant resource consumption and greenhouse gas emissions.

Dogs are also a core component of One Health research and education. Research into canine behaviour, the ways dogs interact with their environment and human-canine relationships provides vital lessons that can be used to promote integrated public health policy regarding interdependence between humans, animals and the environment.

References

Headey B., Grabka M.M. 2007. Pets and human health in Germany and Australia: national longitudinal results. *Social Indicators Research*, 80(2), 297-311. https://doi.org/10.1007/s11205-005-5072-z

Sykes N., Beirne P., Horowitz A., Jones I., Kalof L. *et al.* 2020. Humanity's best friend: a dog-centric approach to addressing global challenges. *Animals*, 10(3), 502. https://doi.org/10.3390/ani10030502

Jendrny P., Twele F., Meller S., Osterhaus A.D.M.E., Schalke E., Volk H.A. 2021. Canine olfactory detection and its relevance to medical detection. *BMC Infectious Diseases*, 21, 838. https://doi.org/10.1186/s12879-021-06523-8

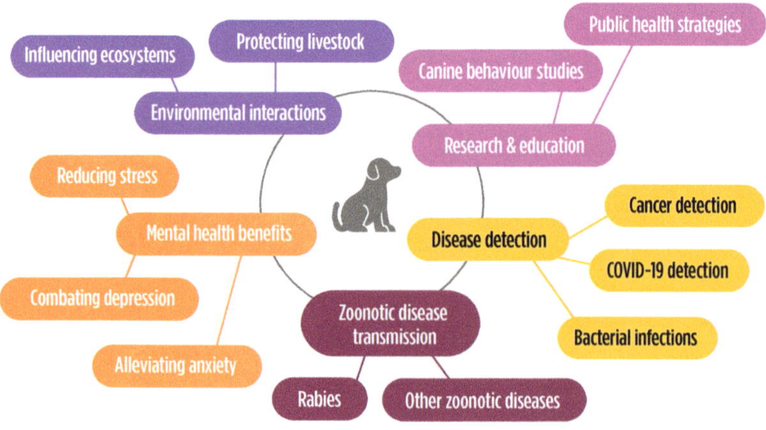

Figure 3. The beneficial and detrimental roles of dogs in the One Health framework.

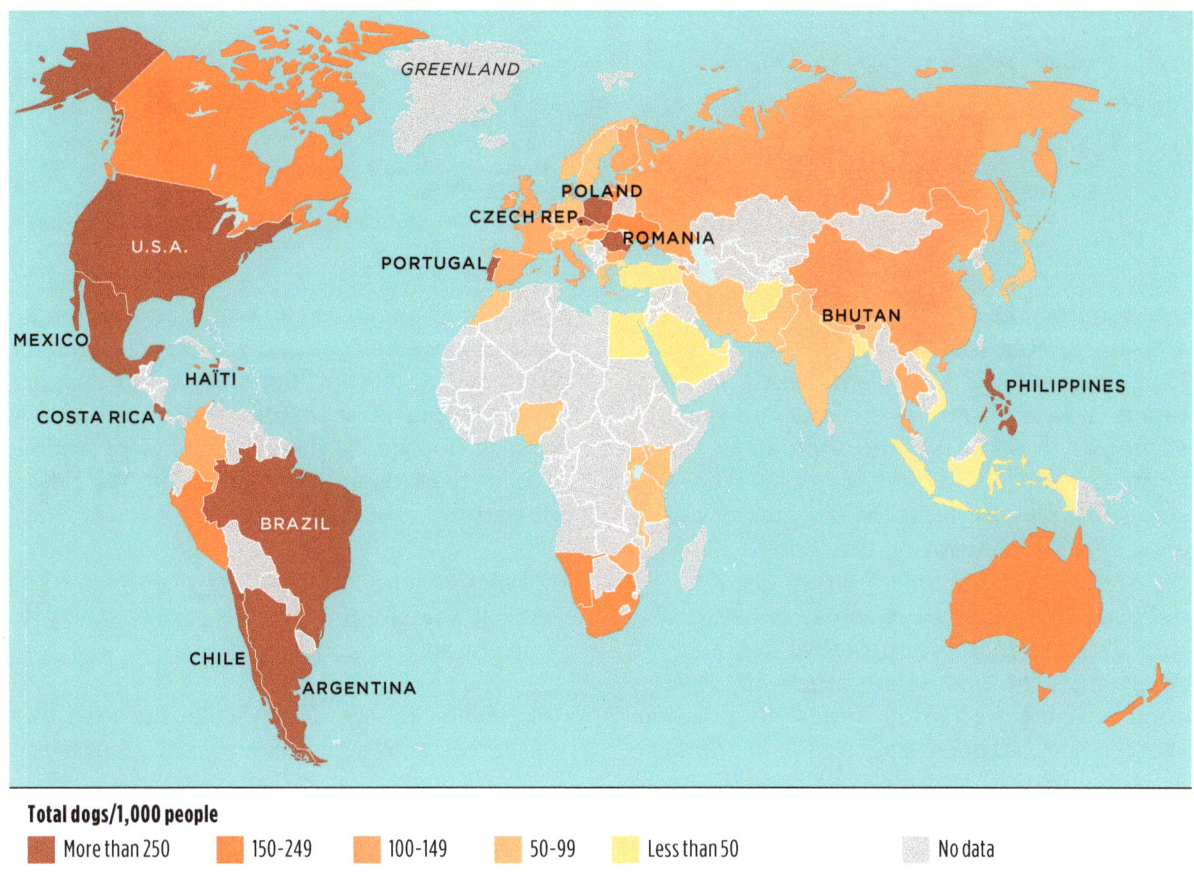

Figure 1. Approximate global per-capita dog population distribution. Source: https://academic.oup.com/icb/article/61/1/154/6262639

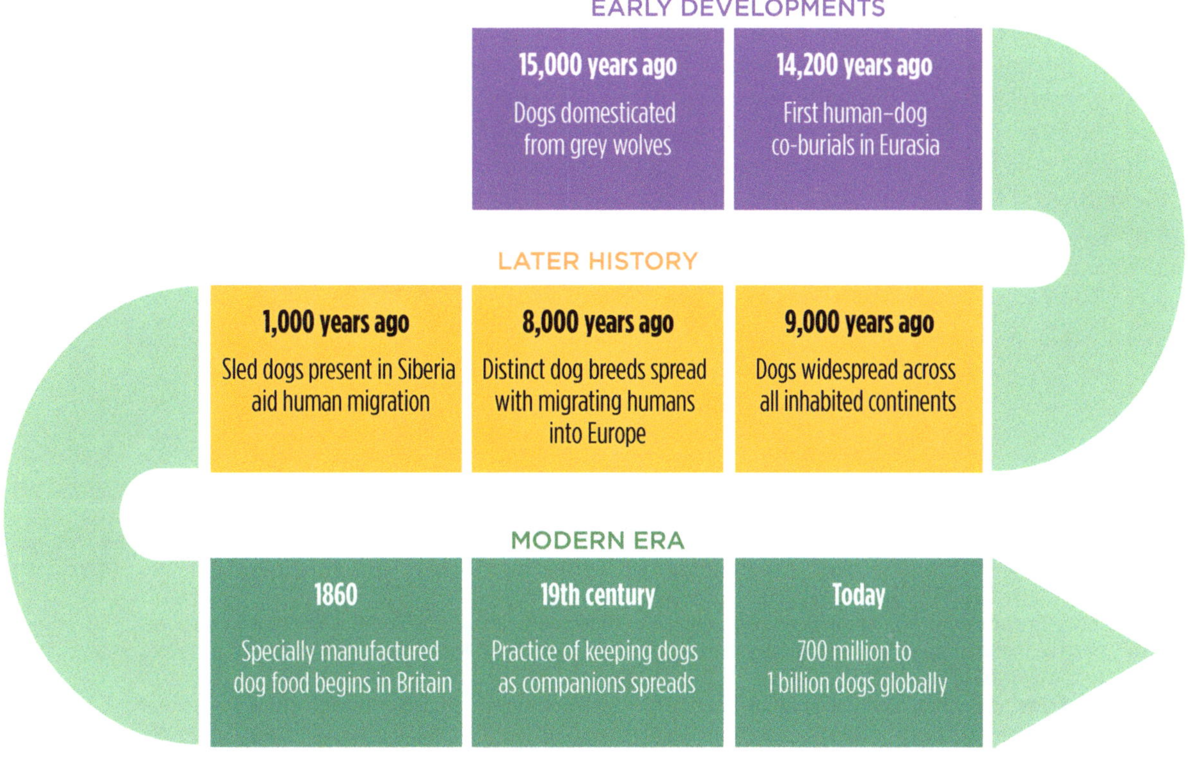

Figure 2. Brief history of dogs and their relationship with humans (based on Sykes *et al.* 2020).

Building a common language for antimicrobial stewardship in One Health

Rebecca Hibbard, Céline Faverjon, Mathilde Paul

Antimicrobial resistance (AMR) is a growing global challenge with implications for human, animal and plant health, necessitating a One Health approach. International strategies to address AMR often advocate an approach including antimicrobial stewardship (AMS). However, there is a lack of clarity around what is meant by the term "antimicrobial stewardship", in no small part because it is defined in different ways across different sectors. This absence of a universal definition of AMS across the One Health spectrum impedes communication on AMS and AMR between sectors and between scientists and the public more broadly, and makes it challenging to identify actions to improve AMS at an operational (field) level.

To develop an inclusive definition of AMS with relevance for the human and animal health sectors, we made use of boundary object theory, an approach from science and technology studies that has been used to theorize methods of cross-sectoral and transdisciplinary collaboration. Boundary objects are concepts, ideas or theories that are weakly structured in general use but strongly structured in specific contexts. AMS can be conceptualized as a boundary object because different definitions of AMS may appear vague when viewed collectively; however, operational definitions used by specific communities within the human and animal health sectors can be functional, meaningful and well-structured for that community. By reviewing existing definitions and descriptions of AMS in an exploratory and iterative process (Figure 1), three elements of antimicrobial stewardship common across the different sectors were identified:

- Collective and temporal responsibility: the notion that AMS implies a responsibility to current and future populations.
- Flexibility in scope and scale: the idea that all actors at all scales can contribute to AMS.
- Contextual contingency: the acknowledgement that what is considered "good" AMS will depend on the context, and implies taking the most appropriate course of action for a given set of circumstances.

These elements were used to construct a definition with relevance across different sectors, designed to function as an intersectoral communicative tool (Figure 2). The proposed definition has the potential to facilitate intersectoral communication and cooperation by providing a coherent explanation of AMS relevant to different actors implicated in One Health AMS interventions and by encouraging more explicit consideration of what AMS means in a One Health context.

References

Dyar O.J., Huttner B., Schouten J., Pulcini C., ESGAP (ESCMID Study Group for Antimicrobial Stewardship). 2017. What is antimicrobial stewardship? *Clinical Microbiology and Infection*, 23(11), 793–798. https://doi.org/10.1016/j.cmi.2017.08.026

Hibbard R., Mendelson M., Page S.W., Ferreira J.P., Pulcini C. *et al.* 2024. Antimicrobial stewardship: a definition with a One Health perspective. *npj Antimicrobials and Resistance*, 2, 1–9. https://doi.org/10.1038/s44259-024-00031-w

Star S.L., Griesemer J.R. 1989. Institutional ecology, "translations" and boundary objects: Amateurs and professionals in Berkeley's Museum of Vertebrate Zoology, 1907–39. *Social Studies of Science*, 19, 387–420.

World Health Organization. 2015. *Global action plan on antimicrobial resistance.* World Health Organization, Geneva. https://www.who.int/publications/i/item/9789241509763

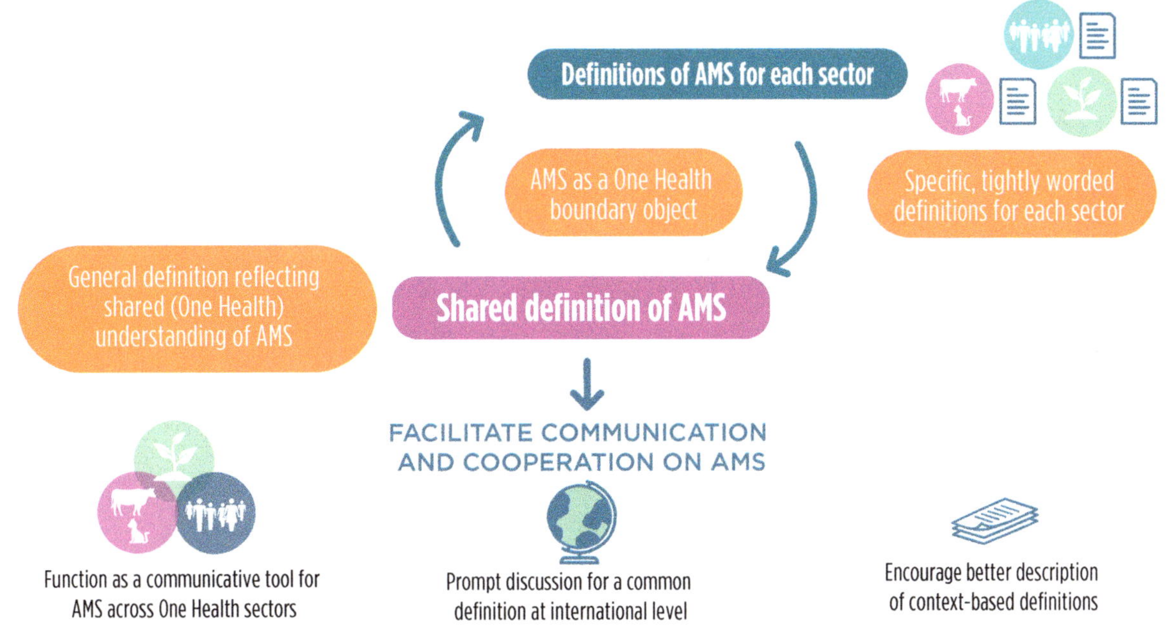

Figure 1. Antimicrobial stewardship (AMS) is depicted as a One Health boundary object, integrating sector-specific definitions (human, animal, and environmental health) into a shared understanding. The bidirectional arrows represent dynamic interactions, enabling mutual learning and adaptation across sectors.

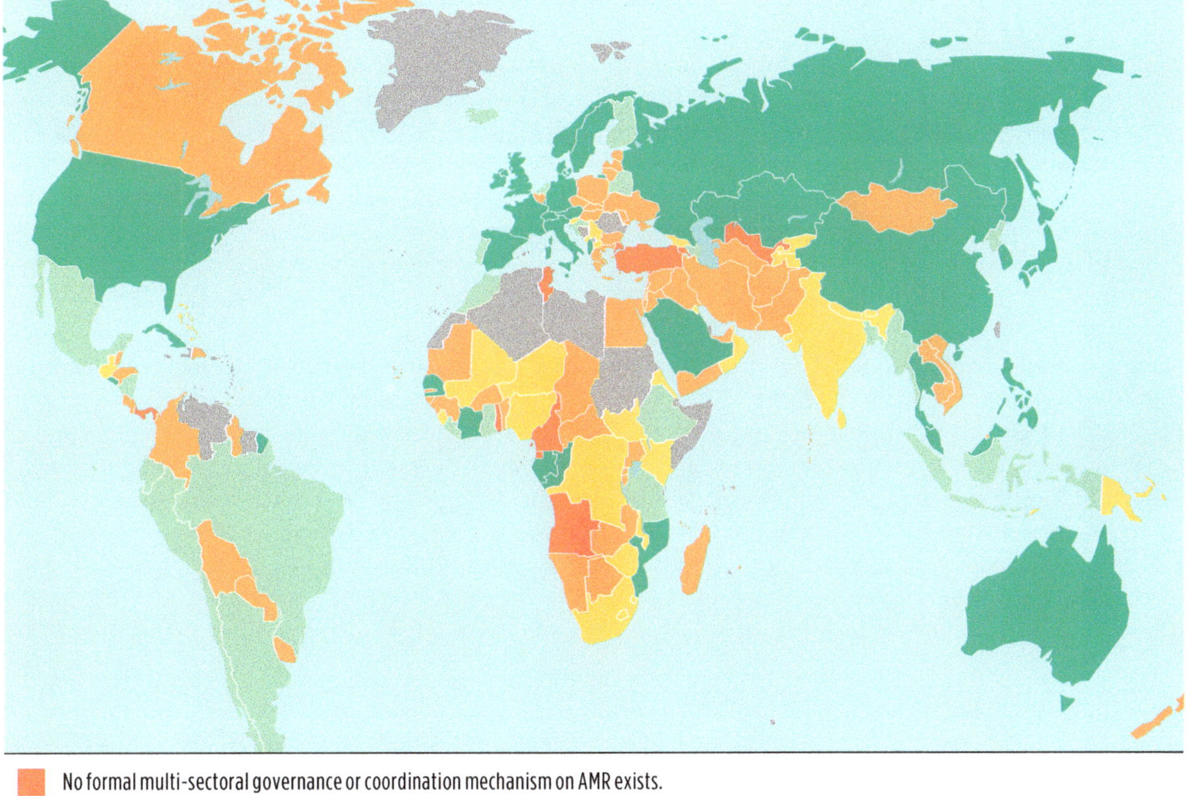

Figure 2. Multi-sector and One Health collaboration/coordination. From TrACSS Database: https://amrcountryprogress.org/#/map-view

One Health at scale: social-ecological system health

Alexandre Caron,
Michel de Garine-Wichatitsky

One Health was initially adopted by major health agencies to promote interdisciplinary collaborations among biomedical scholars and practitioners, and then progressively expanded to those working in the environmental and social sciences, with the aim of establishing a more society-wide responsibility for the health of humans and the whole planetary ecosystem.

This broader concept of One Health engages with the complexity of ecosystems. In the Anthropocene, ecosystems are better framed as social-ecological systems (SES) because the impacts of human activities threaten the sustainability of most ecological functions and services. This destabilization of SES functions negatively impacts the health and well-being of both humans and non-humans.

Thus, the One Health approach to thinking and planning must be embedded in the SES concept to better integrate the multiple dimensions of health risks (e.g. social, economic, genetic). The preparedness and control against sanitary crises such as the COVID-19 pandemic are currently the core target of the One Health approach. However, scaling One Health to SES requires tackling the root causes of these crises that are often associated with environmental and social injustices, driving biodiversity overexploitation and species extinction. This scaling also questions the current dominant human–nature relationship and promotes alternative worldviews that could influence how we protect biodiversity and its functionality and consider progress for human societies.

The SES health framework explicitly defines how health components related to humans/animals/plants/ecosystems interact within SES ("health in") to determine the resilience of the SES components ("health of" the SES). Health in SES encompasses the issues related to human, plant and animal health, including their environmental drivers as defined under the One Health approach, whereas the health of SES is based on the concept of resilience as a property of complex adaptive systems (Figure 1). The multiple dimensions of "health in" depend on various context-dependent parameters (e.g. ecological, social) that must be defined through a transdisciplinary process. Using participatory approaches, SES health places local stakeholders and their knowledge systems at the centre of governance systems for the sustainable use of natural resources and the management of risks (e.g. health risks). SES health promotes healthy and resilient agricultural and natural ecosystems that may sustainably deliver well-being for humans and non-humans (Figure 2).

References

de Garine-Wichatitsky M., Binot A., Ward J., Caron A., Perrotton A., Ross H. *et al.* 2021. "Health in" and "Health of" social-ecological systems: a practical framework for the management of healthy and resilient agricultural and natural ecosystems. *Frontiers in Public Health*, 8, 616328. https://doi.org/10.3389/fpubh.2020.616328

Caron A., Mugabe P., Bourgeois R., Delay E., Bitu F. *et al.* 2022. Social-ecological system health in transfrontier conservation areas to promote the coexistence between people and nature. *One Health Cases*. https://doi.org/10.1079/onehealthcases.2022.0005

de Garine-Wichatitsky M., Binot A., Morand S., Kock R., Roger F. *et al.* 2020. Will the COVID-19 crisis trigger a One Health coming-of-age? *Lancet Planet Health*, 4(9), e377-e378. https://doi.org/10.1016/S2542-5196(20)30179-0

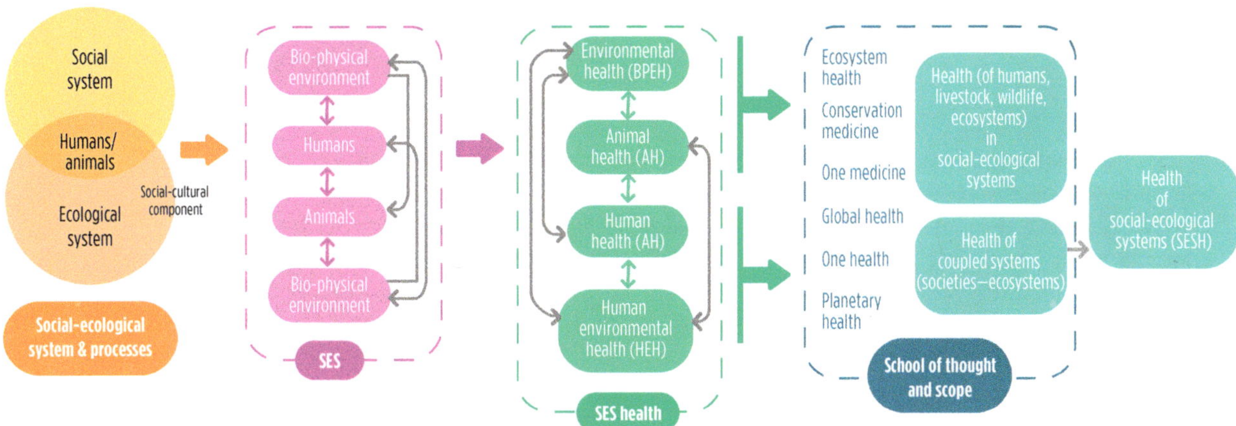

Figure 1. Conceptual model connecting a social-ecological system (SES) template to the health of its components and the health of the whole social-ecological system (left and central parts); main integrated approaches associated with the specific health components and their interactions (right part). From de Garine-Wichatitsky *et al.* 2021.

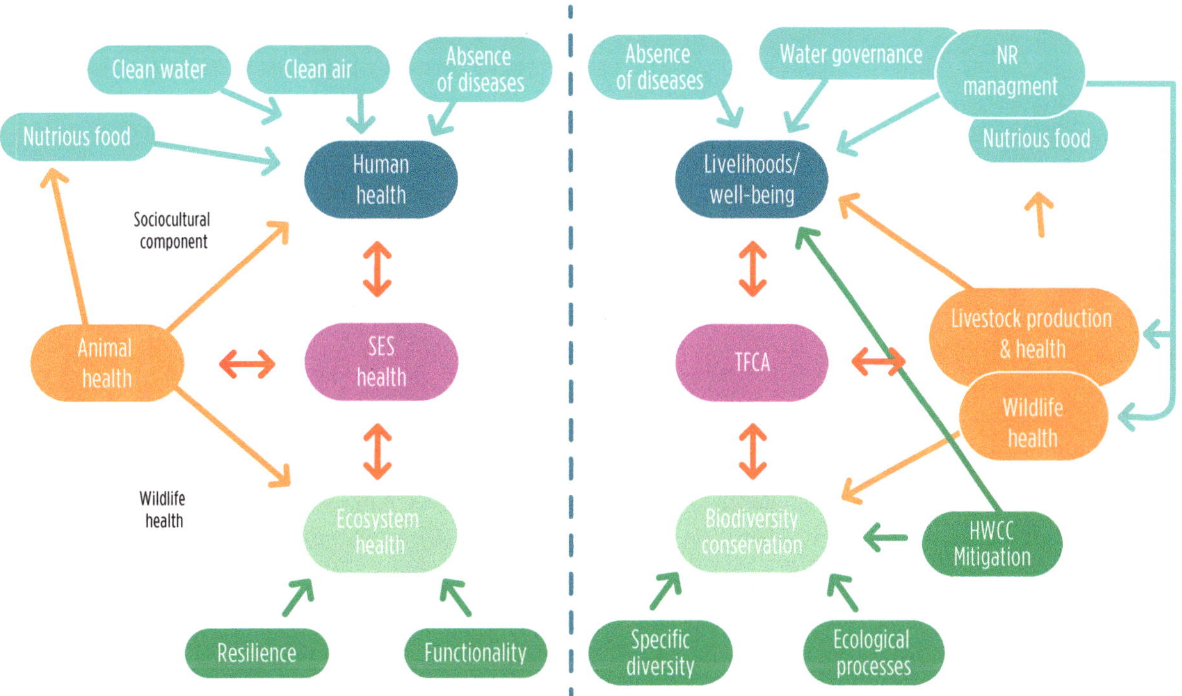

Figure 2. The social-ecological system health concept on the left-hand side translated into the context of transfrontier conservation areas (TFCAs) in Southern Africa on the right. Human health encompasses the absence of disease and mental health and well-being. NR: natural resources; HWCC: human–wildlife conflicts and crimes. From Caron *et al.* 2022.

From bush to fork: managing wild meat value chains to safeguard health and biodiversity

Manon Mispiratceguy,
Sandra Ratiarison

Wild animals are a crucial part of diets, income and culture for millions of people worldwide. For rural communities, wild meat is often the most accessible and affordable protein source, while for wealthier urban dwellers, it can be a luxury good or status symbol. The rising demand for wild meat in urban and other populated areas fuels overhunting in source areas, with profound impacts on both human and ecosystem health. Unsustainable hunting contributes to biodiversity loss, weakening ecosystem functioning and services, while frequent human-wildlife interactions enhance the risk of zoonotic disease emergence and spillover. Wildlife trade and habitat degradation further exacerbate these risks by increasing the frequency and intensity of these interactions.

From hunters to consumers, every stage of the wild meat value chain poses a potential risk for disease transmission, depending on the species consumed, the pathogens they carry, the practices around hunting, handling and cooking, and the behaviour and awareness of the actors involved.

How can we address such a complex and nuanced issue? Blanket bans on wildlife trade and consumption are often ineffective, driving the trade underground with law enforcement agencies unprepared to respond. Without suitable alternatives, bans also threaten the food security and well-being of vulnerable rural communities.

Decades of research and experience point to the need for more inclusive, rights-based and integrated approaches to effectively manage zoonotic risks. Recognizing the deep interconnectedness between human, animal and wild systems, the One Health approach offers a strong framework for developing and implementing cross-sectoral solutions, promoting resilience and sustainability.

For wild meat value chains, the One Health approach is particularly effective in establishing sustainable wildlife management frameworks, with targeted upstream and downstream interventions to prevent zoonotic disease emergence and spillover while securing access to safe food and livelihoods for rural communities. A deep understanding of these complex chains and zoonotic risks is key to designing suitable interventions.

The Sustainable Wildlife Management Programme[1] promotes an integrated "bush to fork" approach in many countries facing ecological, socioeconomic and health challenges related to hunting and wild meat consumption. Following a community rights-based approach, the programme first characterizes value chains and identifies risks and opportunities before engaging with sectoral actors to develop targeted interventions (Figure 1). Grounded in evidence and inclusive collaboration with communities, community-based organizations, public authorities and private sector actors, this step-by-step approach lays a strong foundation for long-term, sustainable impacts for both people and wildlife.

1. https://swm-programme.info/

References

Coad L., Fa J.E., Abernethy K., Van Vliet N., Santamaria C. et al., 2019. Towards a sustainable, participatory and inclusive wild meat sector. Bogor, Indonesia: CIFOR. https://doi.org/10.17528/cifor/007046

IPBES. 2022. Thematic assessment report on the sustainable use of wild species of the Intergovernmental Science-Policy Platform on Biodiversity and Ecosystem Services. IPBES Secretariat, Bonn, Germany. https://doi.org/10.5281/zenodo.10925382

Risk assessments and value chain analyses

- Typology of hunters, consumers and other actors involved in the trade chain
- Volumes and flows of wild meat traded, species concerned
- Hunting, handling, conservation and cooking practices
- Prevalence assessment at different points of the value chain

OH coordination

- Assess and strengthen One Health and health capacities in government and non-government organizations, particularly wild animal health
- Conduct gap analyses to identify weaknesses and opportunities to improve coordination and collaboration among One Health sectors
- Support One Health coordination platforms

DEVELOP TARGETED INTERVENTIONS

ENABLING FACTORS

Sectoral policies and legislation

- Conduct cross-sectoral legal analyses to identify opportunities to strengthen wildlife management, community rights over natural resource access and use, and intersectoral OH collaboration
- Support informed and participatory reform processes and implementation of wildlife and other sectoral laws to promote safe practices

UPSTREAM **DOWNSTREAM**

Sustainable wildlife management

- Promote sustainable hunting and/or conservation practices
- Empower communities to devise their own rules on wildlife use based on customary law and traditional knowledge
- Support nature-friendly enterprises as alternative livelihood to reduce pressure on wildlife
- Diversify local food systems to reduce dependence on wild products
- Mitigate human-wildlife conflicts

Improve food safety and biosecurity

- Include wild meat in food safety and biosecurity services and standards
- Train on safer hygiene and food safety practices
- Facilitate access to equipment, transport and storage infrastructure
- Promote value chain organizations that facilitate control

Behaviour change strategies

- Assess cultural and economic drivers of wild meat consumption
- Promote alternative protein sources
- Organize targeted social marketing campaigns to raise awareness and reduce demand

Urban or peri-urban consumers

International trade

Participatory OH surveillance

- Work with communities and local stakeholders to monitor and report wildlife diseases
- Train local and provincial authorities
- Facilitate linkages with early alert systems (national and international)
- Encourage cross-sector data-sharing to support risk prediction modelling

Commercial hunters/farmers

Transport and processing

Markets/vendors

Restaurants

Subsistence hunters/farmers

Local (rural) consumers

B

GUYANA

- Targets hunters and urban consumers to promote sustainable wild meat harvesting and consumption practices
- Created the One Health Technical Working Group (OHTWG) to enhance cross-sectoral coordination and wildlife health considerations into policies
- Supported the OHTWG in recommending standardized protocols for priority zoonotic risk assessment along wild meat value chains and trained national laboratory agents on these protocols
- Supported the OHTWG in proposing a roadmap for the development of wild meat safety standards by the government of Guyana
- Supported the development of specific regulations under the Fisheries
- Act to manage the wild fish trade

GABON

- Partners with 10 local communities in the Mulundu department to co-develop and implement sustainable hunting plans
- Trained and supported national researchers to conduct zoonotic risk assessments along the wild meat value chain in Mulundu
- Co-developed, trained and piloted with hunters' associations, public health and wildlife services a comprehensive community-based surveillance and rapid alert system, allowing zoonotic disease detection upstream in the wild meat value chain

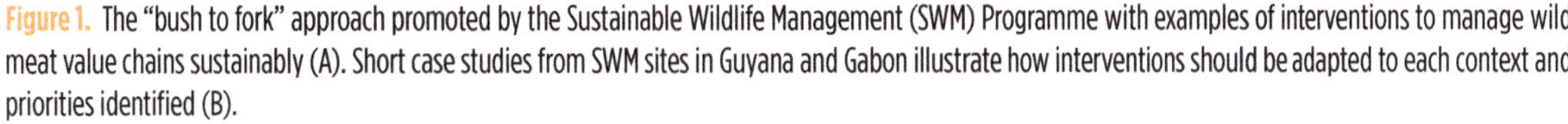

Figure 1. The "bush to fork" approach promoted by the Sustainable Wildlife Management (SWM) Programme with examples of interventions to manage wild meat value chains sustainably (A). Short case studies from SWM sites in Guyana and Gabon illustrate how interventions should be adapted to each context and priorities identified (B).

A systems approach to address climate change, biodiversity loss and health in Canada

Cécile Aenishaenslin

The impacts of climate change on nature, health and the economy affect all regions of the world, and particularly the most vulnerable populations. In our highly urbanized society, natural areas such as nature parks are important for mitigating these impacts. Their benefits include ecosystem services such as temperature reductions and biodiversity conservation, and improved physical and mental health. On the other hand, ecosystem health is also threatened by climate change, extreme fragmentation of natural habitats and other anthropogenic disruptions that affect urbanized countries. In Canada for example, fragmented peri-urban parks have become alternative habitats for white-tailed deer, a species which often becomes overabundant in small natural areas, leading to ecosystem degradation and plant biodiversity loss.

Ecosystem disturbances also increase risks for human and animal health. In Canada, the emergence of tick-borne diseases is a major issue that stems directly from these disturbances (Figure 1). Indeed, climate change and deer abundance favour the establishment and maintenance of *Ixodes* tick populations, which are the main vectors of tick-borne diseases in North America, namely Lyme disease. The northward expansion of tick populations has generated a rapid emergence of the disease in Canada. Other tick-borne diseases, including anaplasmosis, babesiosis and Powassan virus, are also emerging and creating public health concerns across the country.

The measures currently implemented in Canada to prevent tick-borne diseases rely primarily on promoting the adoption of preventive behaviours against tick bites. However, in order to reduce the adverse effects of climate change and other anthropogenic disturbances on park ecosystems and human health in a sustainable way, it is necessary to adopt a systems approach consistent with One Health that takes into account the inherent complexity of this problem. Measures targeting protection of biodiversity, other environmental interventions aiming at reversing ecological disturbances, as well as human behaviours should also be considered (Figure 2).

References

Côté S.D., Rooney T.P., Tremblay J.-P., Dussault C., Waller D.M. 2004. Ecological impacts of deer overabundance. *Annual Review of Ecology, Evolution, and Systematics*, 35 (1), 113-147. https://doi.org/10.1146/annurev.ecolsys.35.021103.105725

Kulkarni M.A., Berrang-Ford L., Buck P.A., Drebot M.A., Lindsay L.R., Ogden N.H. 2015. Major emerging vector-borne zoonotic diseases of public health importance in Canada. *Emerging Microbes and Infections*, 4 (6), e33. https://doi.org/10.1038/emi.2015.33

Potes L., Bouchard C., Rocheleau J.-P., Richard L., Leighton P. *et al.* 2023. Evaluation of a community-based One Health intervention to reduce the risk of Lyme disease in a high-incidence municipality. *CABI One Health*, 2, ohcs202300017. https://doi.org/10.1079/cabionehealth.2023.0017

LYME DISEASE

It is a bacterial infection caused by *Borrelia burgdorferi*, transmitted to humans through the bite of infected Ixodes ticks. Early symptoms typically include fever, fatigue, headache and a characteristic expanding skin rash called erythema migrans. If left untreated, the disease can progress to affect the joints, nervous system and heart. While many wild animals (mainly small mammals) serve as reservoirs for the bacterium, they usually show no clinical signs of illness. Meanwhile, clinical Lyme disease in domestic animals such as dogs is occasionally reported and may involve lameness, fever or lethargy. Prompt diagnosis and antibiotic treatment in humans are key to preventing complications.

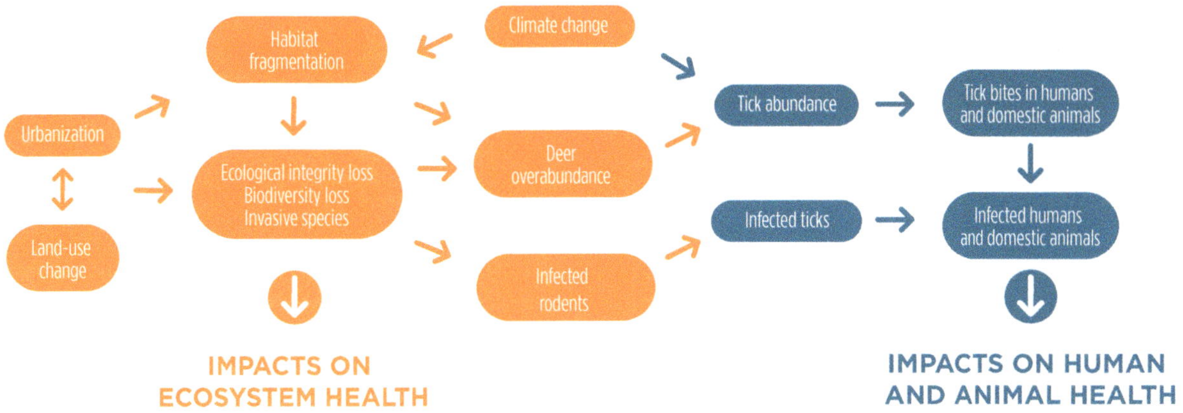

Figure 1. Simplified representation of the dual problems caused by climate and anthropogenic change in peri-urban nature parks.

LYME DISEASE TRANSMISSION	INTERVENTIONS
Drivers of Lyme disease emergence: Land-use change Habitat fragmentation Climate change	**Deep prevention** Restoring natural habitats Ecological corridors Combating climate change
Reservoir hosts infected with *Borrelia*	**Preventing infection of reservoir species** Treating reservoir hosts with acaricides Vaccinating reservoirs
Ticks infected with *Borrelia* Main tick hosts contribute to high tick density	**Controlling tick density** Landscaping Acaricides in tick habitats Controlling deer density Treating tick hosts with acaricides
Human populations/domestic animals at risk of tick bites	**Preventing tick bites** Protective clothing Repellent Tick check Acaricides in domestic animals
Lyme disease in humans and domestic animals	**Preventing *Borrelia* infections** Antibiotic prophylaxis Vaccines

Figure 2. Examples of possible interventions targeting different levels of the Lyme disease transmission system, rather than infection by the infectious agent in humans and/or animals.

The health toll of climate change: why One Health matters

François Roger,
Marisa Peyre

Climate change is intensifying health risks and testing the resilience of health systems. Shifts in temperature, precipitation patterns and extreme weather events fuel the emergence and spread of infectious diseases and contribute to the rise of non-communicable diseases. These changes disrupt animal and plant health, impact food security and destabilize ecosystems. As climate change simultaneously influences agricultural systems, disease propagation, and the biological and social determinants of health, the One Health approach offers a dynamic framework to address these issues (Figure 1).

Climate-sensitive infectious diseases, including vector-borne diseases and zoonoses like avian influenza and rabies, are becoming more frequent as ecological niches shift and host species migrate (Figure 2). Extreme weather events such as floods and droughts also heighten the risks of waterborne diseases and nutritional deficiencies, disproportionately affecting vulnerable populations. Integrating climate data into health surveillance systems is crucial to better predict and mitigate outbreaks. Rising temperatures and flooding promote the spread of resistant bacteria in water, soil and agricultural systems, thus exacerbating antimicrobial resistance. Disruptions in livestock production due to climate stressors drive greater antibiotic use, intensifying selective pressures and fostering the proliferation of resistant pathogens.

Climate change significantly impacts agrifood systems, threatening food security, nutrition and livelihoods. Shifting weather patterns, such as prolonged droughts, erratic rainfall and heatwaves disrupt crop yields and livestock productivity, leading to food shortages and rising prices. Smallholder farmers and low-income populations who rely on subsistence agriculture suffer most and have limited capacity to adapt. Climate-induced changes in pest and disease dynamics further jeopardize food production and safety, increasing the prevalence of mycotoxins and other contaminants in food supplies.

Finally, climate change poses a critical challenge to global health systems, undermining their functionality by increasing demand for healthcare services while simultaneously damaging infrastructure. Extreme weather events, such as hurricanes, floods and heatwaves, can strain already fragile healthcare facilities. The 2024 Lancet Countdown report noted the urgent need to redirect investments from fossil fuels to sustainable alternatives. It calls for structural changes across various sectors to mitigate the health impacts of climate change. One Health is one part of a holistic strategy to navigate these challenges.

References

Mora C., McKenzie T., Gaw I.M., Dean J.M., von Hammerstein H. *et al.* 2022. Over half of known human pathogenic diseases can be aggravated by climate change. *Nature Climate Change*, 12, 869–875. https://doi.org/10.1038/s41558-022-01426-1

Roger F., Bonnet P., Steinmetz P., Salignon P., Peyre M. 2016. The One Health concept to dovetail health and climate change policies. *In* E. Torquebiau (Ed.), *Climate Change and Agriculture Worldwide*, Springer, Éditions Quæ, pp. 239–250. https://doi.org/10.1007/978-94-017-7462-8_18

Shafique M., Khurshid M., Muzammil S., Arshad M.I., Malik I.R. *et al.* 2024. Traversed dynamics of climate change and One Health. *Environmental Sciences Europe*, 36(135). https://doi.org/10.1186/s12302-024-00931-8

Wong C. 2024. Antibiotic resistance is a growing threat — is climate change making it worse? *Nature* https://doi.org/10.1038/d41586-023-04077-0

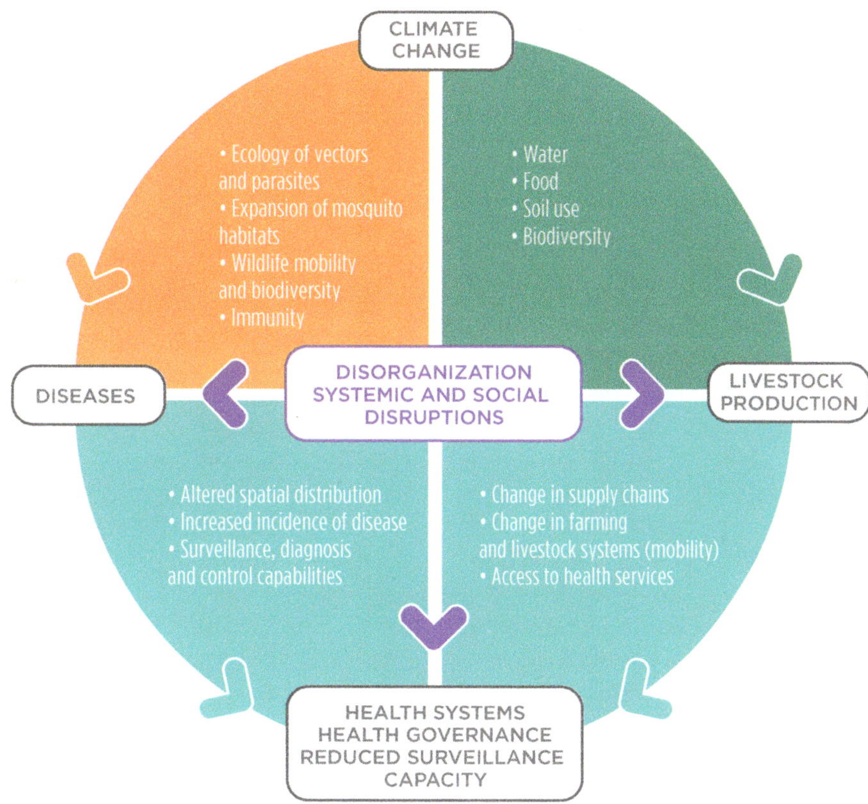

Figure 1. Interconnections between climate change, diseases, livestock production and health systems. This diagram shows how environmental, agricultural and health governance are interlinked, and so require integrated solutions for systemic resilience. From Roger *et al.* 2016.

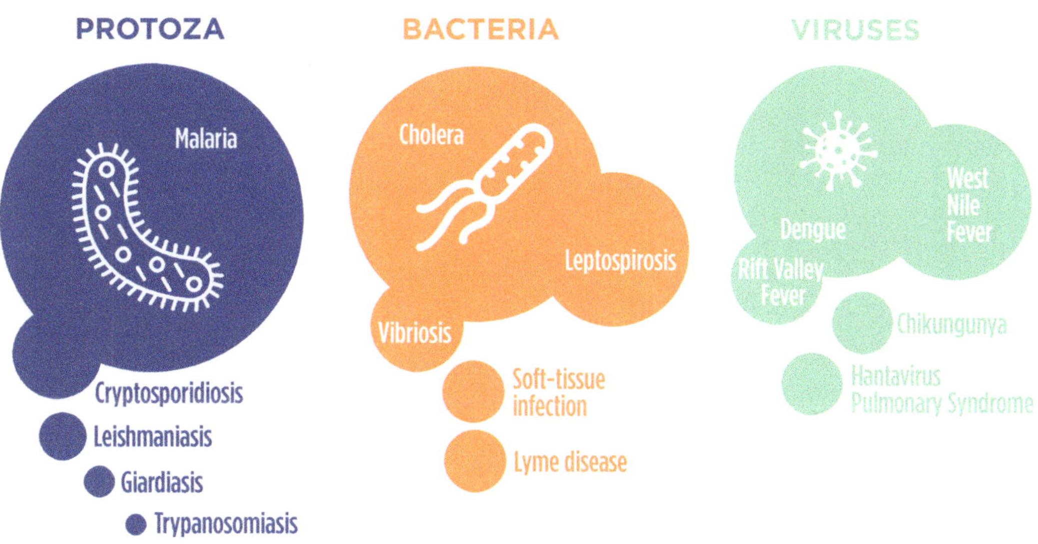

Figure 2. Infectious diseases aggravated by climatic hazards, emphasizing zoonotic diseases (e.g. leptospirosis, Rift Valley fever, and hantavirus pulmonary syndrome) and environmentally linked diseases (e.g. malaria, cholera, and dengue). These diseases underscore the necessity of a One Health approach. From https://www.scidev.net/global/news/truly-scary-climate-change-diseases-study/ and from Mora C. *et al.* 2022.

One Health Alert: A(H5N1) a cross-species threat on the rise

Claire Hautefeuille,
François Roger

Recent avian influenza A(H5N1) virus mutations have prompted global concern, especially given the dangers of the present strain. The FAO, WHO and WOAH reported recent events and dangers based on 2024 data, and while the virus has not yet been shown to spread between humans, the evolution and transmission of A(H5N1), and especially clade 2.3.4.4b, require continual surveillance. Wild birds and mammal species around the world continue to contract A(H5N1) viruses (Figures 1 and 2), but domestic animals are also affected. For example, in 2024, outbreaks occurred in US dairy cattle farms, and infections were also observed in cats after row poultry meat or raw milk consumption. These infections can cause severe illness with neurological symptoms, especially in mammals, leading to concerns about the risks to other species (including humans).

Between 2021 and mid-2024, 35 human cases of A(H5N1) have been reported, mostly from avian or environmental contact, with a few cases in the United States linked to contaminated dairy cattle. Most human cases have been asymptomatic or mild, but severe disease, particularly in Asia, has revealed the virus's possible dangerous effects on human health. In October 2024, this strain was detected in a pig, sparking worries about the risk of viral reassortment, since pigs can act as mixing vessels for influenza viruses. These trends show the need for improved animal and human surveillance, especially in high-exposure areas. Biosecurity, good animal-handling practices and personal protective equipment training for people exposed to sick animals are crucial to prevent the virus from spreading from animals to humans. Although the global danger from A(H5N1) to public health is minimal, a coordinated One Health strategy is strongly recommended to limit threats (Figure 3). This strategy should include cross-sectoral coordination to monitor, manage and lower the risks from A(H5N1) transmission from birds to animals (including livestock) and humans.

Mutations of different avian or swine influenza virus strains or the recombination of these viruses with human seasonal influenza viruses could eventually result in a new zoonotic strain that is able to easily spread between humans. Such a potential emergence crystallizes fears around these viruses and makes the surveillance and control of these serotypes a major public health and One Health issue. The joint FAO/WHO/WOAH review shows that the human, animal and environmental sectors must work together, and the One Health program emphasizes the need for cross-sector data-sharing and collaborative research to assess and manage A(H5N1) risks.

References

Burrough E.R., Magstadt D.R., Petersen B., Timmermans S.J., Gauger P.C. *et al.* 2024. Highly pathogenic avian influenza A(H5N1) clade 2.3.4.4b virus infection in domestic dairy cattle and cats, United States. *Emerging Infectious Diseases*, 30(7), 1335–1343. https://doi.org/10.3201/eid3007.240508

FAO/WHO/WOAH. 2024. Updated joint FAO/WHO/WOAH assessment of recent influenza A(H5N1) virus events in animals and people. Assessment based on data as of 18 July 2024–14 August 2024. www.who.int/publications/m/item/updated-joint-fao-who-woah-assessment-of-recent-influenza-a(h5n1)-virus-events-in-animals-and-people

Plaza P.I., Gamarra-Toledo V., Euguí J.R., Lambertucci S.A. 2024. Recent changes in patterns of mammal infection withhighly pathogenic avian influenza A(H5N1) virus worldwide. *Emerging Infectious Diseases*, 30(3), 444–452. https://doi.org/10.3201/eid3003.231098

Trevennec K., Cowling B.J., Peyre M., Baudon E., Martineau G.-P., Roger F. 2011. Swine influenza surveillance in East and Southeast Asia: A systematic review. *Animal Health Research Reviews*, 12(2), 213–223. https://doi.org/10.1017/S1466252311000181

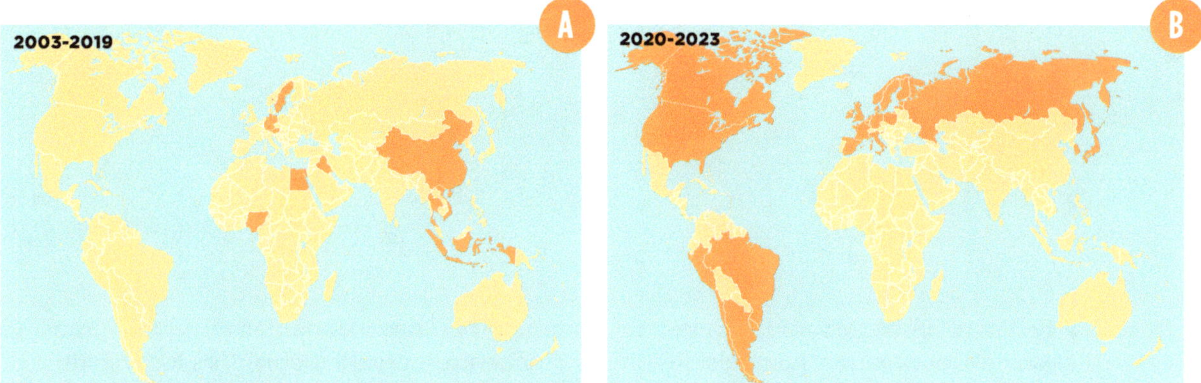

Figure 1. Geographic location of mammal species affected by highly pathogenic influenza virus A(H5N1) in previous waves of infection, 2003–2019 (A), and in the current panzootic, 2020–2023 (B). From Plaza *et al.* 2024.

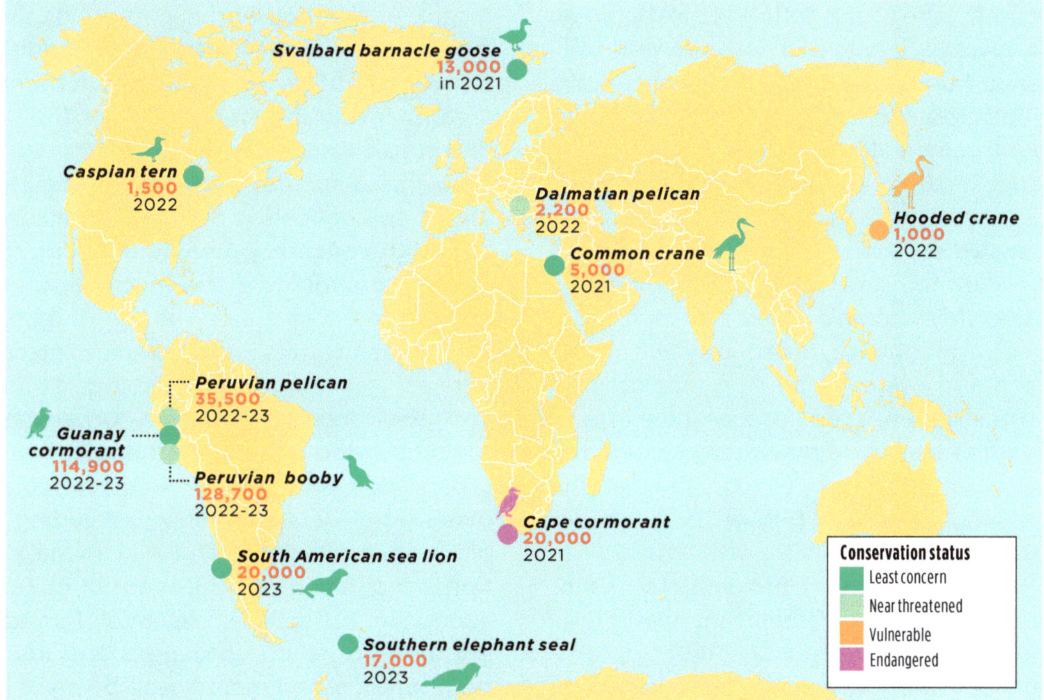

Figure 2. Major mortality events linked to HPAI H5N1 infection in wild animal populations between 2021 and 2024.
The map highlights reported die-offs in marine mammals (e.g. South American sea lions, elephant seals) and wild bird species—including several listed as endangered or vulnerable, such as condors, cormorants, pelicans and cranes. From https://www.theguardian.com/environment/article/2024/sep/04/forgotten-epidemic-with-over-280-million-birds-dead-how-is-the-avian-flu-outbreak-evolving

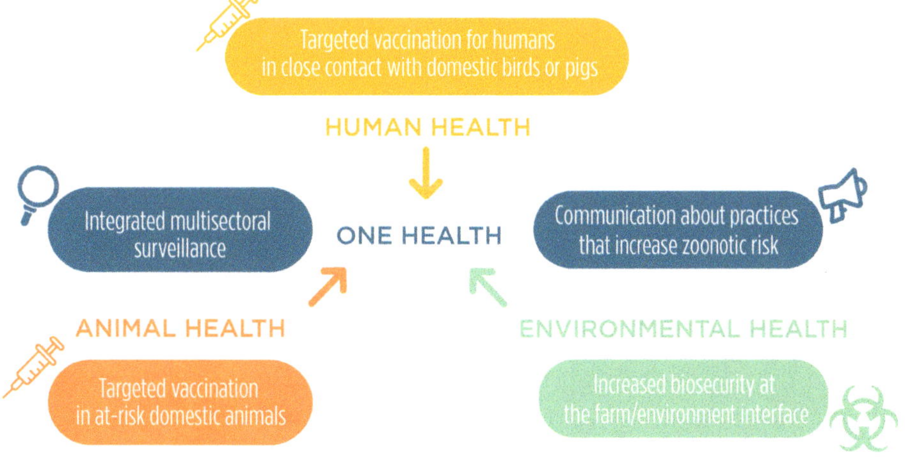

Figure 3. One Health approach to managing zoonotic influenza risks.

One Health in urban settings

Florence Fournet,
Amélie Desvars-Larrive,
Gérard Salem

For the first time in history, more than half of the world's population now resides in cities, and this trend is accelerating. While urban living can improve public health by providing better access to healthcare services, it significantly alters the interactions between people, animals, plants and ecosystems. In cities, dense human populations, air and noise pollution, close human-animal contacts and habitat fragmentation create unique challenges for health and environmental management (Figure 1). Urban areas are also marked by extensive soil artificialization and a reduction in vegetated surfaces. However, cities are increasingly developing green and blue infrastructures to enhance urban dwellers' mental health and overall well-being. Ironically, as cities expand these green spaces, they attract more wildlife, leading to new interactions at the human-wildlife interface, which increases the risk of zoonotic spillover (Figure 2). This situation creates a paradox: while urbanization can bring health benefits (e.g. proximity to healthcare), it also introduces significant health risks by altering disease ecology. These dynamics highlight the need for a comprehensive approach that balances the benefits and risks of urbanization on health. Among these risks is the increased threat of tick-borne diseases, which have emerged as a growing concern in urban settings, particularly due to the rise of tick populations in urban green spaces. These green patches connect peri-urban areas to city centres, creating corridors for wildlife that may carry ticks, which can serve as vectors for disease transmission. Urban small vertebrates, such as rodents and birds, can also act as reservoirs for tick-borne pathogens, increasing the risk to urban residents. In addition, by favouring artificial breeding sites, urbanization creates favourable conditions for the survival of *Aedes* mosquito species and the spread of arboviruses such as dengue, chikungunya and Zika. Leptospirosis is another significant public health concern, primarily affecting tropical regions, though its prevalence is also rising in temperate urban areas. The risk of leptospirosis in cities results from complex interactions between environmental, ecological, climatic and socioeconomic factors. Urbanization is a multifaceted challenge to public health. An urban One Health framework that integrates all dimensions of urbanization—including urban planning, green space and wild animal population management, as well as social, cultural and economic aspects—is essential for addressing modern health challenges and improving both physical and mental well-being of human populations. Adopting a One Health approach in cities—i.e. collaborative, multisectoral and transdisciplinary—can also enhance the resilience and sustainability of urban ecosystems, benefiting humans, animals, plants and the environment.

References

Crits-Christoph A., Levy J.I., Pekar J.E., Goldstein S.A., Singh R. *et al.* 2024. Genetic tracing of market wildlife and viruses at the epicenter of the COVID-19 pandemic. *Cell,* 187(19), 5468–5482.e11. https://doi.org/10.1016/j.cell.2024.08.010

Ellwanger J.H., Byrne L.B., Chies J.A.B. 2022. Examining the paradox of urban disease ecology by linking the perspectives of Urban One Health and Ecology with Cities. *Urban Ecosystems,* 25(6), 1735–1744. https://doi.org/10.1007/s11252-022-01260-5

United Nations - Habitat. 2010. *Hidden Cities, Unmasking and Overcoming Health Inequities in Urban Settings.* UN-HABITAT. https://unhabitat.org/hidden-cities-unmasking-and-overcoming-health-inequities-in-urban-settings

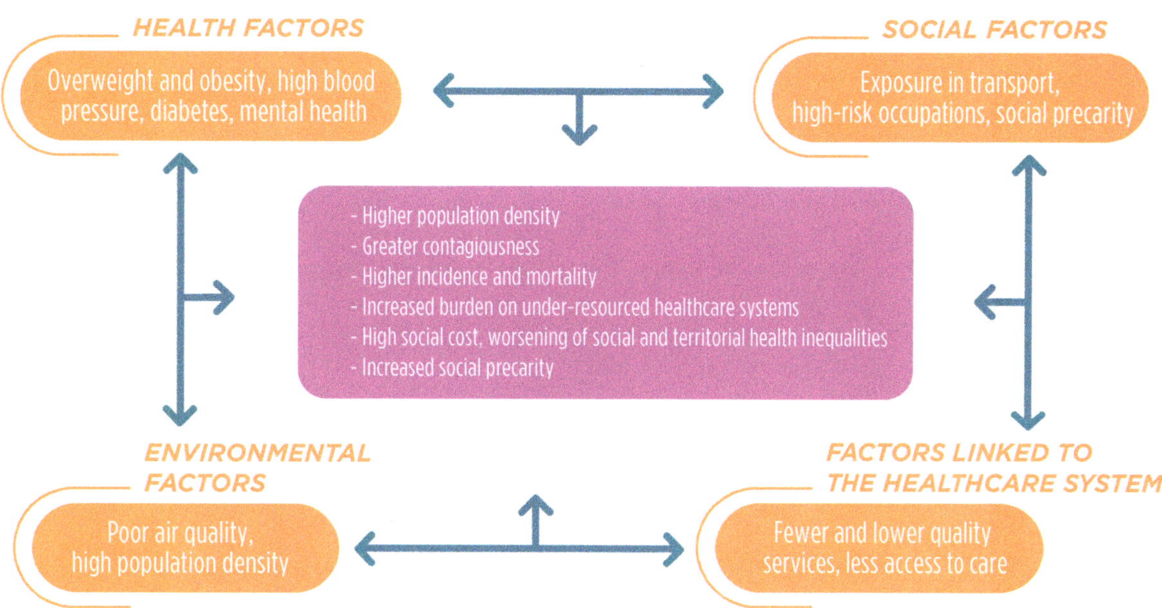

Figure 1. COVID-19 in urban settings: a case study of the greater Paris area (Ile-de-France).

Figure 2. Introduction of zoonotic risks through urban greening. Adapted from Alec Van Landuyt 2022.

Artificial intelligence in One Health: unlocking potential and navigating risks

Written by ChatGPT-4o[1]

Artificial intelligence (AI) presents significant opportunities and risks in the field of One Health, which considers the ways in which human, animal and ecosystem health interact (Figure 1). AI is a transformative technology with the potential to revolutionize research, surveillance and interventions in interconnected health systems. However, its use also raises ethical, technical and operational challenges that must be carefully addressed.

Potentialities

- AI-powered tools can process vast data sets from diverse sources, including social media, clinical records and environmental sensors, to detect emerging health threats. Predictive modelling enables early warning systems for zoonotic disease outbreaks, such as influenza or COVID-19, improving response times and mitigating risks.

- One Health requires the integration of data from various disciplines, such as epidemiology, ecology and veterinary sciences. AI facilitates this integration by harmonizing heterogeneous data sets, uncovering patterns and generating insights that inform cross-sectoral decision-making.

- Machine-learning algorithms can optimize interventions, from vaccine distribution to resource allocation during crises. For example, AI can identify high-risk areas for disease transmission or hotspots for antimicrobial resistance and guide targeted actions.

- Remote sensing and AI-driven analytics can be used to monitor environmental changes such as deforestation and climate variability that influence disease dynamics. These insights support sustainable ecosystem management, a cornerstone of the One Health approach.

Risks

- Collecting and processing sensitive health and environmental data pose risks to privacy. Data breaches or misuse can lead to ethical concerns and loss of trust among stakeholders.

- AI models often reflect biases present in training data sets, which may skew results. This can disproportionately affect vulnerable populations by exacerbating health disparities rather than addressing them.

- Excessive dependence on AI tools risks sidelining traditional expertise and local knowledge. One Health initiatives require the integration of AI alongside participatory approaches to ensure contextual relevance and inclusivity.

- Despite its promises, AI can be siloed within technical disciplines, undermining the holistic perspective that is so essential to One Health. Collaborative frameworks are necessary to ensure AI complements rather than replaces interdisciplinary approaches.

AI has the potential to significantly advance the goals of One Health by improving the understanding, prediction and management of health risks at the human-animal-environment interface. However, for it to be effectively integrated, ethical, technical and social challenges must be addressed. Transparent governance, inclusive collaboration and robust data stewardship are essential to harnessing AI's benefits while mitigating its risks. By balancing innovation with responsibility, AI can become a critical enabler of a healthier, more sustainable future for all.

1. Operating on 11 January 2025 under the supervision of François Roger and Marie-Marie Olive. The text generated by ChatGPT was reviewed by Teri Jones-Villeneuve.

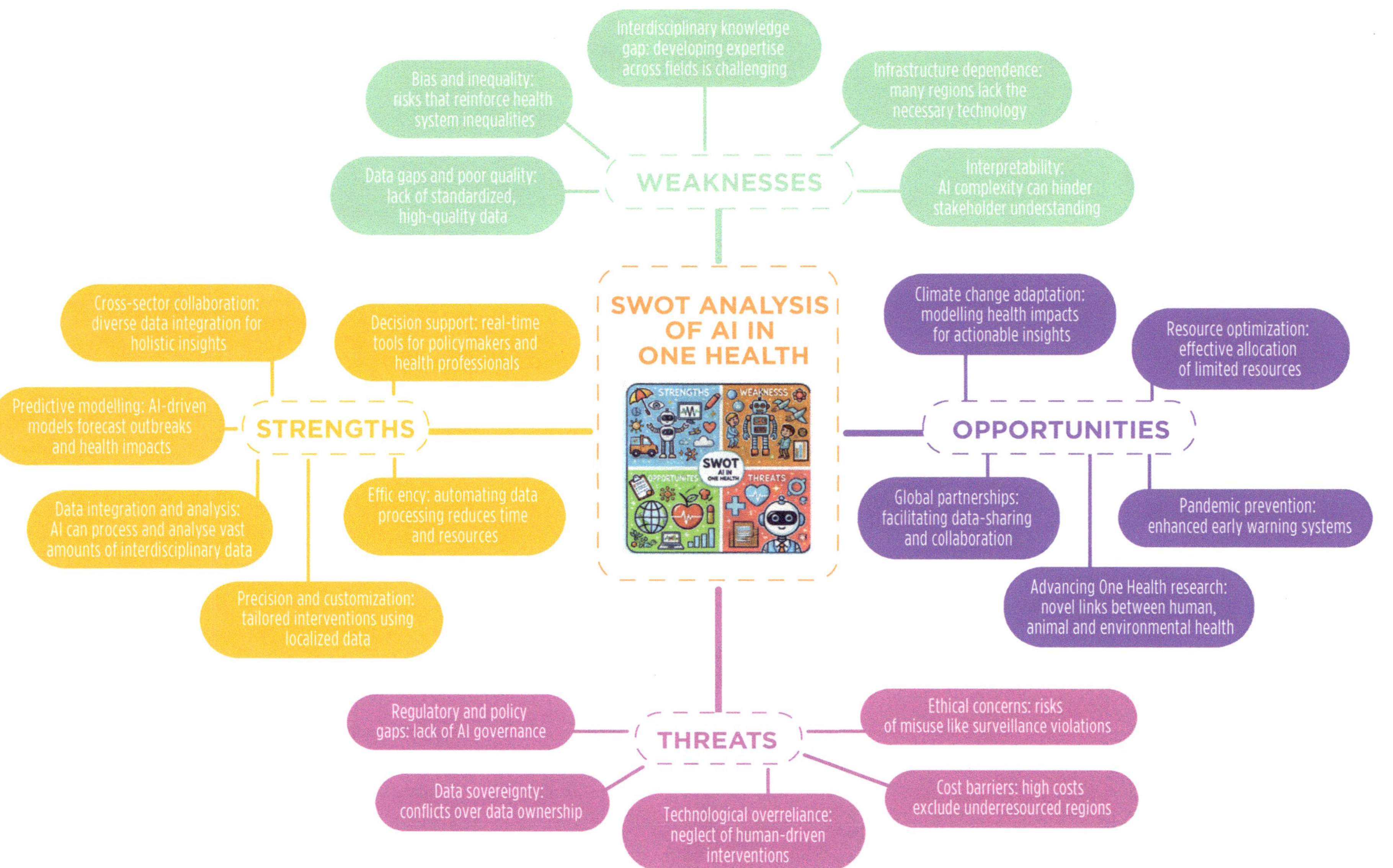

Figure 1. AI-generated SWOT analysis of artificial intelligence in One Health. The central illustration was generated with DALL.E. The entire diagram was adapted by Laetitia Perotin-Meslay. The English was reviewed by Teri Jones-Villeneuve.

Conclusion: One Health Risk governance and the science-policy-society interface

Dirk Pfeiffer

Although the One Health approach has been promoted fairly extensively for about 20 years, it has yet to become institutionalized in most, if any, countries around the world. This failure to act comes despite the SARS outbreak in 2002–04 and the COVID-19 pandemic in 2020–23, as well as the continuing threat of avian influenza and various other zoonotic disease outbreaks (Nipah, Ebola, etc.). Policymakers around the world have either not embraced the One Health approach at all or only to a limited extent to aid their decision-making processes. It is therefore essential to reflect on the utility of the One Health approach for risk governance and how a more widespread adoption can be promoted. Recognising this situation, the European Union published a scientific opinion on this topic that was informed by a commissioned review produced by an expert panel which had been tasked to identify constraints to the effective implementation of One Health governance within the European Union.[1; 2]

In this book chapter, a global perspective is used to provide a general introduction to the different considerations that are important when aiming to enhance the effectiveness of One Health policies. Figure 1 brings these together as a broad conceptual framework for governing One Health risks.

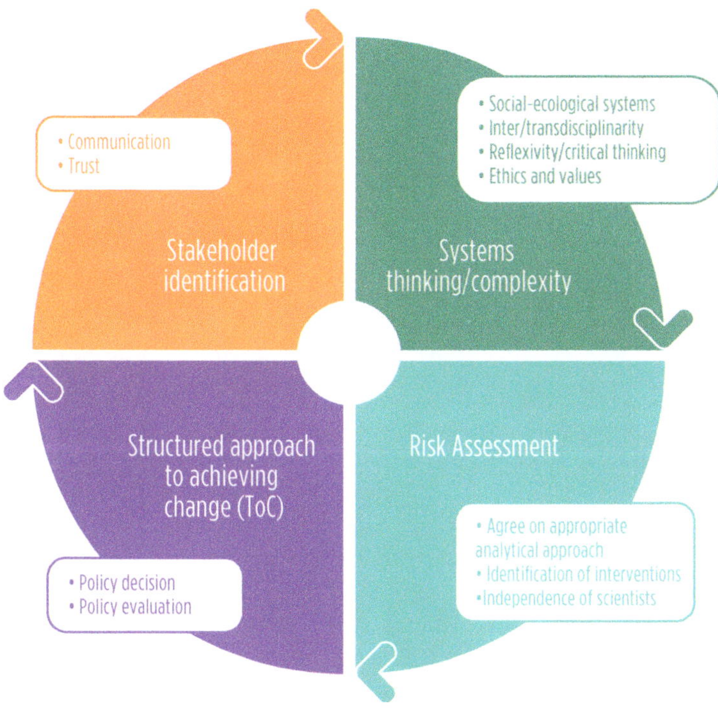

Figure 1. Framework for risk governance within One Health: integrating scientific, social and policy knowledge.

Current state of the science-policy interface

The United Nations Environment Programme (UNEP) concluded in their 2012 foresight report that there is a "broken bridge" between science and policy that needs to be reconnected.[3] The International Network for Governmental Science Advice (INGSA) conducted a survey between August 2019 and March 2020 with 306 respondents consisting of policymakers, scientists and researchers from various Asian countries, including Malaysia, Singapore, Thailand, Indonesia, Pakistan, Cambodia, Myanmar, India, Sri Lanka, China, Korea, Bangladesh, Japan, Vietnam, Philippines, Nepal and Laos.[4] The objective of the survey was to understand the perspectives of policymakers and researchers on the use of scientific knowledge, as well as the role of scientists and researchers in policymaking. The results indicated that the respondents believe there is a gap in collaboration between the scientific community and policymakers. This gap was attributed to differences in jargon used as well as a lack of tradition in collaborating. The survey also found that policymakers believe that researchers do not address their concerns. A recently published report from the UK COVID-19 Inquiry also identified significant issues in science-policy communication.[5] These are only three of many possible examples that indicate an urgent need to improve the effectiveness of the science-policy interface.

The way research is conducted impacts its usefulness for the policy process. Traditional research is performed in disciplinary or sectoral silos, which runs contrary to the fact that our environment is actually a complex whole and that people live their lives in an integrated way.[6] As a consequence, specific disciplinary research output often has limited direct policy relevance, and so policymakers must do the work of integrating scientific research output with other types of knowledge for it to be useful. This frustration among policymakers led to the Bamako call to action in 2008.[7] The One Health approach and the related emphasis on the need for inter- and transdisciplinary research approaches have been an indirect result of that event. The recent COVID-19 pandemic has shown that the One Health approach has yet to be widely adopted, let alone institutionalized, thus severely compromising our preparedness for the next pandemic.[8-10]

Inter- and transdisciplinarity

The emergence of SARS-CoV-2 has demonstrated once again that government policies on the prevention and control of infectious diseases must be informed by the most current scientific knowledge. For this scientific knowledge to remain current, it will very likely have to be continuously updated, even during outbreak situations. Perceptions of what is considered to be useful scientific knowledge or evidence also vary. Typically, knowledge generated by the natural sciences is rated more highly than that produced by the social sciences, which usually translates into major differences in political influence.[11] Furthermore, Western knowledge cultures still dominate, even though they do not adequately reflect the cultures and societal values in other parts of the world. There is now an increasing acceptance that society and nature in different areas of the world require the production of locally relevant knowledge.[12-14] It is also important to reflect on the relationship between knowledge and governance in different cultural and societal settings.[13] Policies designed to manage different types of societal risks are influenced by a wide range of factors and different types of knowledge, not "just" scientific knowledge. As such, it is useful to understand the wider context within which decisions about risk are being made. The risk governance framework was designed to do this, and it expresses the "various ways in which many actors, individuals, and institutions—public and private—deal with risks surrounded by uncertainty, complexity and/or ambiguity".[16] Within this framework, researchers play a role of producing and communicating relevant scientific evidence that contributes, together with other knowledge providers, to the integrated knowledge that should then underpin policy development. Such integration requires the use of an interdisciplinary research approach, and when dealing with One Health risks, it requires the inclusion of the social sciences and humanities.[17] However, experience with the implementation of interdisciplinary approaches in research projects shows a need for significant changes in the attitudes, expectations and communication among researchers from different scientific disciplines working together.[18] Knowledge management plays a central role in risk governance, and it is important to realize that it is not just scientific knowledge that is relevant here. Parkes et al.[19] present different knowledge perspectives based on knowledge types described by Brown.[20] They emphasize the paramount importance of creating links between different knowledge perspectives and among users of that knowledge, which is the essential goal of a transdisciplinary approach.[21] Pohl et al.[22] outline potential approaches for how researchers, policymakers and other participants could work together to co-produce inter- and transdisciplinary knowledge. They suggest starting with problem framing, followed by problem analysis and impact exploration.

Being able to integrate knowledge produced by different disciplines about the various elements and relationships in a system in a meaningful way requires critical thinking and consideration of the biases associated with research hypotheses, data collection through measurement tools or personal observation, and the analysis and interpretation of the results.[23, 24] These principles for dealing with knowledge have been accepted and applied more widely in the social sciences than in the natural sciences, mainly because social scientists are more likely to use qualitative research approaches. This is because these methods are considered to be less "objective" than the quantitative methods and the scientific method, which predominate in the natural sciences. Furthermore, when it comes to considering the utility of scientific evidence for policy development, given that we are living in a world strongly influenced by post-truth politics, reflexive strategies for assessing knowledge need to be used rather than purely relying on trust in particular scientific methodologies.[25]

Systems thinking and complexity

Linear and reductionistic thinking has dominated Western science since the beginning of the scientific revolution. It is fundamental to the scientific method, which has been the basis of the enormous improvements in human well-being in many parts of the world over the last 200 years. We live in a world that is subject to volatility, uncertainty, complexity and ambiguity.[26] While this term is now common in management science, it also applies to the challenges associated with policy development in a One Health context, often referred to as "wicked problems". When facing these types of challenges, policymakers will typically resort to reductionist interventions, just as they did during the recent COVID-19 pandemic, without appropriately considering their effects on the complex societal and ecological systems as a whole.[27]

It is notable that the transition from a linear, reductionist approach to systems thinking happened in the management and business sciences quite some time ago.[28] While systems and complexity science has been recognized for over 100 years,[29, 30] its importance has only recently been emphasized in medical and veterinary science.[31-34] It is now considered a key component of the One Health approach.[35, 36]

Recognizing the importance of systems thinking, the government of the United Kingdom developed a series of guides for civil servants called *How to use systems thinking to drive improved outcomes in complex situations*.[37, 38] Similarly, the European Commission's Knowledge4Policy platform offers a workshop-based tool to make it easier to adopt a systems thinking approach as part of a participatory policy development process.[39] The Organisation for Economic Co-operation and Development (OECD) also promotes systems thinking in policy development, emphasizing the importance of using it to identify tipping points, interconnectedness and resilience within complex systems.[40, 41]

A system comprises elements, interrelationships and a particular function or functions,[42] and is often described as "a whole that is more than the sum of its parts"—the very reason why reductionist interventions are likely to have unexpected effects on elements or interrelationships which had not been previously considered.[43] An important feature of a system is "emergence". Emergence refers to effects that are a consequence of the interactions between a system's elements which cannot be identified or predicted by examining the individual parts by themselves.[33, 44] These interactions arise from positive or negative feedback loops among system elements. An example of such effects is the closure of live bird markets in Shanghai in 2013 in response to human deaths due to avian influenza A(H7N9). Following that policy intervention, poultry value-chain actors responded in various ways to continue to make a living. One effect was that poultry was sold in neighbouring provinces; this resulted in spreading H7N9 further across the country, even though it was effectively controlled in Shanghai, which achieved the provincial government's objective.[45] Examples from other disease outbreaks include situations where farmers may stop raising one particular animal species and switch to another, such as a different domestic animal species or a wild animal species. Both actions will result in system changes and have the potential to generate new, unanticipated risks. Accordingly, when designing an intervention to mitigate a One Health risk, the appropriate approach would be to examine the system as a whole and to identify optimal leverage points, which, if changed, lead to major desired changes without (or at least minimizing) any unwanted effects.[42] The Intergovernmental Science-Policy Platform on Biodiversity and Ecosystem Services (IPBES) used systems thinking when identifying leverage points as part of defining pathways toward a sustainable future.[41] The integrated landscape approach promoted by IPBES and the German Advisory Council on Global Change (WBGU) applies systems thinking based on a social-ecological systems (SES) perspective. Both organizations use it to consider the trade-offs and co-benefits between biodiversity, climate change and human health.[46, 47]

An SES perspective is very useful for considering One Health risks and designing effective interventions.[48] SES are a type of complex adaptive system for which Preiser et al.[49] developed a typology around six organizing principles:

- **Principle 1:** Relationships rather than the elements of the system must be emphasized;
- **Principle 2:** SES adapt and evolve in response to interactions between elements;
- **Principle 3:** System behaviour is dynamic and non-linear;
- **Principle 4:** SES are open, which means that defining their boundaries is extremely challenging;
- **Principle 5:** SES are context dependent, i.e. the system may change in response to external as well as internal factors;
- **Principle 6:** Cause-effect relationships are complex, which also gives rise to emergence.

Stakeholder identification and involvement

The development of policies and their impact are critically influenced by human behaviour and relationships. Kuijper[50] emphasizes the very tight link between a country's economic system and its "ecological, geological, geographical, financial political, legal, ideological, social and cultural systems". In such a context, the political economy describes the power relationships within and between countries that are important for effective risk management. The One Health approach needs to embrace the systems perspective described by Kuijper since doing so will be essential for achieving desired outcomes. The process of producing the required knowledge and translating it into action needs to be based on true inter- and/or transdisciplinarity involving systems thinking. This includes having to identify stakeholders that must be considered when developing and implementing One Health interventions. The role of the different stakeholders with respect to their interests and influence in relation to the impact of specific interventions can be very effectively visualized using XY charts.[51] Mapping relationships between stakeholders can also be useful.[52, 53]

An important part of stakeholder identification is the involvement of industry players. It is worth considering that the environmental, social and corporate governance framework aims to provide industry players with economic incentives to operate ethically, responsibly and sustainably to generate value for all organizational stakeholders (such as employees, customers, suppliers and financiers). As a result, large companies should be increasingly willing to engage in risk management associated with ecosystem health. The importance of engaging with industry players should also be considered in the context of increasing the concentration of power and corporate control in the global food system.[54]

Communication between science, policy and society

Within the risk governance framework, there are many communication interfaces, one of which is the interface between scientists and policymakers. This science-policy interface is crucial for effective policy development, and it must involve a two-way flow of information. This means policymakers and researchers should work together to define research questions based on identified knowledge gaps and, if relevant, to agree on the associated theory of change. To do this, researchers must be able to effectively communicate scientific evidence produced by their research to policymakers. The characteristics of the social relationship between researchers and policymakers are therefore important, since they will strongly influence the effectiveness of the science-policy interface. One important attribute of that social relationship is mutual trust. In 2014, INGSA Secretariat Chair Sir Peter Gluckman published his top ten principles on providing science advice to government.[55] He emphasized the essential role of scientists engaging with the policy community, and noted that trust should be earned by acting as knowledge brokers instead of issue advocates. A High-Level Expert Group of the European Commission established in 2021 was tasked with reviewing the effectiveness of science-policy interfaces in the context of improved food systems governance. One of their conclusions was that "there is a need to move beyond traditional unidirectional science-policy interfaces to [science-policy-society interfaces]" to ensure greater holistic stakeholder engagement and involvement.[56] This shift will also complement the implementation of the systems thinking approach among both scientists and policymakers.

Independence of scientists

The utility of scientific knowledge for policy development depends not only on research quality but also on the trust that other stakeholders have in the scientists who produced the knowledge. A key principle in this context is scientists" independence from stakeholder interests. The World Economic Forum published a Code of Ethics for Researchers which emphasizes seven principles, including the need to pursue the truth, to be accountable and to engage with decision makers and the public.[57] Aven[58] discusses the reasons for risk science that is politically neutral in more detail.

The European Food Safety Authority (EFSA) is an excellent example of how scientific independence can be promoted and ensure transparency throughout every step of a risk assessment process.[59] EFSA was established by the European Parliament as a scientific organization tasked with conducting scientific risk assessments. It is independent of the European Commission, which is the policymaker at the European Union level. Unfortunately, this model of a separate risk assessment agency cannot realistically be replicated by individual countries due to budgetary constraints, and therefore the relationships between scientists and other stakeholders, including policymakers, are often difficult to work out. As a consequence, trust across the diverse spectrum of stakeholders in the process and its outcomes is much more fragile, and may even be completely absent.

Ethics and values

While there is widespread agreement in terms of the need to adopt inter- and transdisciplinary approaches when dealing with One Health challenges, there is less consensus on the importance of including expertise in philosophy and ethics. Since 2007, the German Ethics Council has had the legal mandate in Germany to handle "ethical, social, scientific, medical and legal issues as well as the likely consequences for the individual and society that arise in connection with research and developments, especially in the field of life sciences and their application to humans."[1] The German Ethics Council increased its public profile dramatically during the COVID-19 pandemic by providing independent opinions in the context of public health policy decisions associated with lockdowns and vaccination. They also produced an opinion paper analysing decision-making challenges and issued recommendations on how to handle such situations in future. The paper includes a discussion of the different dimensions of human well-being that should be considered when trying to appropriately balance interests during emergency situations.[60]

One Health risk governance needs to explicitly take into account the different dimensions of human vulnerability, which also means looking at them from an anthropological perspective rather than a strictly natural science perspective. This conclusion led the German Advisory Council on Global Change to develop a normative compass to guide policymakers in their decision-making. At its centre is the central importance of human dignity, and the three points of the compass are (1) sustaining natural life-support systems, (2) ensuring inclusion of all people and (3) ensuring respect for diversity and development possibilities.[61] Effective One Health risk governance should be associated with this type of normative compass that all stakeholders support.

Competencies

Recognizing the need for more effective science-policy engagement, the European Commission has developed two competence frameworks, one for policymakers "for innovative, effective and evidence-informed policymaking" and another for researchers "contributing to policymaking with evidence and advice".[62] The particular strength of these frameworks is that they reflect the need for policymakers and scientists to communicate more effectively with each other and for both to understand the role of science in policymaking. The competences have been identified and defined in educational language, which makes them more helpful. Such competences should be integrated into professional training programmes as well as internships, short courses, etc. Early career researchers also need to be introduced to the interface between science, policy and society (practice) so they can identify where their research sits and how they can achieve impact.[63] Relevant competences from the European Commission's competence framework for researchers should be included in graduate and postgraduate training programmes.

Identifying, assessing and interpreting the risks

Risk is at the core of the risk governance process. The purpose of this process is to manage risk, ideally through effective prevention, and if prevention is not possible, through early detection and effective response. Structured scientific risk assessment is typically used to identify the processes that generate risk, to identify knowledge gaps and to estimate risk in qualitative or quantitative terms. But before conducting a risk assessment, consideration must be given to the specific disciplinary and interdisciplinary perspectives of risk, since the impact of any risk management interventions will ultimately be determined by how those affected by a particular risk view it in terms of acceptability.[64, 65]

1. Der Ethikrat: Gesetz zur Einrichtung des Deutschen Ethikrat (Ethikratgesetz - EthRG) issued on 16 July 2007 (Federal Law Gazette I p. 1385); entered into force on 1 August 2007.

Typically, a technical and natural-science-based approach dominates the process, from identifying risks to assessing and managing those risks. It is used where there is a single, clearly defined outcome of event occurrence, such as animal or human mortality. However, this approach is mainly suitable for "simple" risk problems characterized by a limited degree of complexity, uncertainty and ambiguity.[66] This limitation has been demonstrated during major crises, such as the bovine spongiform encephalopathy (BSE) epidemic over 30 years ago as well as during the recent COVID-19 pandemic.

The decision about the appropriate risk assessment approach should be based on a critical and reflective examination of the key causal relationships identified using the systems thinking process. When discussing and framing the policy challenge, scientists and policymakers need to agree on which analytical approach to use; choices range from linear risk pathway-based qualitative or quantitative risk assessment models to complex qualitative or quantitative simulation models. When assessing risks associated with complex systems, causal loop diagrams are one of several methodologies which have been applied in public health and in different eco-social contexts.[67–69] In the veterinary field, the publication by Matthiessen et al.[34] is an excellent reflective example of what this particular method can and cannot do. More information on systems thinking and analysis methods can be found in a wide range of textbooks.[29, 70–72]

In terms of risk-assessment outcomes, the technical and natural-science-based approach focuses on estimating the likelihood of an adverse event and its consequences. This approach neither considers the wider context nor the subjective or cultural frames which influence perceptions of risk and harm.[65] The German Advisory Council on Global Change therefore came up with a set of nine criteria that capture the wider societal context, which were developed further by Renn.[65] In addition to the probability of occurrence and extent of damage, they include incertitude, ubiquity, persistence, reversibility, delay effect, violation of equity and potential of mobilization. The COVID-19 pandemic has demonstrated the importance of all these criteria when it comes to considering the acceptability of risk and the consequential interventions. Policymakers typically only use the likelihood and consequences of the adverse event to decide whether a particular risk is acceptable, tolerable or intolerable.[73, 74] That decision, in turn, will determine what the desired aim of the risk management policy should be, i.e. elimination, reduction or acceptance of the risk. Interpreting the uncertainty associated with the risk estimates and the risk-generating mechanisms is a major challenge in this communication process, especially – but not only – between scientists and policymakers.[75] The policymaker's decision-making dilemma is to achieve optimum outcomes that all affected stakeholders will consider to be an epidemiologically, socio-politically and environmentally acceptable compromise. In this situation, the normative compass described by the German Advisory Council on Global Change can be used for guidance.[61]

Structured approach to achieving change

A key requirement for effective policy development and implementation is transparent, clear communication and evaluation of the assumptions, knowledge gaps, pathways and processes for achieving the desired outcomes. This issue can be addressed by communicating policy interventions using a theory of change (ToC) approach, as has become common practice in development projects when dealing with complex challenges involving the need for social change, including the emergence of infectious disease risks.[76, 77] The ToC is informed by the outcome of the systems thinking and mapping process. Ideally, the ToC should be developed and agreed upon by all stakeholders, as doing so increases the likelihood that they will provide effective support. For example, the UK-funded research project One Health Poultry Hub developed a ToC for the overall project which was then adapted to the specific needs of each of the four project countries, Bangladesh, India, Sri Lanka and Vietnam (Figure 2).

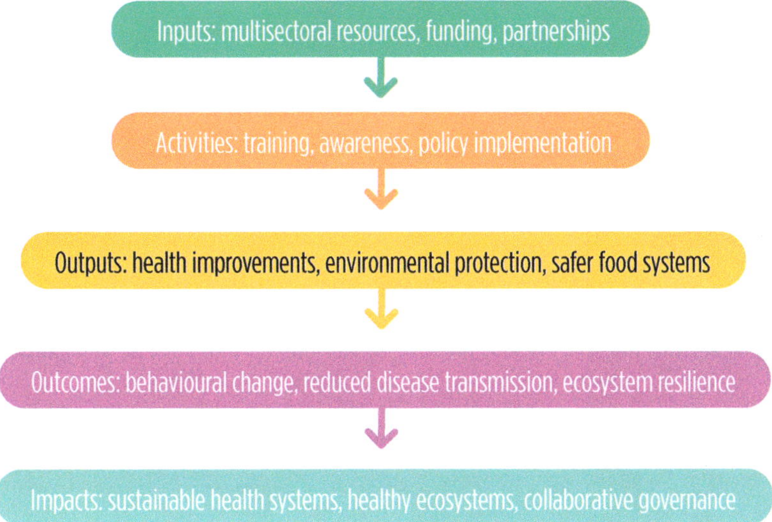

Figure 2. A Theory of Change (ToC) is a planning tool that describes how and why the attainment of targeted intervention or project goals will be achieved through logical sequences of interventions. In the context of the One Health approach, a ToC maps how actions clearly oriented towards human, animal and environmental health can address complex challenges such as zoonotic diseases, antimicrobial resistance or ecosystem degradation.

Conclusions

Effective One Health risk governance must be based around a functional science-policy-society interface. Achieving this requires a clear understanding of the roles of scientists and policymakers, who must be able to meaningfully communicate with each other as well as with society in general. Capacity-building programmes based on the competence frameworks for researchers and decision makers developed by the European Commission would be immensely beneficial for improving the effectiveness of the science-policy interface. They should be adapted to the needs of different eco-social-political contexts and then be integrated into relevant under- and postgraduate courses and continuing professional development.

Figure 1 presents an example of a potential sequence of steps involved in developing effective One Health policies, based on effective communication and collaboration between policymakers, researchers and stakeholders. It is essential that both scientists and policymakers adopt a systems thinking approach to be able to understand the interrelationships between the various elements within social-ecological systems that influence risk and how any intervention will affect the system as a whole. This understanding of the system could inform the development of a ToC, which would allow the comprehensive framing of a policy problem together with the intended interventions and other actions. If used effectively, the ToC also provides a platform for facilitating communication between all stakeholders. This communication includes interactions between scientists and policymakers, since the ToC clearly outlines the pathway from the current situation to the desired future outcomes. It would then be possible to conduct policy-relevant research and, at the same time, for all stakeholders to appreciate the wider socioecological contexts. The ToC needs to go together with the adoption of an interdisciplinary approach to scientific research. This means that the boundaries which have been institutionalized by academic institutions between the different scientific disciplines over the last 200 years must be overcome and that researchers must develop the competences to conduct truly interdisciplinary research that addresses the policy problems presented to them. Researchers in particular, also need to practice reflexivity so that biases are transparent when it comes to communicating between scientists and policymakers.

The challenges for the planet, including humanity, are significant, and new approaches are required to be able to tackle them. It will mean appropriately integrating knowledge—including, but not only, scientific knowledge—produced in different ways. The sciences will have to embrace inter- and transdisciplinarity, but without losing the ability to strive for excellence within their disciplines. The science–policy–society interface has a key role to play in achieving meaningful and sustainable impact. It needs to become the centre of an effective One Health approach, rather than an afterthought.

References

[1] **Scientific Advice for Policy by European Academies (SAPEA).** 2024. One Health governance in the European Union: Evidence review report. *In:* Scientific Advice Mechanism to the European Commission, p. 186.

[2] **European Commission: Directorate-General for Research Innovation and Group of Chief Scientific Advisors.** 2024. One Health Governance in the European Union, Publications Office of the European Union.

[3] **United Nations Environment Programme (UNEP).** 2012. 21 Issues for the 21st Century: Result of the UNEP Foresight Process on Emerging Environmental Issues. *In:* United Nations Environment Programme (UNEP), Nairobi, Kenya, p. 56.

[4] **International Network for Government Science Advice Asia.** 2020. Science in policy making survey. *In:* The International Network for Government Science Advice, p. 34.

[5] **Hallett H.** 2024. UK Covid-19 Inquiry: Module 1 Report – The resilience and preparedness of the United Kingdom. *In:* UK Government, London, United Kingdom, p. 217.

[6] **Falk I., Wallace R.** 2011. Managing plant biosecurity across borders. *In: Managing Biosecurity across Borders* (I. Falk, R. Wallace & M.L. Ndoen, eds), Springer, Heidelberg, pp. 3–19.

[7] ***The Lancet.*** 2008. The Bamako call to action: research for health. *The Lancet*, 372(9653), 1855. https://doi.org/10.1016/S0140-6736(08)61789-4

[8] **Lefrançois T., Malvy D., Atlani-Duault L., Benamouzig D., Druais P.-L. *et al.*** 2023. After 2 years of the COVID-19 pandemic, translating One Health into action is urgent. *The Lancet*, 401(10378), 789–794. https://doi.org/10.1016/S0140-6736(22)01840-2

[9] **Sironi V.A., Inglese S., Lavazza A.** 2022. The "One Health" approach in the face of Covid-19: how radical should it be? *Philosophy, Ethics, and Humanities in Medicine*, 17(1), 3. https://doi.org/10.1186/s13010-022-00116-2

[10] **Mwatondo A., Rahman-Shepherd A., Hollmann L., Chiossi S., Maina J. *et al.*** 2023. A global analysis of One Health Networks and the proliferation of One Health collaborations. *The Lancet*, 401(10376), 605–616. https://doi.org/10.1016/S0140-6736(22)01596-3

[11] **Wright S.** 2006. Reflections on the disciplinary gulf between the natural and social sciences. *Community Genetics*, 9(3), 161-169. https://doi.org/10.1159/000092652

[12] **Nowotny H., Scott P., Gibbons M.** 2001. *Re-thinking Science – Knowledge and the Public in an Age of Uncertainty*, Polity Press, Cambridge, UK.

[13] **Nowotny H., Scott P., Gibbons M.** 2003. Introduction: "Mode 2" revisited: The new production of knowledge. *Minerva (London)*, 41(3), 179–194. https://doi.org/10.1023/A:1025505528250

[14] **Grek S.** 2024. *The New Production of Expert Knowledge: Education, Quantification and Utopia*, 1 edn., Springer Nature, Cham.

[15] **Glückler J., Herrigel G., Handke M.** 2020. On the reflexive relations between knowledge, governance, and space. *In: Knowledge for Governance* (J. Glückler, G. Herrigel & M. Handke, eds), Springer International Publishing, Cham, pp. 1-21.

[16] **Hermans M.A., Fox T., van Asselt M.B.A.** 2012. Risk governance. *In: Handbook of Risk Theory: Epistemology, Decision Theory, Ethics, and Social Implications of Risk* (S. Roeser, R. Hillerbrand, P. Sandin & M. Peterson, eds), Springer Netherlands, Dordrecht, pp. 1093-1117.

[17] **Barry A., Born G.** 2013. Interdisciplinarity: *Reconfigurations of the Social and Natural Sciences*, Routledge, London; New York, NY.

[18] **Barnett T., Pfeiffer D.U., Ahasanul Hoque M., Giasuddin M., Flora M.S. *et al.*** 2020. Practising co-production and interdisciplinarity: Challenges and implications for one health research. *Preventive Veterinary Medicine*, 177, 104949. https://doi.org/10.1016/j.prevetmed.2020.104949

[19] **Parkes M.W., Bienen L., Breilh J., Hsu L.-N., McDonald M. *et al.*** 2005. All hands on deck: Transdisciplinary approaches to emerging infectious disease. *EcoHealth*, 2(4), 258–272. https://doi.org/10.1007/s10393-005-8387-y

[20] **Brown V.A.** 2001. Planners and the planet. *Australian Planner*, 38(2), 67–73. https://doi.org/10.1080/07293682.2001.9657941

[21] **Choi B.C., Pak A.W.** 2006. Multidisciplinarity, interdisciplinarity and transdisciplinarity in health research and policy: 1. Definitions, objectives, and evidence of effectiveness. *Clinical & Investigated Medicine*, 29, 351–364.

[22] **Pohl C., Klein J.T., Hoffmann S., Mitchell C., Fam D.** 2021. Conceptualising transdisciplinary integration as a multidimensional interactive process. *Environmental Science & Policy*, 118, 18–26. https://doi.org/10.1016/j.envsci.2020.12.005

[23] **Jamieson M.K., Govaart G.H., Pownall M.** 2023. Reflexivity in quantitative research: A rationale and beginner's guide. *Social and Personality Psychology Compass*, 17(4), e12735. https://doi.org/10.1111/spc3.12735

[24] **Whitaker E.M., Atkinson P.** 2021. *Reflexivity in Social Research, Springer International Publishing*, Cham, Switzerland.

[25] **Berling T.V., Bueger C.** 2017. Expertise in the age of post-factual politics: An outline of reflexive strategies. *Geoforum*, 84, 332–341. https://doi.org/10.1016/j.geoforum.2017.05.008

[26] **Mack O., Khare A.** 2016. Perspectives on a VUCA World. *In: Managing in a VUCA World* (O. Mack, A. Khare, A. Krämer & T. Burgartz, eds), Springer International Publishing, Cham, pp. 3–19.

[27] **Selin N.E.** 2021. Lessons from a pandemic for systems-oriented sustainability research. *Science Advances*, 7(22), eabd8988. https://doi.org/10.1126/sciadv.abd8988

[28] **Senge P.M.** 2006. *The Fifth Discipline: The Art and Practice of the Learning Organization*. Rev. and updated. edn., Doubleday/Currency, New York.

[29] **Jackson M.C.** 2019. Critical Systems Thinking and the Management of Complexity: Responsible Leadership for a Complex World, Wiley & Sons, Hoboken, NJ.

[30] **Capra F.** 2009. The new facts of life: Connecting the dots on food, health, and the environment. *Public Library Quarterly* (New York, N.Y.), 28(3), 242–248. https://doi.org/10.1080/01616840903110107

[31] **Rutter H., Savona N., Glonti K., Bibby J., Cummins S. *et al.*** 2017. The need for a complex systems model of evidence for public health. *The Lancet*, 390(10112), 2602–2604. https://doi.org/10.1016/S0140-6736(17)31267-9

[32] **Berezowski J., Ruegg S.R., Faverjon C.** 2019. Complex system approaches for animal health surveillance. *Frontiers in Veterinary Science*, 6, 153. https://doi.org/10.3389/fvets.2019.00153

[33] **Galea S., Riddle M., Kaplan G.A.** 2009. Causal thinking and complex system approaches in epidemiology. *International Journal of Epidemiology*, 39(1), 97–106. https://doi.org/10.1093/ije/dyp296

[34] **Matthiessen L.E., Hald T., Vigre H.** 2022. System mapping of antimicrobial resistance to combat a rising global health crisis. *Frontiers in Public Health*, 10. https://doi.org/10.3389/fpubh.2022.816943

[35] **Bunch M.J., Waltner-Toews D.** 2021. Grappling with complexity: the context for One Health and the ecohealth approach. *In: One Health: The Theory and Practice of Integrated Health Approaches* (Z. Jakob, S. Esther, C. Lisa, W. Maxine, T. Marcel & S. Craig, eds), CAB International, Wallingford, Oxfordshire, UK, pp. 39–51.

[36] **Burger P.A.** 2024. Integrating One Health into systems science. *One Health*, 18, 100701. https://doi.org/10.1016/j.onehlt.2024.100701

[37] **UK Government Office for Science.** 2024. Systems thinking for civil servants. https://www.gov.uk/government/publications/systems-thinking-for-civil-servants
[38] **Department for Environment F.a.R.A.D.** 2022. Integrating a systems approach into Defra. https://www.gov.uk/government/publications/integrating-a-systems-approach-into-defra/integrating-a-systems-approach-into-defra#:~:text=This%20section%20brings%20together%20the%20need%20for
[39] **European Commission – Complex System Analyser.** https://knowledge4policy.ec.europa.eu/foresight/topic/complex-system-analyser_en
[40] **OECD, International Institute for Applied Systems Analysis**. 2020. Systemic Thinking for Policy Making, OECD Publishing, Paris.
[41] **Intergovernmental Science-Policy Platform on Biodiversity and Ecosystem Services (IPBES).** 2019. Global assessment report on biodiversity and ecosystem services of the Intergovernmental Science-Policy Platform on Biodiversity and Ecosystem Services (Version 1), IPBES secretariat, Bonn, Germany, p. 1144.
[42] **Meadows D.H., Wright D.** 2008. *Thinking in Systems: a Primer*, Chelsea Green Pub., White River Junction, Vt.
[43] **Calenbuhr V. 2020**. Chapter 11. Complexity Science in the Context of Policymaking. *In: Science for Policy Handbook* (V. Šucha, M. Sienkiewicz, eds), Elsevier, pp. 118–127.
[44] **Jackson M.C.** 2024. *Critical Systems Thinking: a Practitioner's guide*, Wiley, Hoboken, New Jersey.
[45] **Fournie G., Pfeiffer D.U.** 2014. *Can closure of live poultry markets halt the spread of H7N9? Lancet*, 383(9916), 496–497. https://doi.org/10.1016/S0140-6736(13)62109-1
[46] **Pörtner H.-O., Scholes R.J., Agard J., Archer E., Arneth A. *et al.*** 2021. Scientific outcome of IPBES-IPCC co-sponsored workshop on biodiversity and climate change, Secretariat of the Intergovernmental Science-Policy Platform on Biodiversity and Ecosystem Services, Bonn, Germany.
[47] **German Advisory Council on Global Change (WBGU).** 2021. Rethinking Land in the Anthropocene: from Separation to Integration, Wissenschaftlicher Beirat der Bundesregierung Globale Umweltveränderungen (WBGU), Berlin, Germany, p. 366.
[48] **Biggs R., Clements H., de Vos A., Folke C., Manyani A. *et al.*** 2021. What are social-ecological systems and social-ecological systems research? *In: The Routledge Handbook of Research Methods for Social-Ecological Systems* (R. Biggs, A. de Vos, R. Preiser, H. Clements, K. Maciejewski & M. Schlüter, eds), Routledge, p. 24.
[49] **Preiser R., Biggs R., De Vos A., Folke C**. 2018. Social-ecological systems as complex adaptive systems: organizing principles for advancing research methods and approaches. *Ecology and Society*, 23(4), https://doi.org/10.5751/ES-10558-230446
[50] **Kuijper H**. 2022. *Comprehending the Complexity of Countries: The Way Ahead*, Springer Nature Singapore.
[51] **Mills P., Dehnen-Schmutz K., Ilbery B., Jeger M., Jones G *et al.*** 2011. Integrating natural and social science perspectives on plant disease risk, management and policy formulation. *Philosophical Transactions of the Royal Society B Biological Science*, 366(1573), 2035–2044.
[52] **Hayes L., Manyweathers J., Maru Y., Loechel B., Kelly J. *et al.*** 2021. Stakeholder mapping in animal health surveillance: A comparative assessment of networks in intensive dairy cattle and extensive sheep production in Australia. *Preventive Veterinary Medicine*, 190, 105326. https://doi.org/10.1016/j.prevetmed.2021.105326
[53] **Barraclough A.D., Cusens J., Måren I.E.** 2022. Mapping stakeholder networks for the co-production of multiple ecosystem services: A novel mixed-methods approach. *Ecosystem Services*, 56, 101461. https://doi.org/10.1016/j.ecoser.2022.101461
[54] **Keenan L., Monteath T., Wójcik D.** 2023. Hungry for power: financialization and the concentration of corporate control in the global food system. *Geoforum*, 147, 103909. https://doi.org/10.1016/j.geoforum.2023.103909
[55] **Gluckman P.** 2014. Policy: The art of science advice to government. *Nature*, 507(7491), 163–165. https://doi.org/10.1038/507163a
[56] **Commission E., Research D.-G.f. & Innovation**. 2022. Everyone at the table – Transforming food systems by connecting science, policy and society, Publications Office of the European Union.
[57] **Anon.** 2018. A code of ethics to get scientists talking. *Nature*, 555(7694), 5. https://doi.org/10.1038/d41586-018-02516-x
[58] **Aven T.** 2024. Risk science and politics: What is and should be the relationship? *Risk Analysis*, 1–9. https://doi.org/10.1111/risa.16558
[59] **European Food Safety Authority (EFSA).** 2024. EFSA's Independence Policy. *Journal*, https://www.efsa.europa.eu/sites/default/files/2024-06/factsheet-independence-policy.pdf
[60] **German Ethics Council.** 2022. Vulnerability and resilience in a crisis – Ethical criteria for decision-making in a pandemic. *In:* German Ethics Council, Berlin, Germany, p 54.

[61] **German Advisory Council on Global Change (WBGU).** 2023. Healthy living on a healthy planet, Wissenschaftlicher Beirat der Bundesregierung Globale Umweltveränderungen (WBGU), Berlin, Germany, 407.

[62] **Schwendinger F., Topp L., Kovacs V.** 2022. Competences for policymaking – Competence frameworks for policymakers and researchers working on public policy. *In:* Publications Office of the European Union, Luxembourg.

[63] **Evans M.C., Cvitanovic C.** 2018. An introduction to achieving policy impact for early career researchers. *Palgrave Communications*, 4. https://doi.org/10.1057/s41599-018-0144-2

[64] **Renn O.** 2008. Concepts of risk: An interdisciplinary review part 1: Disciplinary risk concepts. *GAIA – Ecological Perspectives for Science and Society*, 17(1), 50–66. https://doi.org/10.14512/gaia.17.1.13

[65] **Renn O.** 2008. Concepts of risk: An interdisciplinary review part 2: Integrative approaches. *GAIA – Ecological Perspectives for Science and Society,* 17(2), 196–204. https://doi.org/10.14512/gaia.17.2.7

[66] **Renn O.** 2012. Risk governance in a complex world. *In: Encyclopedia of Applied Ethics* (R. Chadwick, ed), Academic Press, London, UK, pp. 846–854.

[67] **Kiekens A., Dierckx de Casterlé B., Vandamme A.-M.** 2022. Qualitative systems mapping for complex public health problems: A practical guide. *PLOS ONE*, 17(2), e0264463. https://doi.org/10.1371/journal.pone.0264463

[68] **Cassidy R., Borghi J., Semwanga A.R., Binyaruka P., Singh N.S., Blanchet K.** 2022. How to do (or not to do)…using causal loop diagrams for health system research in low and middle-income settings. *Health Policy and Planning,* 37(10), 1328–1336. https://doi.org/10.1093/heapol/czac064

[69] **Baugh Littlejohns L., Hill C., Neudorf C.** 2021. Diverse approaches to creating and using causal loop diagrams in public health research: Recommendations from a scoping review. *Public Health Reviews*, 42. https://doi.org/10.3389/phrs.2021.1604352

[70] **Salmon P.M., Stanton N.A., Walker G.H., Hulme A., Goode N. *et al.*** 2022. *Handbook of Systems Thinking Methods.* 1st edn., CRC Press.

[71] **Bala B.K., Arshad F.M., Noh K.M.** 2017. *System Dynamics – Modelling and Simulation*, Springer, Singapore.

[72] **Barbrook-Johnson P., Penn A.S.** 2022. *Systems Mapping: How to Build and Use Causal Models of Systems*, Springer International Publishing, Cham.

[73] **International Risk Governance Council (IRGC).** 2017. Introduction to the IRGC risk governance framework, revised version. *In:* EPFL International Risk Governance Centre, Lausanne, Switzerland, p. 48.

[74] **Renn O., Klinke A., van Asselt M.** 2011. Coping with complexity, uncertainty and ambiguity in risk governance: A synthesis. *Ambio*, 40(2), 231–246.

[75] **Stirling A.** 2010. Keep it complex. *Nature*, 468(7327), 1029–1031.

[76] **Serrat O.** 2017. Theories of change. *In: Knowledge Solutions: Tools, Methods, and Approaches to Drive Organizational Performance* (O. Serrat, ed), Springer Singapore, Singapore, pp. 237–243.

[77] **Stein D., Valters C.** *2012. Understanding theory of change in international development.* The Justice and Security Research Programme.

Epilogue

As the journey through this One Health Atlas draws to a close, we recognise that we have reached a critical moment in planetary history, where the interconnected challenges of health, the environment and society call for collaboration across disciplines, sectors and national borders. This book is not merely a compilation of knowledge; it is a call to action. It urges researchers, practitioners and policymakers to forge stronger, more inclusive alliances, recognizing that no single discipline or sector can address the complexities of our time in isolation.

The insights shared in this book aim to inspire innovative solutions and enduring partnerships. From zoonotic disease surveillance to agroecological innovations, from governance frameworks to education and capacity-building, this Atlas highlights the remarkable progress made in applying the One Health approach. Yet it also serves as a reminder of the work that still remains to be done.

Some significant areas not addressed in this Atlas require further attention. One major challenge in One Health is the need to refine evaluation methods, including economic assessments, to effectively measure the impact and sustainability of initiatives. Such analyses are crucial for ensuring informed resource allocation, improving public health outcomes and fostering long-term economic stability, with tools such as the theory of change providing valuable support. A pressing issue is the risk of "One Health washing", where superficial commitment to the approach overshadows meaningful action. Additionally, addressing biases, such as an overreliance on veterinary perspectives, is vital to guarantee a more inclusive and balanced implementation. These gaps underscore the importance of continuous reflection and adaptation as the One Health field continues to evolve.

One Health is more than a framework or an approach; it is a commitment to shared responsibility. It is a recognition that the health of humans, animals and the environment are inextricably linked, and that addressing one without considering the others risks perpetuating cycles of harm. This approach calls for systemic thinking, humility to learn from diverse knowledge systems, and the courage to navigate the complexities of collaboration.

As we look to the future, we must continue to push the boundaries of what One Health can achieve. This Atlas serves as an introduction—a resource to explore the multidimensional aspects of this evolving discipline. However, it is just one step in a much broader journey that continues in laboratories, classrooms, farms, forests and communities across the globe.

François Roger
Marie-Marie Olive
Marisa Peyre
Dirk Pfeiffer
Jakob Zinsstag

Afterword

Real-world lessons from One Health practices

Bangladesh can offer valuable insights into how One Health can be institutionalized effectively and can serve as an important reference point for other countries featured in this Atlas. When we consider Bangladesh's One Health journey, one thing stands out: progress was not the result of a single initiative or institution. Instead, it came from collaboration—across disciplines, sectors and borders. The story of One Health in Bangladesh is a lesson in how a shared vision, persistent effort, partnerships and strong networks can transform a concept into real impact.

Bangladesh recognized the value of One Health early on. In 2007, during the avian influenza outbreak, it became clear that health challenges at the human, animal and environmental interface could not be tackled in silos. This realization brought together professionals—doctors, veterinarians, agriculturists, environmentalists, wildlife experts, ecologists, anthropologists, economists, public health practitioners and activists—who initiated informal discussions that soon gained momentum. An international conference organized in Chattogram in 2008 with the theme "Changing World, Emerging Challenges: A One World One Health Approach" solidified these efforts, leading to the formation of One Health Bangladesh, a civil society platform dedicated to promoting interdisciplinary collaboration.

Over time, One Health Bangladesh evolved into more than just an advocacy group. It became a dynamic community of practice, engaging government agencies, research institutions and international organizations. The establishment of the One Health Secretariat in 2016 and its government funding in 2017 demonstrated institutional buy-in, ensuring One Health became embedded in policy and practice.

One Health in Bangladesh is more than a theoretical framework, or a forum for exchanging ideas—it is a catalyst for real and impactful change. National strategies have led to better surveillance, outbreak investigations and cross-sectoral responses to emerging and zoonotic infectious diseases. Joint initiatives, such as mass rabies vaccination campaigns and antimicrobial resistance (AMR) control programmes, illustrate the power of a multisectoral approach. These activities have been underpinned by enactment into legislation.

The country's One Health approach has also fostered international collaboration. Projects like the One Health Poultry Hub and the Fleming Fund's AMR initiatives have strengthened Bangladesh's research capacity and policy influence. Bangladeshi experts are now recognized on global platforms, shaping international One Health strategies and funding priorities.

Bangladesh is not unique in having developed a national One Health framework, though it has been more successful than most in terms of institutionalization and garnering international support. What can others learn from Bangladesh's One Health journey? First, sustained commitment is essential. Institutionalizing One Health takes time and requires consistent advocacy, research and capacity-building. Second, partnerships matter. The most significant breakthroughs happen when diverse actors—from government agencies to grassroots organizations—work together. Lastly, the approach must be flexible. Emerging health threats evolve, and so must our responses. The adaptability of Bangladesh's One Health movement has been key to its success. These experiences align closely with the broader themes explored in this Atlas, particularly regarding governance mechanisms, multisectoral collaboration and global engagement.

The journey is far from over. As Bangladesh continues to refine its One Health strategies, the lessons learned can guide other nations seeking to integrate human, animal and environmental health. The country's experience shows that when we break down silos and collaborate, we do more than just improve health outcomes—we build resilience for the future.

Nitish Debnath and Md Ahasanul Hoque
Chattogram Veterinary and Animal
Sciences University, Bangladesh

Reference

Debnath N., Rahman M., Sabrina Flora M., Shirin T., Kalam M.A. *et al.* 2024. *Toward the Institutionalization of a One Health Agenda: What the World can Learn from Bangladesh.* CABI One Health Cases. https://doi.org/10.1079/onehealthcases.2024.0018

Glossary

Agroecology

Agroecology is a holistic approach to agriculture that integrates sustainable farming, biodiversity conservation and social equity. It aligns with One Health by promoting healthy ecosystems, reducing disease risks and enhancing food security. By minimizing chemical inputs and fostering resilient communities, agroecology supports the interconnected health of humans, animals and the environment. As a key component of agroecology, **agroforestry** integrates trees, crops and livestock to improve biodiversity, soil health and ecosystem services. Within the One Health framework, it contributes to sustainable food production, ecosystem resilience and climate change mitigation to benefit human, animal and environmental health. By diversifying habitats and reducing land degradation, agroforestry also helps regulate disease dynamics and strengthens the links between agriculture and natural ecosystems.

Antimicrobial resistance (AMR)

Antimicrobial resistance is the ability of microorganisms such as bacteria, viruses, fungi and parasites to withstand the effects of antimicrobial drugs, rendering standard treatments ineffective. AMR is a critical global challenge that threatens the effectiveness of medicines and increases the risks of severe infections, longer hospital stays and higher mortality rates. The One Health approach addresses AMR by promoting responsible **antimicrobial use** (AMU), strengthening surveillance across sectors and reducing environmental contamination.

Disease occurrence

An **outbreak** is the sudden occurrence of a disease in a localized area or population, often involving infectious agents or environmental factors, such as a *Salmonella* outbreak caused by contaminated food. If an outbreak is not contained, it may escalate into an **epidemic**, where the disease occurs at rates significantly higher than expected within a specific region, as seen in the 2014–2016 Ebola epidemic in West Africa. In contrast, an **endemic disease** represents the consistent presence of a disease within a particular geographic area or population, such as malaria in tropical regions, reflecting a stable, baseline level of occurrence. A **pandemic**, the most widespread category, occurs when an epidemic spreads across multiple countries or continents, affecting large populations, as exemplified by the COVID-19 pandemic or the 1918 influenza pandemic. Through a unifying One Health perspective, these terms apply across human, animal and plant health, which highlights the interconnected challenges of managing diseases within ecosystems.

Drivers shaping One Health challenges

Key drivers shaping One Health challenges include biodiversity loss, climate change, land-use change, intensive agriculture, urbanization, globalization, socioeconomic inequalities and gender dynamics. Biodiversity loss disrupts ecosystems, increasing the risk of zoonotic spillovers. Climate change alters disease patterns, vector distributions and food security. Deforestation and habitat destruction bring humans, livestock and wildlife into closer contact, heightening disease transmission risks. Intensive agriculture and antimicrobial use drive antimicrobial resistance, while urbanization and globalization accelerate disease spread through trade and travel. Social determinants, including gender inequalities and marginalization, further exacerbate health disparities and access to resources. Gender roles influence exposure to environmental and health risks, as well as access to healthcare, education and decision-making power. Women, for example, often play critical roles in food production, caregiving and disease prevention but may face barriers in accessing knowledge and resources necessary for resilience. Socioeconomic disparities also shape vulnerability to climate change, infectious diseases and environmental degradation, affecting the ability of communities to respond effectively. Addressing these interconnected drivers through a One Health approach requires integrating environmental, social and economic dimensions to develop inclusive, equitable and sustainable health solutions for humans, animals and ecosystems.

Ecology and health

Ecology is the scientific study of interactions between organisms and their environment, including relationships within ecosystems and the influence of abiotic factors such as climate, soil and water. This discipline plays a crucial role in health sciences, as ecological changes impact human, animal and plant health by influencing disease dynamics, biodiversity and environmental sustainability. **Health ecology** and One Health are two complementary frameworks: while health ecology provides a foundational understanding of the systemic determinants of health, One Health operationalizes this perspective through interdisciplinary collaboration and policy implementation. **EcoHealth** promotes systems thinking, transdisciplinary collaboration and community engagement to develop sustainable solutions. It recognizes that healthy ecosystems are essential for human and animal well-being and seeks to prevent and mitigate health risks arising from environmental degradation.

Environment, ecosystem and One Health

In the field of One Health, the terms "environment" and "ecosystem" are widely used, often interchangeably or imprecisely. This lack of clarity can obscure the understanding of health-environment interactions, even though these dynamics are central to the One Health approach, whose overarching goal is to sustainably balance and optimize the health of people, animals and ecosystems (see the One Health entry in this glossary). The **environment** refers to all abiotic (air, water, soil, climate) and biotic (fauna, flora, microbiota) elements surrounding a living organism. In a One Health context, it also includes human-modified environments—such as agroecosystems, urban areas, landfills and infrastructure—that directly or indirectly affect health through exposure to pathogens, pollutants or global environmental pressures. The concept is often associated with the **exposome**, which encompasses the totality of exposures experienced over a lifetime. An **ecosystem** refers to a functional unit composed of a physical environment (biotope) and the living communities interacting within it (biocenosis). The ecological processes at play—such as nutrient cycling, trophic networks and pathogen dynamics—play a key role in regulating health-related equilibria. Thinking in terms of ecosystems allows us to understand health as the outcome of complex interactions among biological, physical and in some cases anthropogenic factors. It thus helps reposition One Health within an ecological and territorialized perspective.

Epidemiology

Epidemiology is the scientific discipline that studies the distribution, determinants and dynamics of health and disease conditions across human, animal and plant populations. Epidemiologists investigate patterns of disease occurrence, identify risk factors and evaluate the effectiveness of health interventions. By analysing data from various sources, epidemiology can produce essential insights for understanding disease outbreaks, tracking the spread of infections and implementing control measures. It serves as a cornerstone of public health planning, policymaking and the development of strategies for disease prevention and management, including within interdisciplinary approaches such as One Health.

Food system

A food system encompasses all the processes and actors involved in producing, processing, distributing, consuming and disposing of food. It includes biophysical, economic, social and environmental dimensions, shaping how food is grown, accessed and utilized. **Food security** ensures that all people in a particular population, at all times, have access to sufficient, safe and nutritious food for a healthy life. **Food safety** focuses on the proper handling, preparation and storage of food to prevent contamination and foodborne illnesses. Both food security and food safety are integral to the One Health approach. Ensuring a safe and sustainable food system requires interdisciplinary collaboration across agriculture, veterinary science, public health and environmental management.

Global health

Global health is a multidisciplinary field that addresses human health issues and challenges that transcend national boundaries, focusing on improving health equity worldwide. It emphasizes disease prevention and management, efforts to strengthen health systems, and the social, environmental and economic determinants of health. Global health seeks to develop collaborative solutions through research, policymaking and capacity-building across nations and disciplines. It is closely linked with concepts such as One Health and Planetary Health in addressing global challenges such as pandemics, antimicrobial resistance and climate change.

Health

As defined by the World Health Organization (WHO) with regard to humans, "health is a state of complete physical, mental and social well-being, and not merely the absence of disease or infirmity". It encompasses a dynamic balance between an individual's biological, psychological and social conditions and their environment. Health is influenced by a range of factors, including genetics, lifestyle, access to healthcare, socioeconomic status and environmental conditions. In frameworks like One Health, the concept of health extends beyond humans to include the well-being of animals and ecosystems to emphasize the interconnectedness of all living systems.

Health anthropology

Health anthropology is a branch of anthropology that examines health, illness and healthcare systems in cultural and social contexts. It explores how beliefs, practices and social structures influence health outcomes and the ways communities perceive and manage diseases. Within a One Health context, health anthropology bridges gaps between biomedical approaches and local knowledge, fostering culturally sensitive health interventions.

Interdisciplinarity

Interdisciplinarity refers to the integration of concepts, methodologies and knowledge from multiple scientific disciplines to tackle complex problems. An interdisciplinary approach combines diverse perspectives to enable a deeper understanding of challenges like climate change, pandemics and sustainable development.

Microbiology and parasitology

These two scientific fields refer to the study of microorganisms—including bacteria, viruses, fungi and protozoa—and parasitic organisms, such as helminths (parasitic worms) and ectoparasites (ticks, fleas and lice) that can cause disease in humans, animals and plants. Both disciplines play a critical role in understanding infectious diseases, antimicrobial resistance and pathogen transmission. Within the One Health framework, microbiology and parasitology are necessary for identifying and tracking zoonotic pathogens, studying vector-borne diseases and developing vaccines, diagnostic tools and antimicrobial strategies to mitigate health risks across species and ecosystems.

One Health

One Health is an integrated, unifying approach that aims to sustainably balance and optimize the health of people, animals and ecosystems. It recognizes that the health of humans, domestic and wild animals, plants and the wider environment (including ecosystems) are closely linked and interdependent. The approach mobilizes multiple sectors, disciplines and communities at varying societal levels to work together to foster well-being and tackle threats to the health of humans, animals and ecosystems, while taking action on climate change, contributing to sustainable development to address the collective need for healthy food, water, energy and air. This definition is supported by the Quadripartite organizations: the Food and Agriculture Organization of the United Nations (FAO), the United Nations Environment Programme (UNEP), the World Health Organization (WHO), and the World Organisation for Animal Health (WOAH, formerly OIE).

One Health surveillance

Epidemiological surveillance is a systematic approach to collecting, analysing, interpreting and disseminating data on population health to monitor the evolution of diseases and other health indicators. Its primary objectives are to detect health anomalies early, evaluate the effectiveness of interventions and guide public health policies and actions. From a One Health perspective, epidemiological surveillance should integrate data from human, animal and environmental health sectors to identify, monitor and address health challenges that span species and ecosystems.

Participatory science

Participatory science involves the active engagement of non-scientists—such as community members, farmers or local stakeholders—in the research process to generate knowledge, solve problems and drive collective action. In the context of One Health, participatory science plays a key role in integrating local expertise and lived experiences to better address the interconnected challenges of human, animal and environmental health.

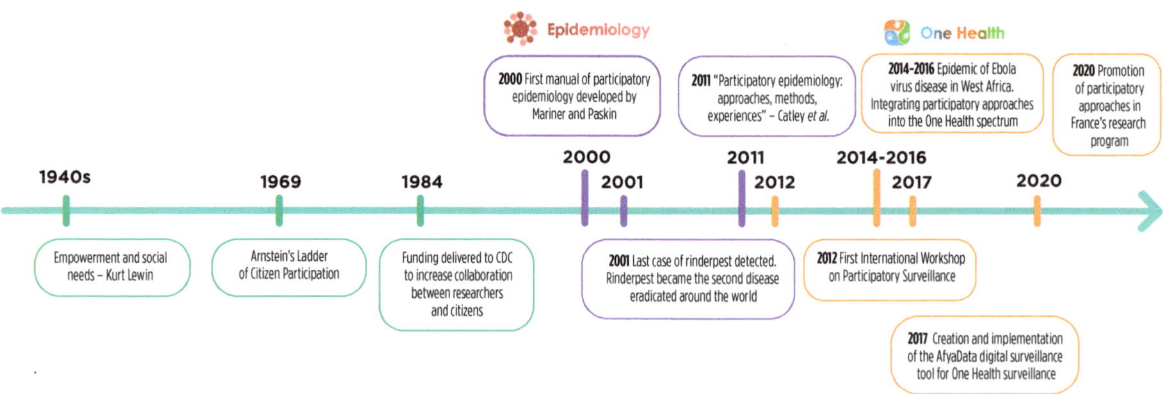

Timeline of the inclusion of the participatory approach within One Health.

Planetary health

Planetary health is an emerging field that examines the links between human health and the Earth's natural systems. One of the key principles of planetary health is sustainable development that stays within the planet's ecological boundaries because human health and well-being depends on a healthy environment.

Social sciences

The social sciences are scientific disciplines studying societies and the social interactions between individuals, groups and their environments. The One Health approach benefits from the social sciences by taking into account the sociocultural, economic and political factors that influence health. Disciplines such as anthropology, sociology, economics and political science support understanding of risk perception, behaviours, governance and equity, facilitating the development and implementation of more effective and inclusive health interventions. Integrating the social sciences into One Health enhances interdisciplinary collaboration and promotes sustainable solutions to global health challenges.

Sustainability science

Sustainability science is a transdisciplinary field that explores how to meet our needs today without endangering the needs of future generations. It addresses the intersection of environmental, social and economic systems to identify pathways for sustainable development.

Systems thinking

Systems thinking is an essential approach for analysing and understanding complex systems, which are characterized by dynamic interactions, feedback loops and emergent behaviours. Unlike a linear and reductionist perspective, it allows us to grasp complexity holistically by considering interdependencies and systemic effects. In a field such as One Health, this approach is crucial for identifying leverage points, preventing unintended consequences and designing more resilient interventions. To apply systems thinking, one maps key interactions, engages stakeholders in a participatory process, and uses modelling tools to test scenarios, predict impacts and guide adaptive decisions.

Transdisciplinarity

Transdisciplinarity is an approach that transcends traditional disciplinary boundaries to integrate scientific knowledge, stakeholder insights and practical expertise. This approach seeks to create holistic solutions to complex problems by engaging diverse perspectives, including those of non-academic actors.

Zoonosis (plural: zoonoses)

A zoonosis is an infectious disease that is transmitted between animals and humans and vice-versa. Zoonoses can result from direct contact with infected animals, consumption of contaminated food or exposure to environments where pathogens thrive. Examples include rabies, avian influenza and COVID-19. Integrated approaches such as One Health are key to understanding zoonoses, which is critical for preventing outbreaks and mitigating their public health and economic impacts.

General references

Agropolis International. 2019. *Global Health: People, Animals, Plants, the Environment: Towards an Integrated Approach to Health* (Agropolis Dossier No. 25). Agropolis International, Montpellier, France. Available in both English and French versions at https://www.agropolis.fr/dossier-agropolis-sante-globale

This Agropolis International dossier reviews the diverse expertise and systemic health approaches developed by the scientific community in the France's Occitanie region. With a critical mass of research in the medical, agricultural, veterinary, ecological, environmental and social sciences, this community promotes the development and implementation of original concepts and methods. The dossier presents numerous examples illustrating North-South partnerships, including health issues related to diets and food production methods.

FAO. 2024. *One Health Training Manual.* Cairo. https://doi.org/10.4060/cc9282en

The FAO developed a training manual as a national reference in Egypt to raise awareness of the One Health approach among the country's health professionals. The document explains some of the joint actions to be implemented by the ministries responsible for human, animal and environmental health. It provides a useful model for other countries looking to adopt a One Health strategy.

Garcia Pinillos R., Huertas Canén S.M., Eds. 2023. *One Welfare: Animal Health and Welfare, Food Security and Sustainability.* CABI. https://doi.org/10.1079/9781789249507.0000

This book combines theory and real-world case studies to examine the One Welfare concept, highlighting the ways animals, people and their environments are connected through livestock farming, food safety, food security and sustainability. The authors are international experts who cover an extensive range of issues tied to animal-related food production systems, touching on economic aspects, welfare indicators, stress and best practices.

Häsler B., Tvarijonaviciute A., Savic S. 2024. *Principles of One Health for a Better Planet.* CABI. https://doi.org/10.1079/9781800623002.0000

This book offers concrete methods to strengthen cooperation across disciplines in order to better anticipate and manage emerging health risks. It outlines tools and approaches to promote sustainable health through an integrated vision of environments, ecosystems and public health. By adopting a systemic perspective, it encourages professionals to embrace an interdisciplinary approach, highlighting the importance of interconnected knowledge and collective action.

Ottinger M.A., Geiselman C.K. 2023. *One Health Meets the Exposome: Human, Wildlife, and Ecosystem Health.* Academic Press. https://doi.org/10.1016/C2020-0-03393-1

This book explores the intersection of One Health and the exposome, emphasizing how environmental exposures impact human, animal and ecosystem health. The book discusses the role of pollutants, climate change and emerging diseases in shaping health outcomes, highlighting interdisciplinary approaches to address these challenges. It provides case studies and frameworks to integrate exposome research into One Health strategies for better risk assessment and policymaking.

Peyre M., Roger F., Goutard F., Eds. 2022. *Principles for Evaluation of One Health Surveillance: The EVA Book.* Springer International Publishing, Cham. https://doi.org/10.1007/978-3-030-82727-4

This book examines the evaluation of epidemiological surveillance systems through a One Health perspective. It introduces key theoretical foundations and highlights the main challenges involved. A broad range of traditional and innovative methods are presented, including economic evaluation approaches. Each method is illustrated with concrete case studies from animal health and One Health surveillance, drawn from both the Global North and the Global South.

Rüegg S.R., Häsler B., Zinsstag J., Eds. 2018. *Integrated Approaches to Health.* Wageningen Academic, Leiden, The Netherlands. https://doi.org/10.3920/978-90-8686-875-9

This handbook, the result of an interdisciplinary effort by the Network for Evaluation of One Health (NEOH), offers a scientific guide for evaluating One Health approaches. It presents a structured and standardized method based on a systems perspective, drawing on concepts from the health sciences, social sciences, economics and ecology. Practical protocols are provided to help plan and carry out effective evaluations. The handbook is intended for researchers, practitioners, evaluators and funders engaged in integrated health initiatives.

SPILF, CMIT, SFMTSI, SMV. 2022. *ePILLY Trop* [online]. 3rd Web Edition. Éditions Alinéa Plus, Paris. https://www.infectiologie.com/fr/pillytrop.html [French]

The 2022 edition of ePILLY Trop is a free, French-language online reference on infectious diseases, with a focus on tropical medicine. It targets physicians and medical students in francophone countries of the Global South, as well as healthcare professionals in the Global North managing infections in migrants and travellers. Practical and educational, it offers recommendations adapted to limited-resource settings. This updated edition includes new epidemiological data and diagnostic and therapeutic advances, as well as chapters on the One Health approach.

Zinsstag J., Schelling E., Crump L., Whittaker M., Tanner M., Stephen C., Eds. 2021. One Health: The Theory and Practice of Integrated Health Approaches (2nd ed.). CABI. https://doi.org/10.1079/9781789242577.0000

French version: Zinsstag J., Schelling E., Waltner-Toews D., Whittaker M., Tanner M. 2017. *One Health, une seule santé.* Théorie et pratique des approches intégrées de la santé. Éditions Quæ, Versailles, France. https://doi.org/10.35690/978-2-7592-3097-6

This book provides a clear and practical introduction to the One Health concept. It explores the evolution of the approach and focuses on zoonotic diseases linked to wildlife, domestic animals and livestock. It provides practical tools for research, including methodological frameworks, data collection and monitoring techniques, study designs and modelling. Key issues such as food security are also addressed. Using case studies to illustrate its main points, the book helps bridge theory and practice, making it a valuable resource for professionals involved in One Health initiatives.

List of acronyms

A

A4NH CGIAR Research Program on Agriculture for Nutrition and Health

ABC Brazilian Cooperation Agency

ACP Agroecological crop protection

ADB Asian Development Bank

ADFD Abu Dhabi Fund for Development

AFD Agence Française de Développement/ French Development Agency

AFF-CF Africa Frontline First Catalytic Fund

AFROHUN Africa One Health University Network

AHF AIDS Healthcare Foundation

AI Artificial intelligence

AMR Antimicrobial resistance

AMS Antimicrobial stewardship

AMU Antimicrobial use

ANSES Agence nationale de sécurité sanitaire de l'alimentation, de l'environnement et du travail/French Agency for Food, Environmental and Occupational Health and Safety

APCOVE Asia Pacific Consortium of Veterinary Epidemiology

ARISE Accelerating Research on Inclusive and Sustainable Epidemics

ASEAN Association of Southeast Asian Nations

AU-IBAR African Union–Interafrican Bureau for Animal Resources

AusAID Australian Agency for International Development

AVSF Agronomes & Vétérinaires Sans Frontières

B

BCOMING Biodiversity Conservation to Mitigate the risks of emerging infectious diseases

BUILD Boosting Uganda's Investments in Livestock Development

C

C19RM COVID-19 Response Mechanism

CAHW Community-based animal health worker

CCM Comitato Collaborazione Medica

CDC Centers for Disease Control and Prevention

CEPI Coalition for Epidemic Preparedness Innovations

CFI Canal France International

CGIAR Consultative Group on International Agricultural Research

CHI Center for Health International

CHW Community-based health worker

CIDCA China International Development Cooperation Agency

CIPARS Canadian Integrated Program for Antimicrobial Resistance Surveillance

CIRAD Centre de coopération internationale en recherche agronomique pour le développement/French Agricultural Research Centre for International Development

CMBP Common Microbial Biotechnology Platform

COHESA Capacitating One Health in Eastern and Southern Africa

CORDS Connecting Organisations for Regional Disease Surveillance

COVARS Comité de Veille et d'Anticipation des Risques Sanitaires/French Committee for Monitoring and Anticipating Health Risks

COVAX COVID-19 Vaccines Global Access

COVID-19 Coronavirus disease 2019

D

DALY Disability-adjusted life year

DRC Democratic Republic of the Congo

DAC Development Assistance Committee

Debt2Health Debt Relief for Health Initiative

DIF Development Impact Fund

E

EAC East African Community

ECTAD Emergency Centre for Transboundary Animal Diseases (FAO)

EFSA European Food Safety Authority

EHESP École des Hautes Études en Santé Publique/Public Health grande école

EID Emerging infectious disease

ESBL-Ec Extended-spectrum β-lactamase (*Escherichia coli*)

EU European Union

F

FAIR Findable, accessible, interoperable and reusable

FAO Food and Agriculture Organization

FAWC UK Farm Animal Welfare Committee (now the Animal Welfare Committee - AWC)

FBLI Field Building Leadership Initiative

FCDO Foreign, Commonwealth & Development Office

FETN Field Epidemiology Training Network

FISONG Facilité d'Innovation Sectorielle ONG/ Sectoral Innovation Fund for NGOs (AFD)

FSC Sustainable Finance Corporation

G

GARC Global Alliance for Rabies Control

GAVI Global Alliance for Vaccines and Immunization (now known as Gavi, the Vaccine Alliance)

GDP Gross domestic product

GHG Greenhouse gas

GHSA Global Health Security Agenda

GIZ German Agency for International Cooperation

Global Fund Global Fund to Fight AIDS, Tuberculosis and Malaria

GREASE Global Research Alliance for Emerging and Animal Diseases

H

HAT Human African trypanosomiasis

HEAL Humans, Environment, Animals and Livelihoods project

HIAF Health impact assessment framework

HORN One Health Regional Network for the Horn of Africa

HPAI Highly pathogenic avian influenza

I

IBRD International Bank for Reconstruction and Development

ICT4Health Information and communication technologies for health

IDA International Development Association

IDRC Canadian International Development Research Centre

IFFIm International Finance Facility for Immunisation

IHD International Health Division

IHR International Health Regulations (WHO)

ILRI International Livestock Research Institute

IMF International Monetary Fund

ImpresS Impact of research in the South

INGSA International Network for Governmental Science Advice

INRAE Institut National de Recherche pour l'Agriculture, l'Alimentation et l'Environnement/ French National Research Institute for Agriculture, Food and Environment

INSERM Institut National de la Santé et de la Recherche Médicale/French National Institute of Health and Medical Research

IPBES Intergovernmental Science-Policy Platform on Biodiversity and Ecosystem Services

IPCC Intergovernmental Panel on Climate Change

IRD Institut de Recherche pour le Développement/French National Research Institute for Sustainable Development

IsDB Islamic Development Bank

IUCEA Inter-University Council for East Africa

J

JICA Japan International Cooperation Agency

K

KFAED Kuwait Fund for Arab Economic Development

M

MBDS Mekong Basin Disease Surveillance network

MCE Multicriteria evaluation

MERS-CoV Middle East respiratory syndrome coronavirus

MTBC Mycobacterium tuberculosis complex

N

NADPM&RCP Bhutan National Accelerated Dog Population Management and Rabies Control Program

NAPRE Indian National Action Plan for Rabies Elimination

NARS National Agricultural Research Systems

NDB New Development Bank (BRICS)

NGO Non-governmental organization

NHL National Health Laboratory

NRCP Indian National Rabies Control Program

NZD Neglected zoonotic diseases

O

OH One Health

OH4HEAL One Health for Humans, Environment, Animals and Livelihoods

OHEJP One Health European Joint Programme

OHHLEP One Health High-Level Expert Panel

OHI CGIAR Initiative on One Health

OHJPA One Health Joint Plan of Action

OHRECA One Health Research, Education and Outreach Centre in Africa

OHTC One Health training catalogue

OIE Office International des Epizooties (now the WOAH)

P

PAZ People, Animals and their Zoonoses

PEF Pandemic Emergency Financing Facility

PEP Post-exposure prophylaxis

PEPFAR US President's Emergency Plan for AIDS Relief

PEPR Priority Research Programme and Equipment

PPM Preventive medicine program

PPP Public–private partnership

PPR Peste des petits ruminants

PREZODE Preventing Zoonotic Disease Emergence

PROPARCO Promotion and Participation for Economic Cooperation

PVS Performance of veterinary services

Q

QFFD Qatar Fund for Development

QGIS Quantum geographic information system

R

RLMF Resilient Livelihoods and Market Fund

Rossotrudnichestvo Russian Federal Agency for International Cooperation

RVF Rift Valley fever

RVFV Rift Valley fever virus

S

SAARC South Asian Association for Regional Cooperation

SACIDS Southern African Centre for Infectious Disease Surveillance

SADPA South African Development Partnership Agency

SafePORK Market-based approaches to improving the safety of pork in Vietnam

SAPRE State Action Plan for Rabies Elimination (individual states in India)

SARS Severe acute respiratory syndrome

SARS-CoV/SARS-CoV-2 Severe acute respiratory syndrome coronavirus

SCAP-RU Community-based early warning and response unit (Niger)

SDC Swiss Agency for Development and Cooperation

SDGs Sustainable Development Goals (UN)

SEAOHUN Southeast Asia One Health University Network

SEFASI Selecting Efficient Farm-level Antimicrobial Stewardship Interventions

SES Social-ecological system

SFD Saudi Fund for Development

SGPE Secrétariat Général à la Planification Écologique/French General Secretariat for Ecological Planning

SPF Santé Publique France/Public Health France

SWM Sustainable Wildlife Management Programme

T

TARTARE Burden of Foodborne Disease Project – Ethiopia

TB Tuberculosis

ToC Theory of change

TZG Tripartite Zoonoses Guide (FAO, WHO, WOAH)

U

UAA Utilized agricultural area

UNAIDS Joint United Nations Programme on HIV/AIDS

UNEP United Nations Environment Programme

UNICEF United Nations International Children's Emergency Fund

Unitaid International initiative for medicine procurement

USAID United States Agency for International Development

USDA United States Department of Agriculture

VSF Vétérinaires Sans Frontières

WAHIS World Animal Health Information System

WASH Water, sanitation and hygiene

WB World Bank

WBGU German Advisory Council on Global Change

WCS Wildlife Conservation Society

WHO World Health Organization

WNF West Nile fever

WNV West Nile virus

WOAH World Organisation for Animal Health (formerly the OIE)

WP Work package

ZACAM Zone Atelier santé-environnement Camargue (France)

ZooLinK Zoonoses in Livestock in Kenya

List of authors

A

Abadie Catherine, CIRAD, Turrialba, Costa Rica

Abila Ronello, WOAH, Bangkok, Thailand

Aenishaenslin Cécile, University of Montreal, Montréal, Canada

Alokit Christine, CABI, Entebbe, Uganda

Andriamandimby Soa-Fy, IPM, Antananarivo, Madagascar

Angot Jean-Luc, MASA, Paris France

Antoine-Moussiaux Nicolas, University of Liège, Liège, Belgium

Arsevska Elena, CIRAD, Montpellier, France

Auplish Aashima, FAO, Rome, Italy

B

de Balogh Katinka, FAO, Rome, Italy

Bart Jean-Mathieu, IRD, Conakry, Guinea

Baufumé Servane, CIRAD, Montpellier, France

Berthe Franck, World Bank, Dakar, Senegal

Bett Bernard, ILRI, Nairobi, Kenya

Bhattacharyya Pranaba Nanda, Department of Botany, Nanda Nath Saikia College, Assam, India

Binot Aurélie, MSH, CIRAD, Montpellier, France

Bonfoh Bassirou, Centre suisse de recherches scientifiques en Côte d'Ivoire, Abidjan, Côte d'Ivoire

Booijink Yoeri, CIRAD, Montpellier, France

Bordier Marion, CIRAD, Dakar, Senegal

Bourgarel Mathieu, CIRAD, Maputo, Mozambique

Boutin Nathalie, CNRS, Montpellier, France

Bucher Alvar, Swiss TPH, Basel, Switzerland

Bucheton Bruno, IRD, Montpellier, France

C

Caffier Camille, University of Montpellier, Montpellier, France

Camara Mamadou, PNLMTN, Conakry, Guinea

Cappelle Julien, CIRAD, Montpellier, France

Cardinale Éric, ANSES, Maisons-Alfort, France

Caron Alexandre, CIRAD, ILRI, Nairobi, Kenya

Chege Florence, CABI, Nairobi, Kenya

Chevalier Véronique, CIRAD, Antananarivo, Madagascar

Chirouze Émilie, CIRAD, Montpellier, France

Chitnis Nakul, Swiss TPH, Basel, Switzerland

Choden Kinley, WOAH, Bangkok, Thailand

Conan Anne, CIRAD, Harare, Zimbabwe

Costis Claire, GRET, Nogent-sur-Marne, France

Crump Lisa, Swiss TPH, Basel, Switzerland

D

Danielsen Solveig, CABI, Leusden, The Netherlands

De Nys Hélène, CIRAD, Harare, Zimbabwe

Debnath Nitish, Chattogram Veterinary and Animal Sciences University, Chittagong, Bangladesh

Debnath Partho Pratim, Chulalongkorn University, Bangkok, Thailand

Deguine Jean-Philippe, CIRAD, Montpellier, France

Delabouglise Alexis, CIRAD, Montpellier, France

Destoumieux-Garzon Delphine, CNRS, Montpellier, France

Desvars-Larrive Amélie, University of Veterinary Medicine & Complexity Science Hub, Vienna, Austria

Dieuzy-Labaye Isabelle, GALVmed, Edinburgh, UK

Dimov Artemiy, Swiss TPH, Basel, Switzerland

Duboz Raphaël, CIRAD, Dakar, Senegal

Dukpa Kinzang, WOAH, Tokyo, Japan

Durand Benoit, ANSES, Maisons-Alfort, France

Duru Michel, INRAE, Honorary Researcher, Toulouse, France

E

Etter Éric, CIRAD, Petit-Bourg, France

F

Fascendini Micol, Amref Health Africa, Rome, Italy

Faverjon Céline, EpiMundi, Lyon, France

Fink Guenther, Swiss TPH, Basel, Switzerland

Fournet Florence, IRD, Montpellier, France

Fournié Guillaume, INRAE, Lyon, France

Furco André, WOAH, Bangkok, Thailand

G

de Garine-Wichatitsky Michel, CIRAD, Montpellier, France

Gay Noellie, freelance consultant, Noumea, New Caledonia

Gonzales Mark Jaypee, WOAH, Bangkok, Thailand

Goumou Souana, CERFIG, Conakry, Guinea

Goutard Flavie, CIRAD, Hanoi, Vietnam

Guenin Marie-Jeanne, CEFE-CNRS, Montpellier, France

Guerrero-Sanchez Sergio, Centre for Applied One Health Research and Policy Advice, City University of Hong Kong, Hong Kong SAR, China

H

Häsler Barbara, FAO, Rome, Italy

Hautefeuille Claire, CIRAD, Montpellier, France

Hinjoy Soawapak, Ministry of Public Health, Nonthaburi, Thailand

Herrmann Laetitia, Deakin University, Geelong, Australia.

Hibbard Rebecca, ENVT-INRAE, Toulouse,France

Hoffman Vivian, IFPRI, Washington, United States

Hoque Md Ahasanul, One Health Institute, Chattogram Veterinary and Animal Sciences University, Chittagong, Bangladesh

Hussain Salman, UNEP-TEEB, Geneva, Switzerland

I

Iraki Bibiana, ISAAA AfriCenter, Nairobi, Kenya

J

Justeau-Allaire Dimitri, IRD, Conakry, Guinea

K

Kabir Kazi Ahmed, CIRAD, Phnom Penh, Cambodia

Kagbadouno Moïse, PNLMTN, Conakry, Guinea

Karembu Margaret, ISAAA AfriCenter, Nairobi, Kenya

Kassié Daouda, CIRAD, Antananarivo, Madagascar

Keatts Lucy, Wildlife Conservation Society, Washington DC, USA

Keita Alpha Kabinet, CERFIG-UGANC, Conakry, Guinea; IRD, Montpellier, France

Knight-Jones Theo, ILRI, Nairobi, Kenya

Kugita Hirofumi, WOAH, Tokyo, Japan

L

Lahellec Brieuc, CFI, Paris, France

Lam Chun Ting Angus, Centre for Applied One Health Research and Policy Advice, Hong Kong, China

Lam Steven, ILRI, Nairobi, Kenya

Lambert Marie-Christine, CIRAD, Montpellier, France

Laury Emmanuel, CIRAD, Montpellier, France

Lee-Cruz Larisa, CIRAD, Montpellier, France

Lefrançois Thierry, CIRAD, Paris, France

Lesueur Didier, CIRAD, Hanoi, Vietnam

Leyens Stéphane, University of Namur, Namur, Belgium

Loire Étienne, CIRAD, Hanoi, Vietnam

M

Mathevet Raphaël, CNRS, Montpellier, France

Meunier Julie, CIRAD, Montpellier, France

Miller Manuelle, AVSF, Lyon, France

Mispiratceguy Manon, FAO, Bangkok, Thailand

Munyeme Musso, University of Zambia, Lusaka, Zambia

Muset Sophie, WOAH, Paris, France

N

Netanyahu Sinaia, WHO, Bonn, Germany

Nguyen Thi Dien, Vietnam National University of Agriculture, Hanoi, Vietnam

Nguyen-Viet Hung, ILRI, Nairobi, Kenya

O

Olive Marie-Marie, CIRAD, Montpellier, France

Onivogui Dobo, GRET, N'Zérékoré, Guinea

P

Parmley Jane, University of Guelph, Ontario, Canada

Paul Mathilde, ENVT-INRAE, Toulouse,France

Peeters Martine, IRD, Montpellier, France

Peyre Marisa, CIRAD, Montpellier, France

Pfeiffer Dirk, CityU, Hong Kong, China; RVC, London, UK

Phu Tran Minh, Can Tho University, Can Tho, Vietnam

Pieretti Isabelle, CIRAD, Montpellier, France

Pruvot Mathieu, University of Calgary, Calgary, Canada

R

Raj Eleanor, FAO, Rome, Italy

Rakotoarison Hobiniaina Anthonio, IPM, Antananarivo, Madagascar

Ratiarison Sandra, FAO, Libreville, Gabon

Ratnadass Alain, CIRAD, Saint-Pierre, France

Raynal Jean-Claude, CNRS, Montpellier, France

Richards Shauna, ILRI, Nairobi, Kenya

Rieux Adrien, CIRAD, Saint-Pierre, France

Roche Mathieu, CIRAD, Montpellier, France

Rodkhum Channarong, Chulalongkorn University, Bangkok, Thailand

Roger François, CIRAD, Hanoi, Vietnam

Rosset Bruno, CIRAD, Montpellier, France

Rüegg Simon, University of Zurich, Zurich, Switzerland

S

Salem Gerard, University of Paris-Nanterre, Paris, France

Sarter Samira, CIRAD, Montpellier, France

Schelling Esther, Swiss TPH, Basel, Switzerland

Seck Papa Serigne, PREZODE, Dakar, Senegal

Sheath Danny, FAO, Rome, Italy

Soubeyran Emmanuelle, WOAH, Paris, France

Soumah Abdoul Karim, CERFIG, Conakry, Guinea

Srivastava Rahul, WOAH, Paris, France

Suit-B Yong Chyna, WOAH, Bangkok, Thailand

T

Talla Mba Mathias, CIRAD, Montpellier, France

Taylor Katrin, FAO, Rome, Italy

Tesch Maxime, CIRAD, Montpellier, France

Thomas Frédéric, CNRS, Montpellier, France

Thompson Lesa, WOAH, Tokyo, Japan

Thys Séverine, CIRAD, Vientiane, Lao PDR

Tiensin Thanawat, FAO, Rome, Italy

Timmermans Eddy, VSF, Brussels, Belgium

Tolno Saa André, CIRAD, Conakry, Guinea; ISSMV, Dalaba, Guinea

Touré Abdoulaye, CERFIG, Conakry, Guinea

Tran Annelise, CIRAD, Montpellier, France

Trevennec Carlène, INRAE, Montpellier, France

Truong Dinh Bao, Nong Lam University, Ho Chi Minh-City, Vietnam

Truong Thi Duyen, CIRAD, Hanoi, Vietnam

Tshering Pasang, WOAH, Thimphu, Bhutan

V

Vink Daan, CIRAD, Montpellier, France

Vittecoq Marion, Tour du Valat, Arles, France

Vu Thi Phuong, Ministry of Agriculture and Environment, OHP, Hanoi, Vietnam

Yang Liuhuaying, Complexity Science Hub, Vienna, Austria

Wacrenier Océane, CIRAD, Montpellier, France

Wannous Chadia, WOAH, Paris, France

Wijayathilaka Tikiri, WOAH, Bangkok, Thailand

Zaw Win Tu Tu, CIRAD, Bangkok, Thailand

Zinsstag Jakob, Swiss TPH, Basel, Switzerland

A Vietnamese peasant woman and her buffalo near Lake Ba Be.

Photo credits

Cover: © noerizki, istockphoto.com

p. 4: Zebu drinking from a pond at the foot of the Allée des Baobabs in Morondava, Madagascar. Cyrille Cornu, © Cirad

p. 12: © frdric, stock.adobe.com

p. 14: African livestock. © birtoiu, stock.adobe.com

p. 18: Monkey family sitting on a temple step with walls in moss. © irinakuz9, stock.adobe.com

p. 46: Floating market. © Chalabala, stock.adobe.com

p. 86: Students conducting fieldwork on a village oil palm plantation, Krabi region, Southern Thailand. Alain Rival, © Cirad

p. 130: Tourists at Prague Old Town Square. © Bits and Splits, stock.adobe.com

p. 174: Beautiful sunset in the background as some cattle graze on grass. © Confidence, stock.adobe.com

p. 188: Shepherding livestock, Shombole area of Kenya. © Overflightstock, stock.adobe.com

p. 190: Rhesus macaque on a wall high above Jaipur, Rajasthan. © schame87, stock.adobe.com

p. 207: Marisa Peyre, © Cirad

Edition: Teri Jones-Villeneuve
Cartography: AFDEC
Layout and cover: Laetitia Perotin-Meslay

Printed in July 2025 by:
SEPEC NUMERIQUE
Z.A des Bruyères
01960 Peronnas
France
Legal deposit: July 2025

www.ingramcontent.com/pod-product-compliance
Lightning Source LLC
LaVergne TN
LVHW070408110826
845147LV00016B/967
* 9 7 8 1 8 3 6 9 9 2 5 5 4 *